ADVANCED CONSTRUCTION TECHNOLOGY

FOURTH EDITION

ROY CHUDLEY   *MCIOB*
and
ROGER GREENO   *BA (HONS), FCIOB, FIPHE, FRSA*

PEARSON
Prentice
Hall

Harlow, England • London • New York • Boston • San Francisco • Toronto • Sydney • Singapore • Hong Kong
Tokyo • Seoul • Taipei • New Delhi • Cape Town • Madrid • Mexico City • Amsterdam • Munich • Paris • Milan

Pearson Education Limited
Edinburgh Gate
Harlow
Essex CM20 2JE
England

and Associated Companies throughout the world

Visit us on the World Wide Web at:
www.pearsoned.co.uk

First published (as *Construction Technology*) 1976 (Volume 3), 1977 (Volume 4)
Second edition 1987
Third edition (published as a single volume, with revisions by Roger Greeno) 1999
Reprinted 1999, 2001, 2002 (twice), 2003, 2004 (twice)
Fourth edition 2006
Reprinted 2007

British Library Cataloguing in Publication Data
A catalogue entry for this title is available from the British Library

ISBN 978-0-13-201985-9

10   9
10   09   08   07

Set by 35 in 10/12pt Ehrhardt
Printed in Malaysia (CTP-VVP)

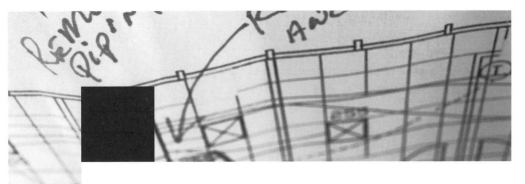

CONTENTS

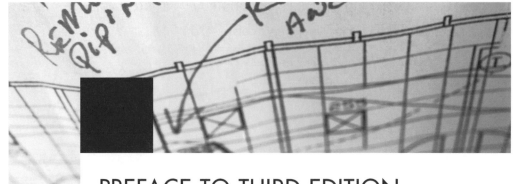

PREFACE TO THIRD EDITION

Roy Chudley's *Construction Technology* was first published in four volumes, between 1973 and 1977. The material has since been continuously updated through numerous reprints and full second editions in 1987. The books have gained a world-wide readership, and their success – and their impact on construction education – is a tribute to Roy Chudley's experience in further and higher education and his talents as a skilled technologist, illustrator and writer.

As a former colleague, it has been a privilege to once again work with Roy, on this occasion revising his original work, and compiling the material into two books: *Construction Technology* and *Advanced Construction Technology*. The content forms a thorough study for all students of building, construction management, architecture, surveying and the many other related disciplines within the diverse construction profession.

The original presentation of comprehensive text matched by extensive illustration is retained. Changes in legislation, such as the Building and Construction Regulations, have been fully incorporated into the text; however, as much of the original work as possible has been purposely retained as it contains many relevant examples of existing construction. Additional material discusses the new developments and concepts of contemporary practice.

The two new volumes are complementary, as many of the topics introduced in *Construction Technology* are further developed here. Together the books provide essential reading for all students aspiring to management, technologist and professional qualifications. They should be read alongside the current local building regulations and national standards, and where possible supplemented by direct experience in the workplace.

Roger Greeno
Guildford 1998

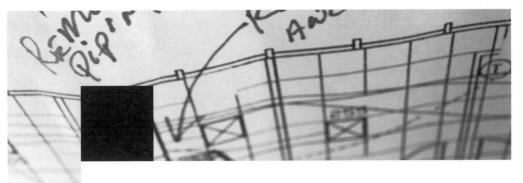

PREFACE TO FOURTH EDITION

Since the previous edition, reprint opportunities have permitted some amendments. These have included new procedures, relative to legislative and practice changes. This revised edition develops these further, with greater attention to information and detail. It also incorporates more recent issues, especially aspects of the Building Regulations that require buildings to be designed and constructed to higher energy-efficient standards. The responsibilities on building designers and owners with regard to human rights are considered in a new chapter outlining the facilities required for the convenience of the less able using buildings other than dwellings.

Notwithstanding contemporary requirements, the book's established construction principles are retained. These provide a useful reference to existing building stock, and, where appropriate, modifications are included to illustrate ongoing change.

The content represents the basic elements of construction practice. The book is neither extensive nor prescriptive, as there is insufficient space in any book to include every possible means for constructing commercial and industrial buildings. However, the content is generally representative, and the reader is encouraged to develop their knowledge through experiential learning, observation in the workplace, and reading manufacturer's literature and technical articles in professional journals. Reference sources for supplementary reading are provided throughout.

In conjunction with this edition's companion volume, *Construction Technology*, the reader should gain an appreciation of the subject material to support progression through any technical, academic or professional qualification study programme that includes construction as core or supplementary modules.

Roger Greeno
Guildford 2006

ACKNOWLEDGEMENTS

This book originated in the 1970s as part of a four-volume series written by Roy Chudley. As a result of its popularity, numerous reprints and a new edition followed. In 1998 the series was rewritten by Roger Greeno as two separate volumes: the initial two volumes formed the basis for the companion title, *Construction Technology*, and the remainder, *Advanced Construction Technology*.

The book's endurance is a tribute to Roy's initial work in representing construction practice with comprehensive illustrative guidance and supporting text. I am particularly grateful to the founding author for allowing me the opportunity to continue this work and to emulate his unique presentation. I am also grateful to the late Colin Bassett as general editor. It was his initiative and enthusiasm that encouraged me to pursue this work.

No book can succeed without a good publisher, and Pearson Education have fulfilled that role with their supportive editorial and production team. In particular, Pauline Gillett has been a constant source of direction and help throughout the preparation of the manuscript.

Roger Greeno
Guildford 2006

We are grateful to the Building Research Establishment and The Stationery Office Ltd for permission to reproduce material from the BRE Digests and various Acts, Regulations and Statutory Instruments.

Extracts from British Standards are reproduced with the permission of BSI. Complete copies can be obtained by post from BSI Customer Services, 389 Chiswick High Road, London W4 4AL.

INTRODUCTION

Advanced Construction Technology is a development of the relatively elementary construction detailed in the associated volume, *Construction Technology*. This volume augments the associated volume with further topics relating to domestic buildings and lightweight-framed structures, in addition to concentrating primarily on complex and specialised forms of construction.

It is designed to supplement a student's lecture notes, projects and research assignments as well as to provide a valuable professional reference. It also complements the associated subjects of science, mathematics, materials technology, design procedures, structural analysis, structural design, services, quantity surveying, facilities management and management studies, and is therefore appropriate for most undergraduate and higher-level construction study programmes.

The format adopted follows that of *Construction Technology*, providing concise notes and generous illustrations to elaborate on the text content. The reader should appreciate that the illustrations are used to emphasise a point of theory and must not be accepted as the only solution. A study of working drawings and details from building appraisals given in the various construction journals will add to background knowledge and comprehension of construction technology.

No textbook or work of reference is ever complete. Therefore readers are recommended to seek out all sources of reference on any particular topic of study, to maximise information and to gain a thorough comprehension of the subject. Construction technology is not purely academic; lectures and textbooks can only provide the necessary theoretical background to the building processes of design and site application. Practical experience and monitoring of work in progress are essential components of any study programme involving the subject of construction technology.

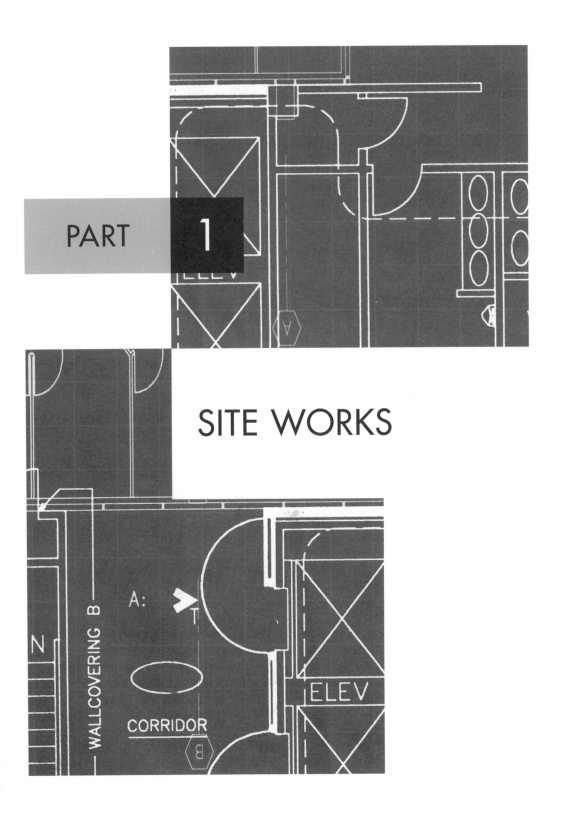

PART 1

SITE WORKS

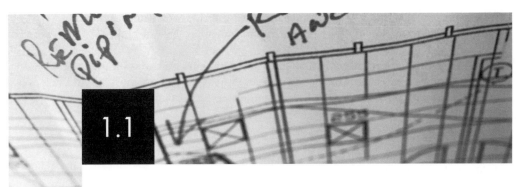

SITE LAYOUT

The construction of a building can be considered as production with a temporary factory, the building site being the 'factory' in which the building contractor will make the product. To enable this activity to take place the builder requires operatives, materials and plant, all of which have to be carefully controlled so that the operatives have the right machines in the most advantageous position, the materials stored so that they are readily available and not interfering with the general site circulation, and adequate storage space and site accommodation.

There is no standard size ratio between the free site space required to construct a building and the total size of the site on which the building is to be erected: therefore each site must be considered as a separate problem in terms of allocating space for operatives, materials and plant. To obtain maximum efficiency there is an optimum way of laying out the site and also a correct amount of expenditure to support the proposed site layout. Any planned layout should be reviewed periodically and adjusted to suit the changing needs of the site activities. If this aspect of building construction is carefully considered, planned and controlled, it will be reflected in the progress and profitability of the contract.

Before any initial planning of the site layout can take place certain preliminary work must be carried out, preferably at the pre-tender stage. The decision to tender will usually be taken by the managing director or, for small works, by the senior estimator up to a contract value laid down by the managing director. With given designs and specifications the best opportunity for the contractor to prepare a competitive and economic tender is in the programming and planning of the construction activities. A thorough study of the bill of quantities will give an indication of the amount and quality of the materials required and also of the various labour resources needed to carry out the contract. A similar study of the drawings, together with the bill of quantities and the specification, will enable the builder to make a preliminary assessment of the size and complexity of the contract, the plant required, and the amount of money that can reasonably be expended on labour-saving items such as concrete mixing and placing alternatives, handling and transporting equipment and off-site fabrication of such items as formwork and reinforcement.

Before the estimator can make a start on calculating unit rates a site investigation should be carried out, preferably by the site manager, who will supervise the contract should the tender be successful. The manager's report should include the following information:

- **Access to site** On- and off-site access, road and rail facilities, distances involved, rights of way restrictions, local authority or police restrictions and bridge weight or height limitations on approach routes.
- **Services** Available power and water supplies, together with rates of payment, nuisance or value of services already on site, diversions required, and the time element involved in carrying out any necessary diversions together with cost implications.
- **Layout** General site conditions such as nature of soil, height of water table, flooding risks, tidal waters, neighbouring properties, preservation orders, trees, demolition problems and special insurance considerations.
- **Staff** Travel distances, availability of local trade contractors, specialist contractors, local rates of pay and facilities to be provided, e.g. site accommodation, catering, health and safety equipment.
- **Security** Local vandalism and pilfering record, security contractors' facilities, need for night security, fencing and hoarding requirements.

With the knowledge and data gained from contract documents, site investigations, and any information gained from the police and local authority sources the following pre-tender work can now be carried out:

- **Pre-tender programme** Usually in a barchart form showing the proposed time allowances for the major activities.
- **Pre-tender health and safety plan** This is prepared by the project coordinator (usually the architect) to enable tendering contractors to consider the practical and cost implications and adequacy of their resources with regard to assessment of risk in safety issues and provision of welfare requirements.
- **Cost implications** Several programmes for comparison should be made to establish possible break-even points giving an indication of required bank loan, possible cash inflow and anticipated profit.
- **Plant schedule** This can be prepared in the form of a barchart and method statements showing requirements and utilisation, which will help in deciding how much site maintenance, equipment and space for plant accommodation will be needed on site. Consideration of whether to purchase or hire plant can be ascertained from these data, although this is only likely to affect smaller items such as specialised tools, as few builders could justify owning large items of plant. However, a balance of buying and hiring will have to be established at this stage.
- **Materials schedule** Basic data can be obtained from the bill of quantities. The buyer's knowledge of the prevailing market conditions and future trends

will enable usage and delivery periods and the amount of site space and/or accommodation required to be predicted.

- **Labour summary** Basic data obtained from the bill of quantities, site investigation report and pre-tender barchart programme to establish number of subcontract trades required. Also the quantity and type of site personnel accommodation required.
- **Site organisation structure** This is a 'family tree' chart showing the relationships and interrelationships between the various members of the site team, and is normally required only on large sites where the areas of responsibility and accountability must be clearly defined.
- **Site layout** Site space allocation for materials storage, working areas, units of accommodation, plant positions and general circulation areas. Access and egress for deliveries and emergency services.
- **Protection** Protection of adjacent buildings, structures (including trees) with preservation orders and provision of fencing/hoarding to prevent trespass and to protect people in the vicinity. Check adequacy of insurances.

■■■ PLANNING SITE LAYOUTS

When planning site layouts the following must be taken into account:

- site activities;
- efficiency;
- movement;
- control;
- facilities for health, safety and welfare provision;
- accommodation for staff and storage of materials.

SITE ACTIVITIES

The time needed for carrying out the principal activities can be estimated from the data obtained previously for preparing the material and labour requirements. With repetitive activities estimates will be required to determine the most economical balance of units that will allow simultaneous construction processes; this in turn will help to establish staff numbers, work areas and material storage requirements. A similar argument can be presented for overlapping activities. If a particular process presents a choice in the way the result can be achieved the alternatives must be considered: for example, the rate of placing concrete will be determined by the output of the mixer and the speed of transporting the mix to the appropriate position. Alternatives that can be considered are:

- more than one mixer;
- regulated supply of ready-mixed concrete;
- on large contracts, pumping the concrete to the placing position.

All alternative methods for any activity will give different requirements for staff numbers, material storage, access facilities and possibly plant types and numbers.

EFFICIENCY

To achieve maximum efficiency the site layout must aim at maintaining the desired output of the planned activities throughout the working day, and this will depend largely upon the following factors:

- Avoidance, as far as practicable, of double handling of materials.
- Proper storekeeping arrangements to ensure that the materials are of the correct type, in the correct quantity, and available when required.
- Walking distances kept to a minimum to reduce the non–productive time spent in covering the distances between working, rest and storage areas without interrupting the general circulation pattern.
- Avoidance of loss by the elements by providing adequate protection for unfixed materials on site, thereby preventing time loss and cost of replacing damaged materials.
- Avoidance of loss by theft and vandalism by providing security arrangements in keeping with the value of the materials being protected and by making the task difficult for the would-be thief or vandal by having adequate hoardings and fences. Also to be avoided is the loss of materials due to pilfering by site staff, who may consider this to be a perquisite of the industry. Such losses can be reduced by having an adequate system of stores' requisition and material checking procedures. Engaging specialist subcontractors on a 'supply and fit' basis may reduce the main contractor's concern.
- Minimising on-site traffic congestion by planning delivery arrivals, having adequate parking facilities for site staff cars and mobile machinery when not in use, and by having sufficient turning circle room for the types of delivery vehicle likely to enter the site.

MOVEMENT

Apart from the circulation problems mentioned above, the biggest problem is one of access. Vehicles delivering materials to the site should be able to do so without difficulty or delay. Many of the contractors' vehicles will be lightweight and will therefore present few or no problems, but the weight and length of suppliers' vehicles should be taken into account. For example, a fully laden ready-mix concrete lorry can weigh 20 tonnes, and lorries used for delivering structural steel can be 18.000 metres long, weigh up to 40 tonnes and require a large turning circle. If it is anticipated that heavy vehicles will be operating on site it will be necessary to consider the road surface required. If the roads and paved areas are part of the contract and will have adequate strength for the weight of the anticipated vehicles it may be advantageous to lay the roads at a very early stage in the contract, but if the specification for the roads is for light traffic it would be advisable to lay only the base hoggin or hardcore layer at the initial stages because of the risk of damage to the completed roads by the heavy vehicles. As an alternative it may be considered a better policy to provide only temporary roadways composed of railway sleepers, metal tracks or mats until a later stage in the contract, especially if such roads will only be required for a short period. See also the Construction (Health, Safety and Welfare) Regulations 15 to 17.

CONTROL

This is concerned mainly with the overall supervision of the contract, including staff, materials, and the movement of both around the site. This control should form the hub of the activities, which logically develops into areas or zones of control radiating from this hub or centre. Which zone is selected for storage, accommodation or specific activities is a matter of conjecture and the conditions prevailing on a particular site, but as a rule the final layout will be one of compromise, with storage and accommodation areas generally receiving priority.

FACILITIES

These must be planned for each individual site, but certain factors will be common to all sites – not least the implications of the Construction (Health, Safety and Welfare) Regulations 1996, the Work at Height Regulations 2005 and the Health and Safety at Work etc. Act 1974. The main contractor is obliged to provide a safe, healthy place of work, and safe systems of work, plant and equipment that are not a risk to health. Equipment for the conduct of work must be provided with adequate information for its safe use and, where appropriate, training in its application. Both regulations are wide ranging and set goals or objectives relating to risk assessment to ensure reasonably practicable steps are taken to ensure safety provision. Prescriptive requirements for such provisions as scaffold guard rail heights and platform widths are scheduled in the Work at Height Regulations.

The principal considerations under the Construction (Health, Safety and Welfare) Regulations can be summarised as follows:

- **Regulation 5**: *Safe places of work*. This requires that people are provided with properly maintained safe surroundings in which to work, along with safe means of accessing and leaving that place of work. It is an overall requirement for reasonable precautions to be taken, with the perceived and varying risk associated with every place of work. Sufficient and suitable working space should be provided with regard to the activity being undertaken.
- **Regulation 7**: *Precautions against falling through fragile material*. This applies mainly to work at heights in excess of 2 m, although potential for falls from any height must be assessed. Requirements for sufficient and adequate means of guarding persons from fragile material must be in place, with a prominence of warning notices displayed in the vicinity (see also Work at Height Regulations).
- **Regulation 8**: *Falling objects*. This requires sufficient and suitable means for preventing injury to persons from falling objects. Provisions may include guard rails, toe boards and protective sheeting to scaffold systems. No material to be tipped or thrown from height (see rubble chutes, Chapter 5.2). Material to be stored or stacked in a stable manner to prevent collapse or unintentional movement.
- **Regulations 9 to 11**: *Work on structures*. A large amount of work associated with the construction of buildings is essentially temporary. Therefore potential for structural collapse, e.g. inadequately supported formwork, is very real and must be recognised and assessed by a competent person. The necessary precautions must be taken to prevent danger. Demolition and dismantling are also high-risk

areas, justifying thorough planning and risk analysis before and as work proceeds.

- **Regulations 12 and 13**: *Excavation, cofferdams and caissons.* Substructural work has an inherent danger of collapse. Suitable provision to prevent collapse of trenches etc. must be designed and installed by competent specialists. Awareness of water-table levels and possible variations, e.g. seasonal and tidal, is essential, as is location of underground cables and other services that could be a danger.
- **Regulation 14**: *Prevention of drowning.* This is not applicable to all sites, but if there is a danger from water or other liquids in any quantity then every practical means possible must be taken to prevent people falling into it. Personal protective and rescue equipment must be available, maintained in good order, and water transport must be provided under the control of a competent person.
- **Regulations 15 to 17**: *Traffic routes, vehicles, doors and gates.* These make provision for segregation of vehicles and pedestrians, with definition of routes. The regulations require adequate construction and maintenance of temporary traffic routes, control of unintended traffic movement, warnings (audible or otherwise) of vehicle movements, prohibition of misuse of vehicles, and safeguards for people using powered doors and guards such as that on hoist facilities.
- **Regulations 18 to 21**: *Prevention and control of emergencies.* These make provision for emergency routes, means of escape, evacuation procedures, adequate signing, firefighting equipment, emergency lighting and associated training for dealing with emergencies.
- **Regulation 22**: *Welfare facilities.* Minimum requirements apply even to the smallest of sites. These include an adequate supply of drinking water, sanitary and washing facilities, means to heat food and boil water, adequate outdoor protection including personal protective equipment (PPE), rest accommodation and facilities to eat meals, first-aid equipment under the control of an appointed person, and accommodation to change and store clothing.
- **Regulations 23 to 27**: *Site-wide issues.* General requirements to ensure fresh air availability at each workplace, reasonable temperatures maintained at internal workplaces, protection against inclement weather, adequate levels of lighting (including emergency lights), reasonably clean and tidy workplaces, well-defined site boundaries, and maintenance of site equipment and plant for safe use by operatives.
- **Regulations 28 to 30**: *Training, inspection and reports.* Specialised elements of work to be undertaken only by those appropriately qualified and/or trained. Supervision of others by those suitably qualified may be acceptable. Places of obvious danger and risk, such as excavations, cofferdams and caissons, to be inspected regularly (at least daily and when changes are effected) by a competent person. Written records/reports to be filed after every inspection.

The principal considerations under the Work at Height Regulations apply to any place at or below ground level as well as above ground. They also include the means of gaining access and egress from that place of work. Measures taken by these regulations are designed to protect a person from injury caused by falling any distance. This may be from plant and machinery or from equipment such as scaffolding, trestles and working platforms, mobile or static. In summary:

- **Regulation 4**: *Organisation and planning*. It is the employer's responsibility to ensure that work at height is planned, supervised and conducted in a safe manner. This includes provisions for emergencies and rescues, and regard for assessing risk to persons working during inclement weather.
- **Regulation 5**: *Competence*. Employer's responsibility to ensure that persons engaged in any activity relating to work at height are competent. Any person being trained to be supervised by a competent person.
- **Regulation 6**: *Risk avoidance*. Re risk assessment under Regulation 3 of the Management of Health and Safety at Work Regulations. This is concerned primarily with appraisal of the work task relative to its situation: that is, work should not be undertaken at a height if it is safer to do it at a lower level, e.g. cutting materials. Provisions to be in place for preventing persons sustaining injury from falling.
- **Regulation 7**: *Work at height equipment*. Further requirements for assessment of risk relative to the selection of plant and equipment suitable for collective rather than individual use.
- **Regulation 8**: *Specific work equipment*.
  Scaffold and working platforms:
  - Top guard rail, min. 950 mm high.
  - Intermediate guard rail, positioned so that no gap between it and top rail or toe board exceeds 470 mm.
  - Toe board, sufficient to prevent persons or materials falling from the working platform. Generally taken as 150 mm min. height. For practical purposes a 225 mm wide scaffold board secured vertically.
  - Stable and sufficiently rigid for the intended purpose.
  - Dimensions adequate for a person to pass along the working platform, unimpeded by plant or materials.
  - No gaps in the working platform.
  - Platform surface resistant to slipping or tripping.
  - Platform designed to resist anticipated loading from personnel, plant and materials.
  - Scaffold frame of sufficient strength and stability.
  - If the scaffold is unconventional in any way, calculations are required to prove its structural integrity.
  - Assembly, use and dismantling plan and instructions to be produced by a competent designer. A standard procedure/plan is acceptable for regular applications.
  - During assembly, alteration, dismantling or non-use, suitable warning signs to be displayed as determined by the Health and Safety (Safety Signs and Signals) Regulations. Means to prevent physical access also required.
  - Assembly, alteration and dismantling under the supervision of a competent person qualified by an approved training scheme.
  Nets, airbags or other safeguards for arresting falls:

Used where it is considered not reasonably practical to use other safer work equipment without it. A safeguard and its means for anchoring must be of adequate strength to arrest and contain persons without injury, where they are liable to fall. Persons suitably trained in the use of this equipment, including rescue procedures, must be available throughout its deployment. Where personal fall protection equipment is considered necessary, it should be correctly fitted to the user, adequately anchored, and designed to prevent unplanned use by the user's normal movements.

Ladders:

- Used solely where a risk assessment indicates that it is inappropriate and unnecessary to install more substantial equipment. Generally, this applies to work of a short duration.
- The upper place of support is to be firm, stable and strong enough to retain the ladder without movement. Position to be secured by rope lashing or other mechanical fixing.
- Inclination is recommended at approximately 75° to the vertical, i.e. in the vertical to horizontal ratio of 4:1.
- A suspended ladder to be secured and attached to prevent displacement and swinging.
- The extension of a ladder beyond a place of landing should be sufficient for safe bodily transfer – normally taken as 1.050 m min. measured vertically.
- Where a ladder ascends 9.000 m or more vertically, landing points to be provided as rest platforms.

- **Regulation 9:** *Fragile surfaces.* It is the employer's responsibility to ensure that no person works on or near a fragile surface. Where it is impossible to avoid, then sufficient protection, e.g. platforms, guard rails etc., are to be provided. Location of fragile surfaces is to be indicated by positioning of prominent warning signs.
- **Regulation 10:** *Falling objects.* See also Regulation 8 under the Construction (Health, Safety and Welfare) Regulations. Suitable provisions, e.g. fan hoardings, are required to prevent persons suffering injury from falling objects or materials. Facilities are to be provided for safe collection and transfer of materials between high and low levels, e.g. chutes. No objects to be thrown. Materials to be stacked with regard to their stability and potential for movement.
- **Regulation 11:** *Dangerous areas.* Areas of work of specific danger, e.g. demolition, to be isolated to ensure that persons not engaged in that particular activity are excluded. Warning signs to be displayed.
- **Regulations 12 and 13:** *Inspection.* These regulations specifically apply to scaffolding, ladders and fall protection equipment. After installation or assembly, no equipment may be used until it has been inspected and documented as safe to use by a competent person. Further inspections are required where conditions may have caused deterioration of equipment, or alterations or changes have been made. Following an interval, every place of work at height should be inspected before work recommences.
- **Regulation 14:** *Personnel duties.* Persons working at height should notify their supervisor of any equipment defect. If required to use personal safety/protective

equipment (PPE), individuals should be adequately trained and instructed in its use.

Under the Health and Safety at Work etc. Act employers must have defined duties, which include providing:

- a safe place of work;
- safe access to and egress from places of work;
- safe systems of work;
- safe items of plant and equipment;
- suitable and adequate training, supervision and instruction in the use of equipment;
- suitable and appropriate PPE applicable to head, hands, feet, eyes and mouth;
- materials and substances that are safe to use (COSHH Regulations 1999);
- a statement of health and safety policy.

Employees and the self-employed have duties to ensure that they do not endanger others while at work. This includes members of the public and other operatives on site. They must cooperate with the health and safety objectives of their employer (the main contractor), not interfere with any plant or equipment provided for their use, other than its intended use, and report any defects to equipment and dangers relating to unsafe conditions of work.

The preceding section on provision of facilities under the Construction Regulations, the Work at Height Regulations and the Health and Safety at Work etc. Act is intended as summary comment for guidance only. For a full appreciation, the reader is advised to consult each specific document. These are published by The Stationery Office, www.tso.co.uk.

ACCOMMODATION

Apart from legislative necessities, the main areas of concern will be sizing, equipping and siting the various units of accommodation.

Mess huts

These are for the purposes of preparing, heating and consuming food, which may require the following services: drainage, light, power, hot and cold water supply. To provide a reasonable degree of comfort a floor area of 2.0 to 2.5 m^2 per person should be allowed. This will provide sufficient circulation space, room for tables and seating, and space for the storage of any utensils. Consideration can also be given to introducing a system of staggered meal breaks, thus reducing space requirements. On large sites where full canteen facilities are being provided this will be subcontracted to a catering firm. Mess huts should be sited so that they do not interfere with the development of the site but are positioned so that travel time is kept to a minimum. On sites that by their very nature are large, it is worthwhile considering a system whereby tea breaks can be taken in the vicinity of the work areas. Siting mess huts next to the main site circulation and access roads is not of major importance. It is the principal contractor's responsibility to ensure that reasonable welfare facilities are available on site, although they do not necessarily have to provide these. It may be part of subcontractors' conditions of engagement that they provide their own.

Drying rooms

Used for the purposes of depositing and drying wet clothes. Drying rooms generally require a lighting and power supply, and lockers or racks for deposited clothes. A floor area of 0.6 m^2 per person should provide sufficient space for equipment and circulation. Drying rooms should be sited near or adjacent to the mess room.

Toilets

Contractors are required to provide at least adequate washing and sanitary facilities as set out in Regulation 22 of the Construction (Health, Safety and Welfare) Regulations. All these facilities will require light, water and drainage services. If it is not possible or practicable to make a permanent or temporary connection to a drainage system, the use of chemical methods of disposal should be considered. Sizing of toilet units is governed by the facilities being provided, and if female staff are employed on site separate toilet facilities must be provided. Toilets should be located in a position that is convenient to both offices and mess rooms, which may mean providing more than one location on large sites.

First-aid rooms

Only required on large sites as a specific facility, otherwise a reasonably equipped mess room will suffice. The first-aid room should be sited in a position that is conveniently accessible from the working areas, and must be of such a size as to allow for the necessary equipment and adequate circulation, which would indicate a minimum floor area of 6 m^2. First-aid equipment must be under the charge of a suitably trained, appointed person, with responsibility for accounting for the contents and their use.

Before the proposed site layout is planned and drawn, the contracts manager and the proposed site agent should visit the site to familiarise themselves with the prevailing conditions. During this visit the position and condition of any existing roads should be noted, and the siting of any temporary roads considered necessary should be planned. Information regarding the soil conditions, height of water table, and local weather patterns should be obtained by observation, site investigation, soil investigation, local knowledge or from the local authority. The amount of money that can be expended on this exercise will depend upon the size of the proposed contract and possibly upon how competitive the tenders are likely to be for the contract under consideration.

Figure 1.1.1 shows a typical small-scale general arrangement drawing, and needs to be read in conjunction with Fig. 1.1.2, which shows the proposed site layout.

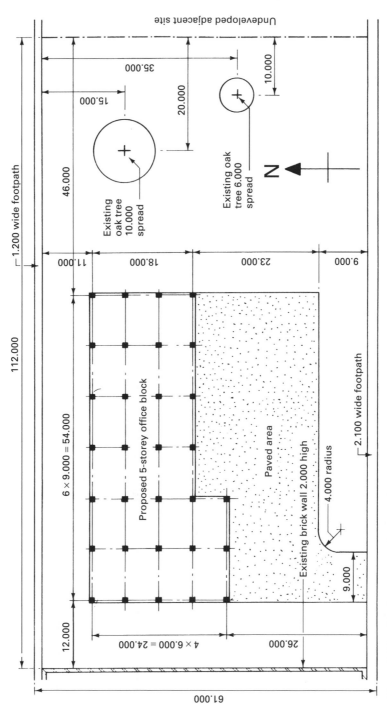

Figure 1.1.1 Site layout example: general arrangement

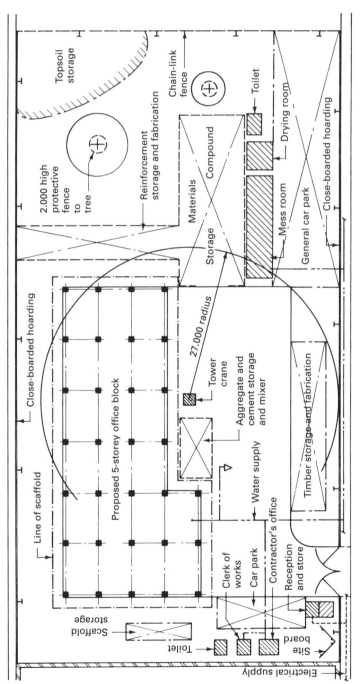

Figure 1.1.2   Site layout example: proposed layout of accommodation and storage

The following data have been collected from a study of the contract documents and by carrying out a site investigation:

- Site is in a typical urban district within easy reach of the contractor's head office and therefore will present no transport or staffing problems.
- Subsoil is a firm sandy clay with a water table at a depth that should give no constructional problems.
- Possession of site is to be at the end of April, and the contract period is 18 months. The work can be programmed to enable the foundation and substructure work to be completed before adverse winter weather conditions prevail.
- Development consists of a single five-storey office block with an *in-situ* reinforced concrete structural frame, *in-situ* reinforced concrete floors and roof, precast concrete stairs, and infill brick panels to the structural frame with large hardwood timber frames fixed into openings formed by the bricklayers. Reduced-level dig is not excessive, but the topsoil is to be retained for landscaping upon completion of the building contract by a separate contractor; the paved area in front of the office block, however, forms part of the main contract. The existing oak trees in the north–east corner of the site are to be retained and are to be protected during the contract period.
- Estimated maximum number of staff on site at any one time is 40, in the ratio of 1 supervisory staff to 10 operatives plus a resident clerk of works.
- Main site requirements are as follows:
  1 office for 3 supervisory staff.
  1 office for resident clerk of works.
  1 office for reception and materials checker/assistant site agent.
  1 hutment as lock-up store.
  1 mess room for 36 operatives.
  1 drying room for 36 operatives.
  Toilets.
  Storage compound for major materials.
  Timber store and formwork fabrication area.
  Reinforcement store and fabrication area.
  Scaffold store.
  Car parking areas.
  1 tower crane and area for concrete deliveries, sand and cement storage, and site mixer.

Sizing and location of main site requirements can be considered in the following manner:

- **Offices for contractor's supervisory staff**   Area required = $3 \times 3.7 \text{ m}^2 = 11.1 \text{ m}^2$. Using plastic-coated galvanised steel prefabricated cabins based on a 2.400 m wide module gives a length requirement of $11.1 \div 2.4 = 4.625$ m: therefore use a hutment 2.400 m wide $\times$ 4.800 m long, giving an area of 11.52 m^2. Other standard internal widths are 2.7 and 3.0 m, and standard internal lengths range from 2.4 to 10.8 m in 1.2 m increments.

- **Office for resident clerk of works** Allowing for one visitor area required = 2×3.7 m^2 = 7.4 m^2. Using same width module as for contractor's office length required = $7.4 \div 2.4 = 3.08$ m, therefore using a 2.400 m wide $\times$ 3.600 m long cabin will give an area of 8.64 m^2. The contractor's office and that for the clerk of works need to be sited in a position that is easily and quickly found by visitors to the site and yet at the same time will give a good view of the site operations. Two positions on the site in question seem to meet these requirements: one is immediately to the south of the paved area and the other is immediately to the west of it. The second position has been chosen for both offices because there is also room to accommodate visitors' cars in front of the offices without disturbing the circulation space given by the paved area.

- **Office for reception and materials checker** A hut based on the requirements set out above for the clerk of works would be satisfactory. The office needs to be positioned near to the site entrance so that materials being delivered can be checked, directed to the correct unloading point, and – most important – checked before leaving to see that the delivery has been completed. It also needs to be easily accessible for site visitors, thus preventing unsupervised wandering onto the site.

- **Lock-up store** This needs to be fitted with racks and storage bins to house valuable items, and a small unit of plan size 2.400 m $\times$ 2.400 m has been allocated. Consideration must be given to security, and in this context it has been decided to combine the lock-up store and the site manager's/materials checker's office, giving a total floor plan of 2.400 m $\times$ 6.000 m. This will enable the issue of stores only against an authorised and signed requisition to be carefully controlled, the assistant site agent fulfilling the function of storekeeper.

- **Mess room** Area required = 36×2.5 m^2 = 90 m^2, using a width module of 3.000 m, length required = $90 \div 3 = 30$ m: therefore a number of combinations based on the standard lengths listed are possible. Perhaps 3 modules of 10.8 m = 32.4 m total length (97.2 m^2) or 5 modules of 6 m length = 30 m total length (90 m^2), the choice depending to some extent on the disposition of the site staff, number and size of subcontractors involved. The mess room needs to be sited in a fairly central position to all the areas of activity, and the east end of the paved area has been selected.

- **Drying room** Area required = 36×0.6 m^2 = 21.6 m^2, using a width module of 3.000 m, length required = $21.6 \div 3 = 7.2$ m: therefore select a single length or 2 modules of 3.600 m. The drying room needs to be in close proximity to the mess room and has therefore been placed at the east end of the mess room. Consideration could be given to combining the mess room and drying room into one unit.

- **Toilets** On this site it is assumed that connection can be made to existing drains. If this is not convenient, temporary (or preferably permanent) drain branches can be connected to a main sewer. Two such units are considered to be adequate, one to be sited near to the mess room and the other to be sited near to the office complex. Adequate sanitary conveniences are required in the Construction (Health, Safety and Welfare) Regulations 1996. For the mess toilet unit catering for 36 operatives two conveniences are considered minimum, but a three-convenience toilet unit will be used, having a plan size of 2.400 m × 3.600 m. Similarly, although only one convenience is required for the office toilet unit, a two-convenience unit will be used with a plan size of 2.400 m × 2.400 m.

■ **Materials storage compound**   An area to be defined by a temporary plywood hoarding 2.400 m high and sited at the east end of the paved area giving good access for deliveries and within reach of the crane. Plan size to be allocated 12.000 m wide × 30.000 m long.

■ **Timber storage**   Timber is to be stored in top-covered but open-sided racks made from framed standard scaffold tubulars. Maximum length of timber to be ordered is unlikely to exceed 6.000 m in length: therefore, allowing for removal, cutting and fabricating into formwork units, a total plan size of 6.000 m wide × 36.000 m long has been allocated. This area has been sited to the south of the paved area, giving good access for delivery and within the reach of the crane.

■ **Reinforcement storage**   The bars are to be delivered cut to length, bent and labelled, and will be stored in racks as described above for timber storage. Maximum bar length to be ordered assumed not to exceed 12.000 m: therefore a storage and fabrication plan size of 6.000 m wide × 30.000 m long has been allocated. This area has been sited to the north of the storage compound, giving reasonable delivery access and within reach of the crane.

■ **Scaffold storage**   Tube lengths to be stored in racks as described for timber storage, with bins provided for the various types of coupler. Assuming a maximum tube length of 6.000 m, a plan size of 3.000 m wide × 12.000 m long. This storage area has been positioned alongside the west face of the proposed structure, giving reasonable delivery access and within reach of the crane if needed. The scaffold to be erected will be of an independent type around the entire perimeter positioned 200 mm clear of the building face and of five-board width, giving a total minimum width of $200 + (5 \times 225) = 1.325$, say 1.400 m total width.

■ **Tower crane**   To be sited on the paved area in front of the proposed building alongside the mixer and aggregate storage position. A crane with a jib length of 27.000 m, having a lifting capacity of 1.25 tonnes at its extreme position, has been chosen so that the crane's maximum radius will cover all the storage areas, thus making maximum utilisation of the crane possible.

■ **Car parking**   Assume 20 car parking spaces are required for operatives, needing a space per car of 2.300 m wide × 5.500 m long, giving a total length of $2.3 \times 20 = 46.000$ m and, allowing 6.000 m clearance for manoeuvring, a width of $5.5 + 6.0 = 11.500$ m will be required. This area can be provided to the south of the mess room and drying room complex. Staff car-parking space can be sited in front of the office hutments, giving space for the parking of seven cars, which will require a total width of $7 \times 2.3 = 16.100$ m.

■ **Fencing**   The north and south sides of the site both face onto public footpaths and highways. Therefore a close-boarded or sheet hoarding in accordance with the licence issued by the local authority will be provided. A lockable double gate is to be included in the south-side hoarding to give access to the site. The east side of the site faces an undeveloped site, and the contract calls for a 2.000 m high concrete post and chain-link fence to this boundary. This fence will be erected at an early stage in the contract to act as a security fence during the construction period as well as providing the permanent fencing. The west side of the site has a 2.000 m high brick wall, which is in a good structural condition, and therefore no action is needed on this boundary.

- **Services**   It has been decided that permanent connections to the foul drains will be made for convenient site use, thus necessitating early planning of the drain-laying activities. The permanent water supply to the proposed office block is to be laid at an early stage, and this run is to be tapped to provide the supplies required to the mixer position and the office complex. A temporary connection is to be made to supply the water service to the mess room complex, because a temporary supply from the permanent service would mean running the temporary supply for an unacceptable distance. An electrical supply is to be taken onto site, with a supply incoming unit housed in the reception office along with the main distribution unit. The subject of electrical supplies to building sites is dealt with in Chapter 1.2. Telephones will be required to the contractor's and clerk of works' offices. It has been decided that a gas supply is not required.
- **Site identification**   A V-shaped board bearing the contractor's name and company symbol is to be erected in the south-west corner of the site in such a manner that it can be clearly seen above the hoarding by traffic travelling in both directions, enabling the site to be clearly identified. The board will also advertise the company's name and possibly provide some revenue by including on it the names of participating subcontractors. As a further public relations exercise it might be worthwhile considering the possibility of including public viewing panels in the hoarding on the north and south sides of the site.
- **Health and safety**   Attached to the hoarding at the site entrance is a board displaying the employer's policy for corporate site safety. Some examples of the standard images that could be used are shown in Fig. 1.1.3.

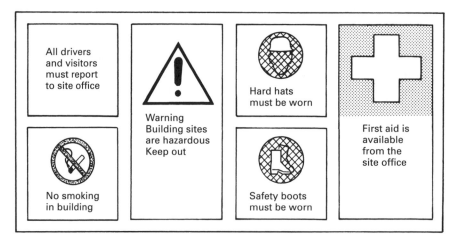

Figure 1.1.3   Site safety board

Note:

Sign colours	Geometric shape	Indication
Red on white background, black image	Circular with a diagonal line	Prohibition
Yellow with black border, black image	Triangular	Warning
Blue with white image	Circular	Mandatory
Green with white image	Oblong or square	Safe condition

References:
Health and Safety at Work etc. Act.
Health and Safety (Safety Signs and Signals) Regulations.
Management of Health and Safety at Work Regulations.
European Directive 92/58 EEC.

The extent to which the above exercise in planning a site layout would be carried out in practice will depend upon various factors, such as the time and money that can reasonably be expended and the benefits that could accrue in terms of maximum efficiency compared with the amount of the capital outlay. The need for careful site layout and site organisation planning becomes more relevant as the size and complexity of the operation increase. This is particularly true for contracts where spare site space is very limited.

1.2

ELECTRICITY ON BUILDING SITES

A supply of electricity is usually required on construction sites to provide lighting to the various units of accommodation. It may also be needed to provide the power to drive small and large items of plant. Two sources of electrical supply to the site are possible:

- portable self-powered generators;
- metered supply from the local section of the national grid distribution network.

As a supply of electricity will be required in the final structure the second source is usually adopted, because it is generally possible to connect a permanent supply cable to the proposed development for construction operations, thus saving the cost of laying a temporary supply cable to the site.

To obtain a metered temporary supply of electricity a contract must be signed between the main contractor and a local area electricity marketing company. They will require the following information:

- Address of site.
- Site location plan.
- Maximum anticipated load demand in kW for the construction period. A reasonable method of estimating this demand is to allow for a loading of 10 W/m^2 for the total floor area of the finished structure and to add for any high-load equipment such as cranes, pumps and drying-out heaters that are to be used.
- Final load demand of the completed building to ensure that the correct rating of cable is laid for the permanent supply.
- Date on which temporary supply will be required.
- Name, address and telephone number of the building owner or their agent, and of the main contractor.

To ensure that the supply and installation are available when required by the builder it is essential that an application for a temporary supply of electricity is made at the earliest possible date.

On any construction site it is possible that there may be existing electricity cables, which can be advantageous or may constitute a hazard or nuisance. Overhead cables will be visible, whereas the routes and depths of underground cables can be ascertained only from the records and maps kept by local area supply companies. Overhead cable voltages should be checked with the local area suppliers, because these cables are usually uninsulated and are therefore classed as a hazard due mainly to their ability to arc over a distance of several metres. High-voltage cables of over 11 kV rating will need special care, and any of the following actions could be taken to reduce or eliminate the danger:

- Apply to the local area supplier to have the cables re-routed at a safe distance or height.
- Apply to have the cable taken out of service.
- Erect warning barriers to keep site operatives and machines at a safe distance. These barriers must be clearly identified as to their intention, and they may be required to indicate the safe distance in both the horizontal and vertical directions. The local area supplier will advise on suitable safe distances according to the type of cable and the load it is transmitting.

The position and depth of underground cables given by electricity suppliers must be treated as being only approximate, because historical records show only the data regarding the condition as laid, and since then changes in site levels may have taken place. When excavating in the locality of an underground cable extreme caution must be taken, which may even involve careful hand excavation to expose the cable. Exposed cables should be adequately supported, and suitable barriers with warning notices should be erected. Any damage, however minor, must be reported to the electricity supplier for the necessary remedial action. It is worth noting that if a contractor damages an underground electric cable that was known to be present, and possibly caused a loss of supply to surrounding properties, the contractor can be liable for negligence, trespass to goods and damages.

■■■ SUPPLY AND INSTALLATION

In Great Britain electrical installations on construction sites are subject to the requirements of the Electricity Supply Regulations 1988 and the Electricity at Work Regulations 1989. These impose duties and expectations on employers, employees and the self-employed, for health and safety responsibilities with regard to the use of electricity. Risk assessment and suitable precautions relating to particular hazards, such as overhead lines and underground cables encountered on site, are contained by the Health and Safety at Work etc. Act 1974 and the Construction (Health, Safety and Welfare) Regulations 1996. Installations should follow rules given in BS 7671: *Requirements for electrical installations* (Institution of Electrical Engineers Wiring Regulations). Section 604 details provision for temporary installations and installations on construction sites. See also, BS 7375:

Code of practice for distribution of electricity on construction and building sites. The
supply distribution assemblies used in the installation should comply with the
recommendations of BS 4363: *Specification for distribution assemblies for reduced
low voltage electricity supplies for construction and building sites.* This covers the
equipment suitable for the control and distribution of electricity from a three-phase
four-wire a.c. system up to a voltage of 400 V with a maximum capacity of 300
A per phase. BS EN 60309-2 specifies plugs, socket outlets and cable couplers for
the varying voltages recommended for use on construction sites.

The appliances and wiring used in temporary installations on construction sites
may be subject to extreme abuse and adverse conditions: therefore correct circuit
protection, earthing and frequent inspection are most important, and this work,
including the initial installation, should be entrusted to a qualified electrician or
to a specialist electrical contractor.

Electrical distribution cables contain three line wires and one neutral, which
can give either a 400 V three-phase supply or a 230 V single-phase supply. Records
of accidents involving electricity show that the highest risk is encountered when
electrical power is used in wet or damp conditions, which are often present on
construction sites. It is therefore generally recommended that wherever possible
the distribution voltage on building sites should be 110 V. This is a compromise
between safety and efficiency, but it cannot be overstressed that a supply of this
voltage can still be dangerous and lethal.

The recommended voltages for use on construction sites are given below:

Mains voltage
400 V three-phase:

- supply to transformer unit, heavy plant such as cranes and movable plant fed via
  a trailing cable;
- hoists and plant powered by electric motors in excess of a 3.75 kW rating.

230 V single-phase:

- supply to transformer unit;
- supply to distribution unit;
- installations in site accommodation buildings;
- fixed floodlighting;
- small static machines.

Reduced voltage
110 V three-phase:

- portable and hand-held tools;
- small mobile plant up to 3.75 kW.

110 V single-phase:

- portable and hand-held tools;
- small items of plant;
- site floodlighting other than fixed floodlighting;

- portable hand-lamps;
- local lighting up to 2 kW.

50 V single-phase and 25 V single-phase:

- as listed for 110 V single-phase but being used in confined and damp situations.

It is worth considering the use of 50 or 25 V battery-supplied hand-lamps if damp situations are present on site. All supply cables must be earthed, and in particular 110 V supplies should be centre-point earthed so that the nominal voltage to earth is not more than 65 V on a three-phase circuit and not more than 55 V on a single-phase circuit.

Protection to a circuit can be given by using bridge fuses, cartridge fuses and circuit breakers. Adequate protection should be given to all main and sub circuits against any short-circuit current, overload current and earth faults.

Protection through earthing may be attained in two distinct ways:

- Provision of a path of low impedence to ensure over-current device will operate in a short space of time.
- Insertion in the supply of a circuit-breaker with an operating coil that trips the breaker when the current due to earth leakage exceeds a predetermined value.

BS 4363 and BS EN 60309-2 recommend that plug and socket outlets are identified by a colour coding as an additional safety precaution to prevent incorrect connections being made. The recommended colours are:

25 V   – violet
50 V   – white
110 V   – yellow
230 V   – blue
400 V   – red
500/650 V – black

The equipment that can be used to distribute an electrical supply around a construction site is as follows:

- **Incoming site assembly (ISA)**   Supply, control and distribution of mains supply on site – accommodates supply company's equipment and has one outgoing circuit.
- **Main distribution assembly (MDA)**   Control and distribution of mains supply for circuits of 400 V three-phase and 230 V single-phase.
- **Incoming site distribution assembly (ISDA)**   A combined ISA and MDA for use on sites where it is possible to locate these units together.
- **Transformer assembly**   Transforms and distributes electricity at a reduced voltage: can be for single-phase, three-phase or both phases and is abbreviated TA/1, TA/3 or TA/1/3 accordingly.
- **Socket outlet assembly (SOA)**   Connection, protection and distribution of final subcircuits at a voltage lower than the incoming supply.
- **Extension outlet assembly (EOA)**   Similar to outlet assembly except that outlets are not protected.

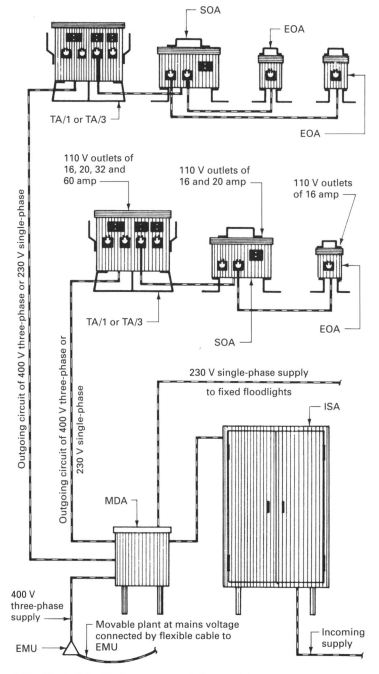

Figure 1.2.1   Typical distribution sequence of site electricity

■ **Earth monitor unit (EMU)**   Flexible cables supplying power at mains voltage from the MDA to movable plant incorporate a separate pilot conductor in addition to the main earth continuity conductor. A very-low-voltage current passes along these conductors between the portable plant and the fixed EMU. A failure of the earth continuity conductor will interrupt the current flow, which will be detected by the EMU, and this device will automatically isolate the main circuit.

The cubicles or units must be of robust construction, strong, durable, rain resistant and rigid to resist any damage that could be caused by transportation, site handling or impact shocks likely to be encountered on a construction site. All access doors or panels must have adequate weather seals. Figure 1.2.1 shows a typical supply and distribution system for a construction site.

The routeing of the supply and distribution cables around the construction site should be carefully planned. Cables should not be allowed to trail along the ground unless suitably encased in a tube or conduit, and even this method should be used only for short periods of time. Overhead cables should be supported by hangers attached to a straining wire and suitably marked with 'flags' or similar visual warning. Recommended minimum height clearances for overhead cables are:

■ 5.200 m in positions inaccessible to vehicles;
■ 5.800 m where cable crosses an access road or any part of the site accessible to vehicles.

Cables that are likely to be in position for a long time, such as the supply to a crane, should preferably be sited underground at a minimum depth of 500 mm and protected by tiles, or alternatively housed in clayware or similar pipes.

In the interest of safety, and to enable first-aid treatment to be given in cases of accident, all contractors using a supply of electricity on a construction site for any purpose must display, in a prominent position, extracts from the Electricity at Work Regulations. Pictographic safety signs for caution of the risk of electric shock are applicable under the Health and Safety (Safety Signs and Signals) Regulations 1996. Suitable placards giving instructions for emergency first-aid treatment for persons suffering from electrical shock and/or burns are obtainable from RoSPA, the St John Ambulance Association, and stockists of custom-made signs.

1.3

LIGHTING BUILDING SITES

Inadequate light accounts for more than 50% of the loss of production on UK construction sites between the months of November and February. Inadequate lighting also increases the risks of accidents and lowers the security of the site. The initial costs of installing a system of artificial lighting for both internal and external activities can usually be offset by higher output, better-quality work, a more secure site, and apportioning the costs over a number of contracts on a use and reuse basis.

The reasons for installing a system of artificial lighting on a construction site are as follows:

- Inclement weather, particularly in winter, when a reduction of natural daylight is such that the carrying out of work becomes impracticable.
- Without adequate light, all activities on construction sites carry an increased risk of accident and injury.
- By enabling work to proceed, losses in productivity can be reduced.
- Reduces the wastage of labour and materials that often results from working in poor light.
- Avoids short-time working due to the inability to see clearly enough for accurate and safe working.
- Improves the general security of the site.

The following benefits may be obtained by installing and using a system of artificial lighting on a construction site:

- Site activities will be independent of the availability of natural daylight, and therefore the activities can be arranged to suit the needs of the contract, the availability of materials, and the personnel involved.
- Overtime and extra shifts can be worked to overcome delays that might occur from any cause.

- Deliveries and collection of materials or plant can be made outside normal site working hours, thus helping to avoid delays and/or congestion.
- The amount of spoilt material and the consequent rectification caused by working under inadequate light can be reduced.
- It provides an effective deterrent to the would-be trespasser or pilferer.
- Contractual relationships will be improved by ensuring regular working hours and thus regular earnings.

Planning the lighting requirements depends on site layout, size of site, shape of site, geographical location, availability of an electrical supply and the planned activities for the winter period. Figure 1.3.1 shows two charts covering various regions in the UK, giving an indication of the periods when external and internal artificial lighting may be required on a construction site under normal conditions. Any form of temporary artificial site lighting should be easy to install and modify as needs change, and should be easy to remove while works are still in progress.

The supply and distribution of an electrical service to a construction site has already been covered in the previous chapter, and it is therefore necessary only to stress again the need for a safe, reliable installation, designed and installed by a specialist contractor.

▪▪▪ ILLUMINATION

Illumination can be considered as the measure of light or illuminance falling on a surface. It is expressed in **lux**, which is one lumen of light falling on 1 m^2 of surface, and this can be measured with a small portable lightmeter, which consists of a light-sensitive cell generating a small current proportional to the light falling on it. The level of illuminance at which an operative can work in safety and carry out tasks to an acceptable standard, in terms of both speed and quality, is quite low, because the human eye is very adaptable and efficient. Although the amount of illuminance required to enable a particular activity to be carried out is a subjective measure, depending largely upon the task, and the age and state of health of the

Table 1.3.1   Typical service values of illuminance

Activities	Illuminance (lux)
External lighting	
Materials handling	200
Open circulation areas	100
Internal lighting	
Circulation	100
Working areas	200
Reinforcing and concreting	300
Joinery, bricklaying and plastering	500
Painting and decorating	500
Fine craft work	1,000
Site offices	500
Drawing board positions	750

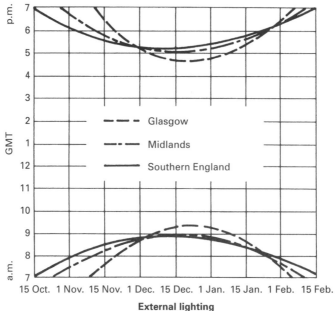

External lighting

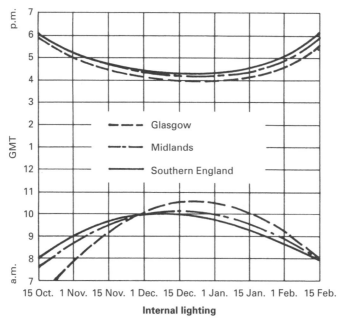

Internal lighting

Figure 1.3.1 Approximate times for site lighting

operative concerned, Table 1.3.1 presents typical service values of illuminance. The values shown in the table do not allow for deterioration, dirt, bad conditions or shadow effects. Therefore in calculating the illuminance required for any particular situation a target value of twice the service value should be used.

When deciding on the type of installation to be used, two factors need to be considered:

- type of lamp to be used;
- nature and type of area under consideration.

The properties of the various types of lamp available should be examined to establish the most appropriate for any particular site requirement.

■■■ LAMPS

- **Tungsten filament lamp**   Ideal for short periods, such as a total of 200 hours during the winter period; main recommended uses are for general interior lighting and low-level external movement. They are cheap to buy but are relatively expensive to run.
- **Tungsten halogen lamp**   Compact fitting with high light output, and suitable for all general area floodlighting. They are easy to mount, and have a more effective focused beam than the filament lamp. These lamps generally have a life of twice that of filament lamps, and quartz lamps have a higher degree of resistance to thermal shock than glass filament lamps. They are dearer than filament lamps and are still relatively expensive to run but should be considered if the running time is in the region of 1,500 hours annually.
- **Mercury tungsten lamps**   Compact, efficient, with a good lamp life, and do not need the expensive starting gear of the vapour discharge lamps. They can be used for internal and external area lighting where lamps are not mounted above 9.000 m high. These are high-cost lamps but are cheap to run.
- **Mercury discharge lamps**   High-efficiency lamps with a long life; can be used for area lighting where lamps are mounted above 9.000 m high. Costs for lamps and control gear are high but the running costs are low.
- **Tubular fluorescent lamps**   Uniformly bright in all directions; used when a great concentration of light is not required; efficient, with a range of colour values. These lamps have a long life and are economical to run.
- **High-pressure sodium discharge lamps**   Compact, efficient, with a long life. For the best coverage without glare they should be mounted above 13.500 m high. Cost for lamp and control gear is high but running costs are low, which makes them suitable for area lighting.

Apart from the cost of the lamps and the running charges, consideration must be given to the cost of cables, controlling equipment, mounting poles, masts or towers. A single high tower may well give an overall saving against using a number of individual poles or masts in spite of the high initial cost for the tower. Consideration can also be given to using the scaffold, incomplete structure or the mast of a tower crane for lamp-mounting purposes, each subject to earthing.

■■■ SITE LIGHTING INSTALLATIONS

When deciding upon the type and installation layout for construction site lighting, consideration must be given to the nature of the area and work to be lit, and also to the type or types of lamp to be used. These aspects can be considered under the following headings:

- external and large circulation areas;
- beam floodlighting;
- walkway lighting;
- local lighting.

EXTERNAL AND LARGE CIRCULATION AREAS

These areas may be illuminated by using mounted lamps situated around the perimeter of the site or in the corners of the site; alternatively, overhead illumination using dispersive fittings can be used. The main objectives of area lighting are to enable staff and machinery to move around the site in safety and to give greater security to the site. Areas of local danger such as excavations and obstructions should, however, be marked separately with red warning lights or amber flashing lamps. Tungsten filament, mercury vapour or tungsten halogen lamps can be used, and these should be mounted on poles, masts or towers according to the lamp type and wattage. Typical mounting heights for various lamps and wattages are shown in Table 1.3.2.

Large areas are generally illuminated by using large, high-mounted lamps, whereas small areas and narrow sites use a greater number of smaller fittings. By mounting the lamps as high as practicable above the working level glare is reduced, and by lighting the site from at least two directions the formation of dense shadows is also reduced. The spacing of the lamps is also important if under-lit and over-lit areas are to be avoided. Figures 1.3.2 and 1.3.3 show typical lamps and the recommended spacing ratios.

Dispersive lighting is similar to an ordinary internal overhead lighting system, and is suitable for both exterior and interior area lighting where overhead

Table 1.3.2   Mounting heights

Lamp type	Watts	Minimum height (m)
Tungsten filament	200	4.500
	300	6.000
	750	9.000
Mercury vapour	400	9.000
	1,000	15.000
	2,000	18.000
Tungsten halogen	500	7.500
	1,000	9.000
	2,000	15.000

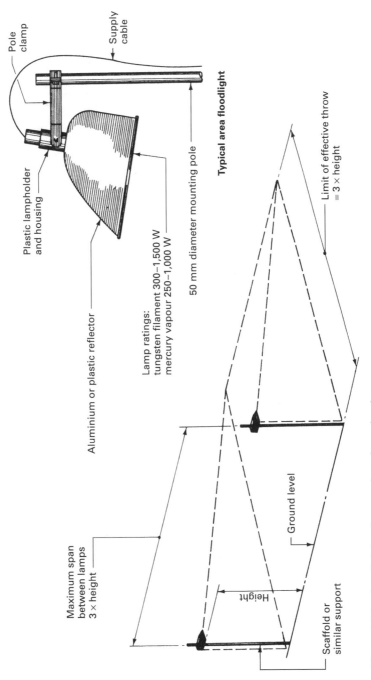

Pole clamp

Supply cable

Plastic lampholder and housing

Aluminium or plastic reflector

Lamp ratings:
tungsten filament 300–1,500 W
mercury vapour 250–1,000 W

50 mm diameter mounting pole

Typical area floodlight

Limit of effective throw
= 3 × height

Maximum span between lamps
3 × height

Ground level

Height

Scaffold or similar support

Figure 1.3.2   Area lighting: lamps and spacing ratios: 1

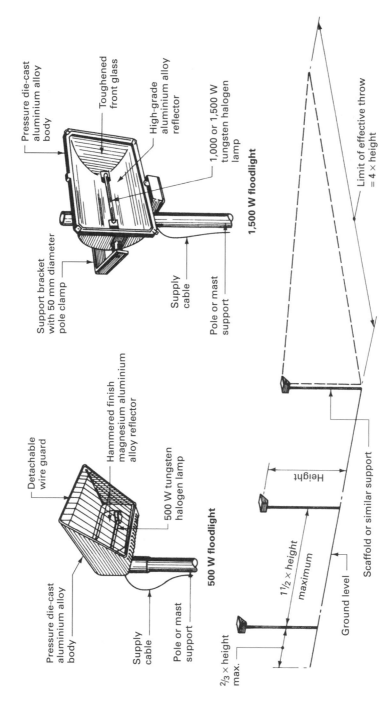

Figure 1.3.3   Area lighting: lamps and spacing ratios: 2

Table 1.3.3   Suspension heights

Lamp type	Watts	Minimum height (m)
Tungsten filament	200	2.500
	300	3.000
	750	6.000
Mercury vapour	250	6.000
	400	7.500
	700	9.000
Fluorescent trough	40 to 125	2.500

suspension is possible. Ordinary industrial fittings should not be used because of the adverse conditions that normally prevail on construction sites. The fittings selected should therefore be protected against rust, corrosion and water penetration. To obtain a reasonable spread of light the lamps should be suspended evenly over the area to be illuminated, as shown diagrammatically in Fig. 1.3.4. Tungsten filament, mercury vapour and fluorescent trough fittings are suitable and should be suspended at a minimum height according to their type and wattage. Typical suspension heights are shown in Table 1.3.3.

Most manufacturers provide guidance as to the choice of lamps or combination of lamps, but a simple method of calculating lamp requirements is as follows:

1. Decide upon the service illuminance required, and double this figure to obtain the target value.

2. Calculate total lumens required $= \dfrac{\text{area (m}^2) \times \text{target value (lux)}}{0.23}$

3. Choose lamp type.

   Number of lamps required $= \dfrac{\text{total lumens required}}{\text{lumen output of chosen lamp}}$

4. Repeat stage 3 for different lamp types to obtain the most practicable and economic arrangement.

5. Consider possible arrangements, remembering that:
   - larger lamps give more lumens per watt and are generally more economic to run;
   - fewer supports simplify wiring and aid overall economy;
   - corner siting arrangements are possible;
   - clusters of lamps are possible.

The calculations when using dispersive lighting are similar to those given above for mounted area lighting except for the formula in stage 2, which has a utilisation factor of 0.27 instead of 0.23.

BEAM FLOODLIGHTING

Tungsten filament or mercury vapour lamps can be used, but this technique is limited in application on construction sites to supplementing other forms of lighting. Beam floodlights are used to illuminate areas from a great distance. The

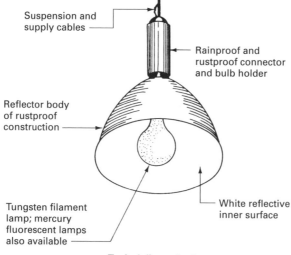

Suspension and supply cables

Rainproof and rustproof connector and bulb holder

Reflector body of rustproof construction

Tungsten filament lamp; mercury fluorescent lamps also available

White reflective inner surface

Typical dispersive lamp

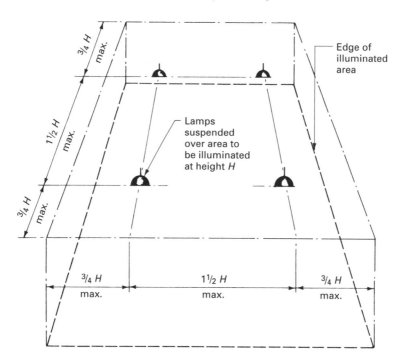

³/₄ H max.

1¹/₂ H max.

³/₄ H max.

Edge of illuminated area

Lamps suspended over area to be illuminated at height *H*

³/₄ H max.

1¹/₂ H max.

³/₄ H max.

Figure 1.3.4 Area lighting: overhead dispersive lamps

beam of light is intense, producing high glare, and it should therefore be installed to point downwards towards the working areas. Generally the lamps are selected direct from the manufacturer's catalogue without calculations.

WALKWAY LIGHTING

Tungsten filament and fluorescent lamps can be used to illuminate access routes such as stairs, corridors and scaffolds. Bulkhead fittings that can be safely installed with adequate protection to the wiring can be run off a mains voltage of 230 V single-phase, but if they are in a position where they can be handled a reduced voltage of 110 V single-phase should be used. Festoon lighting, in which the ready-wired lampholders are moulded to the cable itself, can also be used. A standard festoon cable would be 100.000 m long with rainproof lampholders and protective shades or guards at 3.000 m or 5.000 m centres using 40 W or 60 W tungsten filament bulbs for the respective centres. See Fig. 1.3.5. For lighting to scaffolds of four- or five-board width 60 W lamps should be used, placed at not more than 6.000 m centres and preferably at least 2.400 to 3.000 m above the working platform either to the wall side or centrally over the scaffold.

LOCAL LIGHTING

Clusters of pressed glass reflector floodlamps, tungsten filament lamps, festoons and adjustable fluorescents can be used to increase the surface illumination at local points, particularly where finishing trades are involved. These fittings must be portable so that shadow casting can be reduced or eliminated from the working plane: therefore it is imperative that these lights are operated off a reduced voltage of 110 V single-phase. Fluorescent tubes do not usually work at a reduced voltage, so special fittings working off a 110 V single-phase supply that internally increase the voltage are used. Typical examples of suitable lamps are shown in Fig. 1.3.6.

As an alternative to a system of static site lighting connected to the site mains, electrical supply mobile lighting sets are available. These consist of a diesel-engine-driven generator and a telescopic tower with a cluster of tungsten iodine lamps. These are generally cheaper to run than lamps operating off a mains supply. Small two-stroke generator sets with a single lamp attachment suitable for small, isolated positions are also available.

Another system that can be used for local lighting is flame lamps, which normally use propane gas as the fuel. The 'bulb' consists of a mantle, and a reflector completes the lamp fitting, which is attached to the fuel bottle by a flexible tube. These lamps produce a great deal of local heat and water vapour; the latter may have the effect of slowing down the drying out of the building. An alternative fuel to propane gas is butane gas, but this fuel will not usually vaporise at temperatures below −1 °C.

Whichever method of illumination is used on a construction site it is always advisable to remember the axiom 'A workman can only be safe and work well when he can see where he is going and what he is doing.'

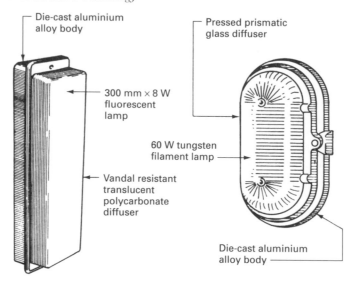

Die-cast aluminium alloy body

Pressed prismatic glass diffuser

300 mm × 8 W fluorescent lamp

60 W tungsten filament lamp

Vandal resistant translucent polycarbonate diffuser

Die-cast aluminium alloy body

Ceiling- or wall-mounted bulkhead lamp fittings

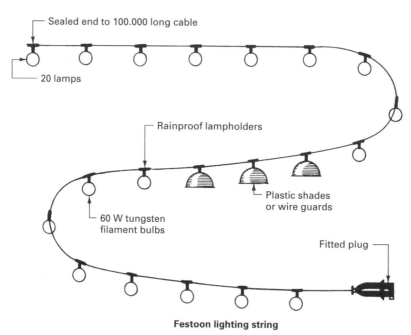

Sealed end to 100.000 long cable

20 lamps

Rainproof lampholders

Plastic shades or wire guards

60 W tungsten filament bulbs

Fitted plug

Festoon lighting string

Figure 1.3.5 Typical walkway lighting fittings

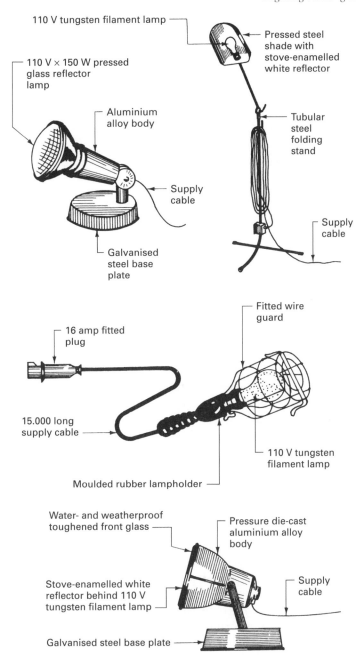

110 V tungsten filament lamp

110 V × 150 W pressed glass reflector lamp

Aluminium alloy body

Pressed steel shade with stove-enamelled white reflector

Tubular steel folding stand

Supply cable

Galvanised steel base plate

Supply cable

16 amp fitted plug

Fitted wire guard

15.000 long supply cable

110 V tungsten filament lamp

Moulded rubber lampholder

Water- and weatherproof toughened front glass

Pressure die-cast aluminium alloy body

Stove-enamelled white reflector behind 110 V tungsten filament lamp

Supply cable

Galvanised steel base plate

Figure 1.3.6 Local lighting: suitable lamps and fittings

1.4

WINTER BUILDING

Approximately one-fifth of the working force of the United Kingdom is employed either directly or indirectly by the building industry. Therefore any fluctuation in productivity will affect a large number of people. The general loss in output in the construction industry in normal circumstances during the winter period is about 10%, which can result in the underemployment of staff, plant and fixed assets together with the loss of good trading relations with suppliers due to goods ordered not being called forward for delivery. A severe winter can treble the typical loss of output quoted above, resulting in loss of cash flow to the main contractor and subcontractors, plus reduced pay to employees. There will also be lower profits, or a profit loss for contractors, and many skilled operatives may leave the industry in search of more secure occupations. The building owner also suffers by the delay in completing the building, which could necessitate extending the borrowing period for the capital to finance the project or the loss of a prospective tenant or buyer.

The major factor in determining the progress of works on site during the winter period is the weather. Guidance as to the likely winter weather conditions for various areas of the United Kingdom can be obtained from maps, charts and statistical data issued by the Meteorological Office, and this is useful for long-term planning, whereas in the short term reliance is placed upon local knowledge, daily forecasts and the short-term monthly weather forecasts. The uncertain nature of the climate in the United Kingdom often discourages building contractors from investing in plant and equipment for winter-building techniques and protective measures that may prove to be unnecessary. Contractors must therefore assess the total cost of possible delays against the capital outlay required for plant and equipment to enable them to maintain full or near-full production during the winter period.

▪▪▪ EFFECTS OF WEATHER

Weather conditions that can have a delaying effect on building activities are rain, high winds, low temperatures, snow and poor daylight levels; the worst effects obviously occur when more than one of the above conditions occur simultaneously.

Rain

Rain affects site access and movement, which in turn increases site hazards, particularly those associated with excavations and earth-moving works. It also causes discomfort to operatives, thus reducing their productivity rate. Delays with most external operations, such as bricklaying and concreting, are usually experienced, particularly during periods of heavy rainfall. Damage can be caused to unprotected materials stored on site and in many cases to newly fixed materials or finished surfaces. The higher moisture content of the atmosphere will also delay the drying out of buildings. If high winds and rain occur together, rain penetration and site hazards are considerably increased.

High winds

Apart from the discomfort felt by operatives, high winds can also make activities such as frame erection and the fixing of sheet cladding very hazardous. They can also limit the operations that can be carried out by certain items of plant such as tower cranes and suspended cradles. Positive and negative wind pressures can also cause damage to partially fixed claddings, incomplete structures and materials stored on site.

Low temperatures

As the air temperature approaches freezing point many site activities are slowed down. These include excavating, bricklaying, concreting, plastering and painting, until they cease altogether at subzero temperatures; also, mechanical plant can be difficult to start, and stockpiles of materials can become frozen and difficult to move. General movement and circulation around the site becomes hazardous, creating with the low temperatures general discomfort and danger for site personnel. When high winds are experienced with low temperatures they will aggravate the above-mentioned effects.

Snow

This is one of the most variable factors in British weather, ranging from an average of five days a year on which snow falls on low ground in the extreme south-west to 35 days in the north-east of Scotland. Snow will impair the movement of labour, plant and materials, as well as create uncomfortable working conditions. Externally stored materials will become covered with a layer of snow, making identification difficult in some cases. This blanket of snow will also add to the load to be carried by all horizontal surfaces. High winds encountered with falling snow can cause drifting, which could increase the site hazards and personal discomfort, and decrease general movement around the site.

It should be appreciated that the adverse conditions described above could have an adverse effect on site productivity, even if they are not present on the actual site, by delaying the movement of materials to the site from suppliers outside the immediate vicinity.

■ ■ ■ WINTER-BUILDING TECHNIQUES

The major aim of any winter-building method or technique is to maintain an acceptable rate of productivity. Inclement weather conditions can have a very quick reaction on the transportation aspect of site operations, movement of vehicles around the site and, indeed, off the site, which will be impaired or even brought to a complete standstill unless firm access roads or routes are provided, maintained and kept free of snow. These access roads should extend right up to the discharge points to avoid the need for unnecessary double handling of materials. If the access roads and hardstandings form part of the contract and are suitable, these could be constructed at an early stage in the contract before the winter period. If the permanent road system is not suitable in layout for contractual purposes, temporary roads of bulk timbers, timber or concrete sleepers, compacted hardcore or proprietary metal tracks could be laid.

Frozen ground can present problems with all excavating activities. Most excavating plant can operate in frozen ground up to a depth of 300 mm, but at a reduced rate of output: this is particularly true when using machines having a small bucket capacity. Prevention is always better than cure. Therefore if frost is anticipated it is a wise precaution to protect the areas to be excavated by covering with fibre mats enclosed in a polythene envelope, insulating quilts of mineral wool or glass fibre incorporating electric heating elements for severe conditions. Similar precautions can be taken in the case of newly excavated areas to prevent them freezing and giving rise to frost-heave conditions. If it is necessary to defrost ground to enable excavating works to be carried out, this can usually be achieved by using flame throwers, steam jet pipes or coils. Care must be taken to ensure that defrosting is complete and that precautions are taken to avoid subsequent re-freezing.

Water supplies should be laid below ground at such a depth as to avoid the possibility of freezing. The actual depth will vary according to the locality of the site, with a minimum depth of 750 mm for any area. If the water supply is temporary and above ground the pipes should be well insulated and laid to falls so that they can be drained at the end of the day through a drain cock incorporated into the service.

Electrical supplies can fail in adverse weather conditions because the vulnerable parts such as contacts become affected by moisture, frost or ice. These components should be fully protected in the manner advised by their manufacturer.

Items of plant that are normally kept uncovered on site – such as mixers, dumper trucks, bulldozers and generators – should be protected as recommended by the manufacturer to avoid morning starting problems. These precautions will include selecting and using the correct grades of oils, lubricants and antifreeze, as well as covering engines and electrical systems, draining radiators where necessary, and parking wheeled or tracked vehicles on timber runners to prevent them freezing to the ground.

Under the Construction (Health, Safety and Welfare) Regulations, operatives will also need protection from adverse winter conditions if an acceptable level of production is to be maintained. Materials too will require shelter. Such protection can be of one or more of the following types:

- temporary shelters;
- framed enclosures;
- air-supported structures;
- protective clothing.

TEMPORARY SHELTERS

These are the cheapest and simplest method of giving protection to the working areas. They consist of a screen of reinforced or unreinforced polythene sheeting of suitable gauge fixed to the outside of the scaffold to form a windbreak. The sheeting must be attached firmly to the scaffold standards so that it does not flap or tear (a suitable method is shown in Fig. 1.4.2). To gain the maximum amount of use and reuse out of the sheeting used to form the windbreaks the edges should be hemmed or reinforced with a suitable adhesive tape with metal eyelets incorporated at the tying positions. Eyelets can be made on site using a special kit; alternatively the sheet can be supplied with prepared edges.

FRAMED ENCLOSURES

These consist of a purpose-made frame having a curved roof clad with a corrugated material and polythene-sheeted sides; alternatively a frame enclosing the whole of the proposed structure can be constructed from standard tubular scaffolding components (see Fig. 1.4.1). Framed enclosures should be clad from the windward end to avoid a build-up of pressure inside the enclosure. It is also advantageous to load the working platform before sheeting in the sides of the enclosure, because loading at a later stage is more difficult. The frame must be rigid enough to take the extra loading of the coverings and any imposed loading such as wind loadings. Anchorage to the ground of the entire framing is also of great importance, and this can be achieved by using a screw-type ground anchor as shown in Fig. 1.4.2.

AIR-SUPPORTED STRUCTURES

Sometimes called **air domes**, these are being increasingly used on building sites as a protective enclosure for works in progress and for covered material storage areas. Two forms are available:

- internally supported dome;
- air rib dome.

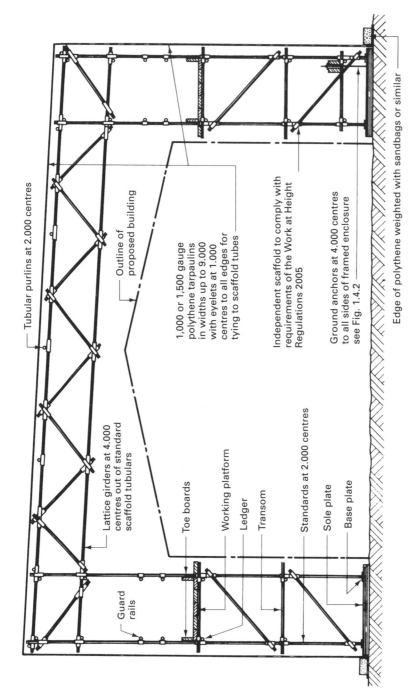

Figure 1.4.1   Protective screens and enclosures: typical framed enclosure

Tubular purlins at 2.000 centres

Lattice girders at 4.000 centres out of standard scaffold tubulars

Outline of proposed building

1,000 or 1,500 gauge polythene tarpaulins in widths up to 9.000 with eyelets at 1.000 centres to all edges for tying to scaffold tubes

Independent scaffold to comply with requirements of the Work at Height Regulations 2005

Ground anchors at 4.000 centres to all sides of framed enclosure see Fig. 1.4.2

Edge of polythene weighted with sandbags or similar

Guard rails

Toe boards

Working platform

Ledger

Transom

Standards at 2.000 centres

Sole plate

Base plate

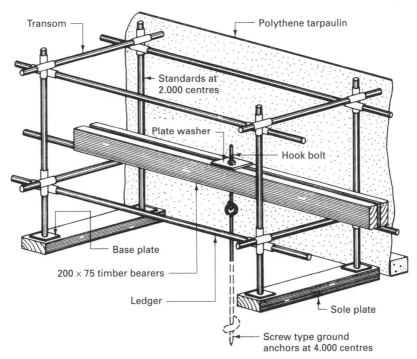

Transom

Polythene tarpaulin

Standards at 2.000 centres

Plate washer

Hook bolt

Base plate

200 × 75 timber bearers

Ledger

Sole plate

Screw type ground anchors at 4.000 centres

Ground anchor for framed enclosure

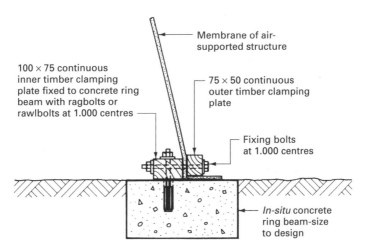

Membrane of air-supported structure

100 × 75 continuous inner timber clamping plate fixed to concrete ring beam with ragbolts or rawlbolts at 1.000 centres

75 × 50 continuous outer timber clamping plate

Fixing bolts at 1.000 centres

In-situ concrete ring beam-size to design

Ring beam anchorage for air-supported structure

Figure 1.4.2   Protective screens and enclosures: anchorages

The internally supported dome is held up by internal air pressure acting against the covering membrane of some form of PVC-coated nylon or rayon with access through an air lock or air curtain door, whereas in the air rib dome the membrane is supported by air-inflated ribs to which the covering membrane is attached. The usual shape for an air-supported structure is semi-cylindrical with rounded ends through which daylight can be introduced by having a translucent membrane over the crown of the structure. The advantages of air-supported structures are that they are low cost, light in weight and reusable; only a small amount of labour is required to erect and dismantle them, and with only a low internal pressure (approximately 150 N/m^2 above atmospheric pressure) workers inside are not affected. Disadvantages are the need to have at least one fan in continuous operation to maintain the internal air pressure, the provision of an air lock or curtain entrance, which will impede or restrict the general site circulation, and the height limitation, which is usually in the region of 45% of the overall span of the structure. The anchorage and sealing of the air-supported structure are also very important, and this can be achieved by using a concrete ring beam as shown in Fig. 1.4.2.

OTHER PROTECTIVE MEASURES

Personal protective clothing/equipment (PPE) is another very important aspect of winter-building techniques. It must be provided to satisfy the health and safety objectives of the Construction (Health, Safety and Welfare) Regulations. Statistics show that the incidence of ailments usually associated with inclement weather, such as rheumatism and bronchitis, is some 20% higher in the building industry when compared with British industry as a whole. Labour costs on building sites are such that maximum utilisation of all labour resources in all weathers must be the ultimate aim, and therefore capital expended in providing protective clothing can be a worthwhile investment. The ideal protective clothing consists of a suit in the form of jacket and trousers made from a lightweight polyurethane-proofed nylon with a removable jacket liner for extreme conditions. A strong lightweight safety helmet and a strong pair of steel-toe-capped rubber boots with a good-grip tread would complete the protective clothing outfit.

Heating equipment may be needed on a building site to offset the effects of cold weather on building operations, particularly where wet trades are involved when the materials used could become frozen at an early stage in the curing process, producing an unacceptable component or member. The main types of heater in use are convectors, radiant heaters and forced-air heaters using electricity, gas or oil as fuel. Electricity is usually too expensive in the context of building operations except for the operation of pumps and fans included in many of the heating appliances using the main other fuels. Gas and oil are therefore the most used fuels, and these must be considered in the context of capital and running costs.

Convector heaters using propane or electricity with an output of 3 to 7 kW are suitable only for small volumes such as site huts and drying rooms. Gas-burning appliances will require a suitable flue to convey the combustion fumes to the outside air.

Radiant heaters using propane are an alternative to the convector heater: they have the advantage of being mobile, making them suitable for site hut heating or for local heating where personnel are working.

Forced-air heaters have high outputs of 30 to 70 kW: they are efficient, versatile and mobile. Two forms of heater are available: the direct forced-hot-air heater, which discharges water vapour into the space being heated, and the vented forced-air heater, where the combustion gases are discharged through a flue to the outside air. Both forms of heater require a fan operated by electricity, propane or a petrol engine. The usual fuels for the heater are propane, paraffin or light fuel oil, which produces sufficient heat for general area heating of buildings under construction and for providing heat to several parts of a building using plastic ducting to distribute the hot air. When using a ducted distribution system it is essential that the air velocity at the nozzle is sufficient to distribute the hot air throughout the system.

The natural drying out of buildings after the wet trades is considerably slower in the winter months, and therefore to meet production targets this drying-out process needs the help of suitable plant. The heaters described above are often used for this purpose, but they are generally very inefficient. Effective drying out or the removal of excess moisture from a building requires the use of a dehumidifier and a vented space heater. Direct-fired heaters are generally unsuitable, because moisture is contained in the combustion gases discharged into the space being dried out. The dehumidifier is used in conjunction with a heater to eliminate the need to heat the cold air that would replace the warm moist air discharged to the outside if only a vented heater was employed. Two basic types of dehumidifier are available:

- **Refrigeration types,** where the air is drawn over refrigeration coils that lower the dewpoint, causing the excess moisture in the air to condense and discharge into a container. According to the model used the extraction rate can be from 10 to 150 litres per day.
- **Chemical desiccant types,** where the moist air is drawn over trays or drums containing hygroscopic chemicals that attract the moisture from the air; the extracted moisture is transferred to a bucket or other suitable container. The advantages of these types are low capital and running costs and their overall efficiency, with an extraction rate between 45 and 110 litres per day according to the model being used. The only real disadvantage is the need periodically to replenish the chemicals in the dehumidifier trays or drums.

The use of steam generators for defrosting plant and materials or heating water is also a possible winter-building technique. The steam generator can be connected to steam coils, which can be inserted into stockpiles of materials; alternatively the steam generator can be connected to a hand-held lance.

The use of artificial lighting on sites to aid production and increase safety on building sites during the winter period when natural daylighting is often inadequate has been fully covered in the previous chapter.

■■■ WORKING WITH TRADITIONAL MATERIALS

CONCRETE

This can be damaged by rain, sleet, snow, freezing temperatures and cooling winds before it has matured, by slowing down the rate at which concrete hardens, or by increasing to an unacceptable level the rate at which the water evaporates. The following precautions should ensure that no detrimental effects occur when mixing and placing in winter conditions:

- Storage of cement under cover and in perfectly dry conditions to prevent air setting.
- Defrosting of aggregates.
- Minimum of delay between mixing and placing.
- Minimum temperature of concrete when placed ideally should be 10 °C.
- Newly placed concrete to be kept at a temperature of more than 5 °C for at least three days, because the rate at which concrete sets below this temperature is almost negligible. It may be necessary to employ the use of covers with heating elements to maintain this minimum temperature, or a polythene tent to provide about 500 mm air space underneath. Warm air can be blown under the tent to create an insulative layer.
- If special additives such as calcium chloride or calcium formate are applied to mass concrete, the manufacturer's specifications should be strictly observed. Calcium chloride should not be used in concrete mixes containing steel-reinforcing bars, as it will encourage corrosion. It may also contribute to the shrinkage of concrete.
- Do not use antifreeze solutions; these are incompatible with concrete mixes.
- Follow the recommendations of BS 8110: *Structural use of concrete*, for guidance on the use of formwork and also the chloride ion content in different types of concrete.
- Air-entraining mixtures (see BS EN's 480 and 934: *Admixtures for concrete and grout*) are derived from wood resins, fatty acids and various other chemicals to create minute air bubbles that adhere to the cement particles. Resistance to frost is achieved, as the bubbles act as expansion chambers for ice formations. Concrete design must allow for the effect of porosity, which can be as much as 15% strength loss.
- If special cements are used, the manufacturer's instructions should be strictly adhered to. Rapid-hardening Portland cement (RHPC) sets at the same rate as ordinary Portland cement (OPC), but because of the extra fineness of particles has greater surface area for the hydration reaction. This provides a high rate of heat evolution to help prevent frost damage and effect faster strength gain.

BRICKWORK

Brickwork can be affected by the same climatic conditions as given above for concrete, resulting in damage to the mortar joints and possibly spalling of the bricks. The following precautions should be taken when bricklaying under winter weather conditions:

- Bricks to be kept reasonably dry.
- Use a 1:1:6 mortar or a 1:5–6 cement: sand mortar with an air-entraining plasticiser to form microscopic air spaces, which can act as expansion chambers for the minute ice particles that may form at low temperatures.
- Cover up the brickwork upon completion with a protective insulating quilt or similar covering for at least three days.
- During periods of heavy frost use a heated mortar by mixing with heated water (approximately 50 °C) to form a mortar with a temperature of between 15 and 25 °C.
- Do not use additives such as calcium chloride, because part of the solution could be absorbed by the absorbent bricks, resulting in damage to the finished work. It can also corrode metal wall ties.

PLASTERING

Precautions that can be taken when plastering internally during cold periods are simple: they consist of closing windows and doors or covering the openings with polythene or similar sheeting, maintaining an internal temperature above freezing, and if necessary using heated (maximum 50 °C) mixing water. Lightweight plasters and aggregates are less susceptible to damage by frost than ordinary gypsum plasters and should be specified whenever possible. External plastering or rendering should be carried out only in dry weather when the air temperature is above freezing.

TIMBER

If the correct storage procedure of a covered but ventilated rack has been followed and the moisture content of the fixed joinery is maintained at the correct level few or no problems should be encountered when using this material during winter weather conditions.

If a contractor's output falls below the planned target figures then fixed charges for overheads, site on-costs and plant costs are being wasted to a proportionate extent. Profits will also be less because of the loss in turnover: therefore it is usually worthwhile expending an equivalent amount on winter-building precautions and techniques to restore full production. In general terms, a decrease of about 10% in productivity is experienced by builders during the winter period, and by allocating 10% of fixed charges and profit to winter-building techniques a builder can bring production back to normal. Cost analysis will show that the more exceptional the inclement weather becomes the greater is the financial reward for expended money on winter-building techniques and precautions.

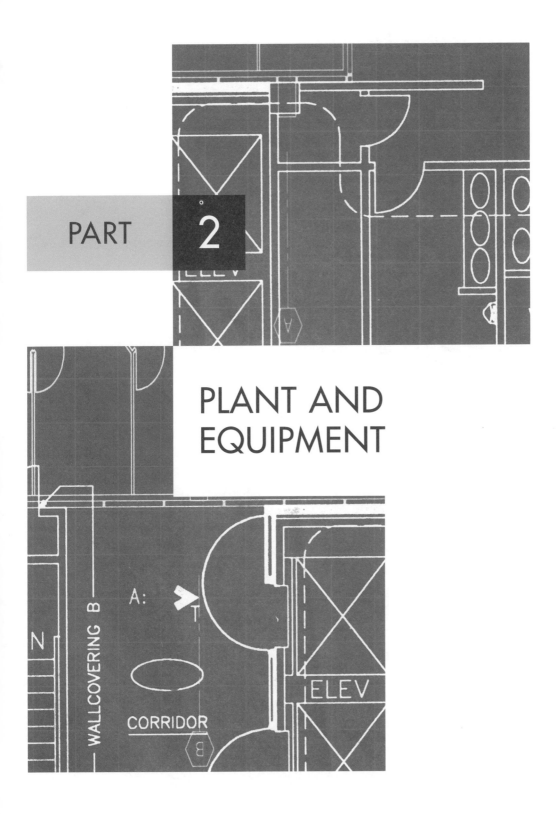

PART 2

PLANT AND
EQUIPMENT

BUILDERS' PLANT: GENERAL CONSIDERATIONS

The general aim of building is to produce a structure at reasonable cost and of sound workmanship within an acceptable time period. To achieve this time period and in many cases to overcome a shortage of suitable manpower the mechanisation of many building activities must be considered. The range of plant now available to building contractors is very extensive, ranging from simple hand tools to very expensive equipment that undertakes tasks beyond the capabilities of manual labour. In a text of this nature it is possible only to consider the general classes of plant and their uses; for a full analysis of the many variations, with the different classifications, readers are advised to consult the many textbooks and catalogues devoted entirely to contractors' plant.

The main reasons for electing to use items of plant are as follows:

- Increase rate of output.
- Reduce overall building costs.
- Carry out activities that cannot be done manually or do them more economically.
- Eliminate heavy manual work, thus reducing fatigue and increasing the productivity of manual workers.
- Maintain a planned rate of production where there is a shortage of either skilled or unskilled operatives.
- Maintain the high standards often required by present-day designs and specifications, especially when concerned with structural engineering works.

It must not be assumed that the introduction of plant to a contract will always reduce costs. This may well be true with large contracts, but when carrying out small contracts such as a traditionally built one-off house it is usually cheaper to carry out the constructional operations by traditional manual methods: the main exception to this is the mixing of concrete and cement mortar using a small mobile batch mixer.

The type of plant to be considered for selection will depend upon the tasks involved, the time element, and the staff available. The person who selects the plant must be competent, the plant operator must be a trained person to obtain maximum efficiency, the manufacturer's recommended maintenance schedule for the plant must be followed and, above all, the site layout and organisation must be planned with a knowledge of the capabilities and requirements of the plant.

Having taken the decision to use plant and equipment, the contractor now has the choice of buying, hiring or combining the two. The advantages of buying plant are as follows:

- Plant is available when required.
- The cost of idle time caused by inclement weather, work being behind planned programme or delay in deliveries of materials will generally be less on owned plant than on hired plant.
- The builder can apportion the plant costs to the various contracts using the plant, by his/her own chosen method.
- There is some resale value.

The advantages of hiring plant are as follows:

- Plant can be hired as required and for short periods.
- Hire firms are responsible for maintenance, repairs and replacements.
- The contractor is not left with unused expensive plant items after completion of the contract.
- Hire rates can include operator, fuel and oil.

Quotations and conditions for hiring plant can be obtained from a plant-hiring company. A full list of companies belonging to the Construction Plant-Hire Association is contained in the *CPA Annual Handbook*. This Association was formed in 1941 to represent the plant-hire industry in the United Kingdom to negotiate general terms and conditions of hiring plant, to give advice, and to promote a high standard of efficiency in the services given by its members. Against plant-hire rates and conditions a contractor will have to compare the cost of buying and owning a similar piece of plant. A comparison of costs can be calculated by the simple straight-line method or by treating the plant as an investment and charging interest on the capital outlay.

Straight-line method

Capital cost of plant	£45,000.00
Expected useful life	5 years
Yearly working, say 75% of total year's working hours	$= 50 \text{ weeks} \times 40 \text{ hours} \times \dfrac{75}{100}$
	$= 1,500 \text{ hours per year}$

Assuming a resale value after 5 years of £7,500.00

Annual depreciation $= \dfrac{45,000 - 7,500}{5} = £7,500$

Therefore hourly depreciation $= \dfrac{7,500.00}{1,500.00} = £5.00$

Net cost per hour	=	5.00
Add 2% for insurance, etc.	=	0.10
Add 10% for maintenance	=	0.50
		£5.60 per hour

To the above costs must be added the running costs, which would include fuel, operator's wages and overheads.

Interest on capital outlay method

Capital cost of plant	£45,000.00
Compound interest on capital @ 6% for 5 years	15,220.00
	£60,220.00
Deduct resale value	7,500.00
	£52,720.00
Add 2% of capital cost for insurance, etc.	900.00
Add 10% of capital cost for maintenance	4,500.00
	£58,120.00

Therefore cost per hour $= \dfrac{£58,120}{5 \times 1,500} = £7.75$

To the above hourly rate must be added the running costs as given for the straight-line method. This second method gives a more accurate figure for the actual cost of owning an item of plant than the straight-line method, and for this reason it is widely used.

Vehicles such as lorries and vans are usually costed on a straight annual depreciation of the yearly book value thus:

Cost of vehicle	£25,000.00
Estimated useful life, 5 years with an annual depreciation of 30%	
Capital cost	£25,000.00
30% depreciation	7,500.00
Value after 1st year	17,500.00
30% depreciation	5,250.00
Value after 2nd year	12,250.00
30% depreciation	3,675.00
Value after 3rd year	8,575.00
30% depreciation	2,572.50
Value after 4th year	6,002.50
30% depreciation	1,800.75
Value at 5th year	£4,201.75

The percentages of insurance and maintenance together with the running costs can now be added on a yearly basis, taking the new book values for each year; alternatively the above additional costs could be averaged over the five-year period.

To be an economic proposition large items of plant need to be employed continuously and not left idle for considerable periods of time. Careful maintenance of all forms of plant is of the utmost importance. This not only increases the working life of a piece of plant, but if a plant failure occurs on site it can cause serious delays and disruptions of the programme, and this in turn can affect the company's future planning. To reduce the risk of plant breakdown a trained and skilled operator should be employed to be responsible for the running, cleaning and daily maintenance of any form of machinery. Time for the machine operator to carry out these tasks must be allowed for in the site programme and the daily work schedules.

On a large contract where a number of machines are to be employed, a full-time skilled mechanic could be engaged to be responsible for running repairs and recommended preventive maintenance. Such tasks would include:

- checking oil levels – daily;
- greasing – daily or after each shift;
- checking engine sump levels – after 100 hours' running time;
- checking gearbox levels – after 1,200 to 1,500 hours' running time;
- checking tyre pressures – daily;
- inspecting chains and ropes – daily.

As soon as a particular item of plant has finished its work on site it should be returned to the company's main plant yard so that it can be re-allocated to another contract. This is equally important for hired plant, which must be returned to the hire company's depot; otherwise excess charges may apply. On its return to the main plant yard or hire depot an item of plant should be inspected and tested so that any necessary repairs, replacements and maintenance can be carried out before it is re-employed on another site. A record of the machine's history should be accurately kept and should accompany the machine wherever it is employed so that the record can be kept up to date.

The soil conditions and modes of access to a site will often influence the choice of plant items that could be considered for a particular task. Congested town sites may severely limit the use of many types of machinery and/or plant. If the proposed structure occupies the whole of the site, component and material deliveries in modules or prefabricated format will eliminate the need and space for fabrication and assembly plant. Concrete-mixing plant will also be impractical: therefore supplies will be delivered ready mixed. Wet sites usually require plant equipped with caterpillar tracks, whereas dry sites are suitable for tracked and wheeled vehicles or power units. On housing sites it is common practice to construct the estate road sublayers at an early stage in the contract to provide firm access routes for mobile plant and hardstanding for static plant such as cement mixers. Sloping sites are usually unsuitable for rail-mounted cranes, but these cranes operating on the perimeter of a building are more versatile than the static cranes. The heights and proximity of adjacent structures or buildings may limit the use of a horizontal jib crane and may even dictate the use of a crane with a luffing jib.

For accurate pricing of a bill of quantities careful consideration of all plant requirements must be undertaken with the preparation of a plant schedule at the pre-tender stage. This should take into account plant types, plant numbers and personnel needed. If the tender is successful a detailed programme should be prepared in liaison with all those to be concerned in the supervision of the contract so that the correct sequence of operations is planned and an economic balance of labour and machines is obtained.

Apart from the factors previously discussed, consideration must also be given to safety and noise emission requirements when selecting items of plant for use on a particular contract. The aspects of safety that must be legally provided are contained in various Acts of Parliament and Statutory Instruments such as the Health and Safety at Work etc. Act 1974, the Construction (Health, Safety and Welfare) Regulations 1996 and the Noise at Work Regulations 1989. These requirements will be considered in the following chapters devoted to various classifications of plant.

Although reaction to noise is basically subjective, excessive noise can damage a person's health and/or hearing; it can also cause disturbance to working and living environments. Under the Health and Safety at Work etc. Act 1974 and its subsidiary regulations provision is made for the protection of workers against noise. If the maximum safe daily noise limits for the unprotected human ear are likely to be exceeded, the remedies available are the issue of suitable ear protectors to the workers, housing the plant in a sound-insulated compartment, or the use of quieter plant or processes.

Local authorities have powers under the Control of Pollution Act 1974 to protect the community against noise. Part III of this Act gives the local authority the power to specify its own requirements as to the limit of construction noise acceptable by serving a notice, which may specify:

- plant or machinery that is, or is not, to be used;
- hours during which the works may be carried out;
- level of noise emitted from the premises in question or from any specific point on these premises, or the level of noise that may be emitted during specified hours.

The local authority is also allowed to make provisions in these notices for any change of circumstance.

A contractor will therefore need to know the local authority requirements as to noise restrictions at the pre-tender stage to enable selection of the right plant and/or processes to be employed. Methods for predicting site noise are given in BS 5228: *Noise and vibration control on construction and open sites*: see Parts 1, 2 and 4.

Part 1 recommends methods of noise control and provides guidance on:

- methods for predicting noise;
- measuring noise;
- minimising the impact noise on people in the vicinity of site work.

Part 2 provides guidance on noise control legislation. Part 4 provides recommendations for noise and vibration control measures specifically for piling work.

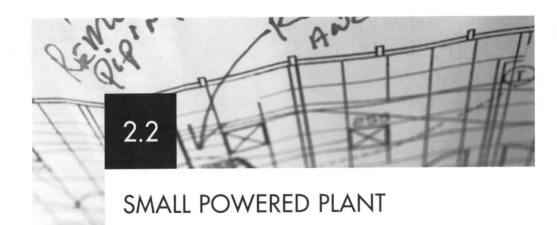

2.2

SMALL POWERED PLANT

A precise definition of builders' small powered plant is not possible, because the term 'small' is relative. For example, it is possible to have small cranes, but these, when compared with a hand-held electric drill, would be considered as large plant items. Generally, therefore, small plant can be considered to be hand-held or operated power tools with their attendant power sources such as a compressor for pneumatic tools; one of the main exceptions is the small static pumps used in conjunction with shallow excavations.

Most hand-held power tools are operated by electricity or compressed air, either to rotate the tool or drive it by percussion. Some of these tools are also designed to act as rotary/percussion tools. Generally, the pneumatic tools are used for the heavier work and have the advantage that they will not burn out if a rotary tool stalls under load. Electrically driven tools are, however, relatively quiet because there is no exhaust noise, and they can be used in confined spaces because there are no exhaust fumes.

▨▦▨ ELECTRIC HAND TOOLS

The most common hand-held tool is the **electric drill** for boring holes into timber, masonry and metals using twist drills. A wide range of twist drill capacities are available with single- or two-speed motors for general-purpose work, the low speed being used for boring into or through timber. Chuck capacities generally available range from 6 to 30 mm for twist drills suitable for metals. Dual-purpose electric drills are very versatile in that the rotary motion can be combined with or converted into a powerful but rapid percussion motion, making the tool suitable for boring into concrete provided that special tungsten-carbide-tipped drills are used (see Fig. 2.2.1).

A variation of the basic electric motor power units is the **electric hammer**, which is used for cutting and chasing work, where the hammer delivers powerful

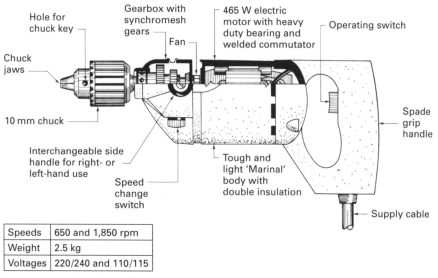

Speeds	650 and 1,850 rpm
Weight	2.5 kg
Voltages	220/240 and 110/115

Percussion version also available giving
18,000 and 31,500 impacts per min.

Figure 2.2.1 Typical electric drill (*courtesy*: Stanley-Bridges)

blows at a slower rate than the percussion drill described. Another variation is the
electric screwdriver, which has an adjustable and sensitive clutch that will operate
only when the screwdriver bit is in contact with the screw head and will slip when a
predetermined tension has been reached when the screw has been driven in. Various
portable electric woodworking tools such as circular saws, jigsaws, sanders, planes
and routers are also available and suitable for site use.

Electric hand-held tools should preferably operate off a reduced voltage supply
of 110 V and should conform to the recommendations of BS 2769 and BS EN
60745: *Hand-held motor-operated electric tools*. All these tools should be earthed
unless they bear the 'two squares' symbol indicating that they are 'all insulated' or
'double insulated' and therefore have their own built-in safety system. Plugs and
couplers should comply with BS EN 60309-2: *Plugs, socket-outlets and couplers for
industrial purposes*, so that they are not interchangeable and cannot be connected
to the wrong voltage supply. Protective guards and any recommended protective
clothing such as goggles and ear protection should be used as instructed by the
manufacturers and as laid down in the Construction (Health, Safety and Welfare)
Regulations 1996, the Personal Protective Equipment at Work Regulations 1992
and the Noise at Work Regulations 1989. Electric power tools must never be
switched off while under load, because this could cause the motor to become
overstrained and burn out. If electrical equipment is being used on a site, the
Electricity at Work Regulations and relevant first-aid placards should be displayed
in a prominent position.

■■■ PNEUMATIC TOOLS

These tools need a supply of compressed air as their power source, and on building sites this is generally in the form of a mobile compressor powered by a diesel, petrol or electric motor, the most common power unit being the diesel engine. Compressors for building works are usually of the piston or reciprocating type, where the air is drawn into a cylinder and compressed by a single stroke of the piston; alternatively two-stage compressors are available where the air is compressed to an intermediate level in a large cylinder before passing to a smaller cylinder where the final pressure is obtained. Air receivers are usually incorporated with and mounted on the chassis to provide a constant source of pressure for the air lines and to minimise losses due to pressure fluctuations of the compressor and/or frictional losses due to the air pulsating through the distribution hoses.

One of the most common pneumatic tools used in building is the **breaker**, which is basically intended for breaking up hard surfaces such as roads. These breakers vary in weight, ranging from 15 to 40 kg with air consumptions of 10 to 20 m^3/min. A variety of breaker points or cutters can be fitted into the end of the breaker tool to tackle different types of surface (see Fig. 2.2.2). **Chipping hammers** are a small lightweight version of the breaker described above, having a low air consumption of 0.5 m^3/min or less. **Backfill tampers** are used to compact the loose spoil returned as backfill in small excavations, and weigh approximately 23 kg with an air consumption of 10 m^3/min. Compressors to supply air to these tools are usually specified by the number of air hoses that can be attached and by the volume of compressed air that can be delivered per minute. Other equipment that can be operated by compressed air includes vibrators for consolidating concrete, small trench sheeting or sheet pile-driving hammers, concrete-spraying equipment, paint sprayers and hand-held rotary tools such as drills, grinders and saws.

Pneumatic tools are generally very noisy, and in view of the legal requirements of the Health and Safety at Work etc. Act 1974 and the Control of Pollution Act 1974 these tools should be fitted with a suitable muffler or silencer. Models are being designed and produced with built-in silencers fitted not only on the tool holder but also on the compressor unit. The mufflers that can be fitted to pneumatic tools such as the breaker have no loss of power effect on the tool provided the correct muffler is used.

The risk of vibration damage to operators of hand-controlled machinery is limited by the requirements of the Control of Vibration at Work Regulations 2005. Several factors combine to define the limits of use. These include the extent of exposure, the degree of exposure, and the operating direction or axis of operation. Formulated values relate to equipment manufacturers' measured vibration levels in metres per second squared (m/s^2) against daily usage time (hours).

■■■ CARTRIDGE HAMMERS

Cartridge hammers or guns are used for the quick fixing together of components or for firing into a surface a pin with a threaded head to act as a bolt fixing. The

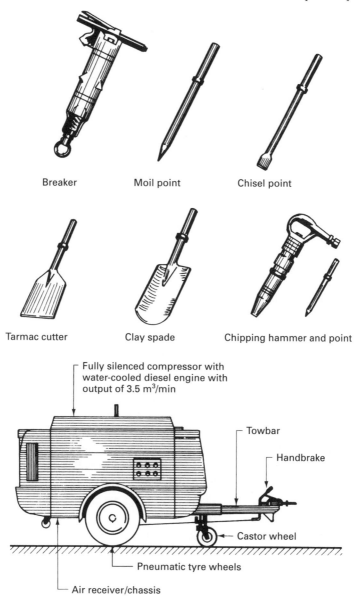

Typical two-tool silenced compressor

Figure 2.2.2 Typical pneumatic tools and compressor

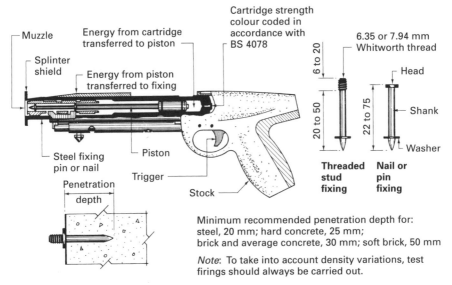

Figure 2.2.3   Typical cartridge hammer (*courtesy*: Douglas Kane Group)

gun actuates a 0.22 cartridge that drives a hardened austempered steel pin into an unprepared surface. Holding power is basically mechanical, caused by the compression of the material penetrated against the shank of the pin, although with concrete the heat generated by the penetrating force of the pin causes the silicates in the concrete to fuse into glass, giving a chemical bond as well as the mechanical holding power. BS 4078: *Powder actuated fixing systems. Code of practice for safe use*, specifies the design, construction, safety and performance requirements for cartridge-operated fixing tools, and this standard also defines two basic types of tool:

- **Direct-acting tools,** in which the driving force on the pin comes directly from the compressed gases from the cartridge. These tools are high–velocity guns with high muzzle energy.
- **Indirect-acting tools,** in which the driving force is transmitted to the pin by means of an intervening piston with limited axial movement. This common form of cartridge-fixing tool is trigger operated, having a relatively low velocity and muzzle energy. A typical example is shown in Fig. 2.2.3.

Hammer-actuated fixing tools working on the piston principle are also available, the tool being hand held and struck with a hammer to fire the cartridge.

The cartridges and pins are designed for use with a particular model of gun, and should in no circumstances be interchanged between models or different makes of gun. Cartridges are produced in seven strengths, ranging from extra low to extra high, and for identification purposes are colour coded in accordance with BS 4078. A low-strength cartridge should be used first for a test firing, gradually increasing the strength until a satisfactory result is obtained. If an over-strength cartridge is used it could cause the pin to pass right through the base material. Fixing pins near to edges can be dangerous because the pin is deflected towards the free edge by following the path of low resistance. Minimum edge distances recommended for concrete are 100 mm using a direct-acting tool and 50 mm using an indirect-acting tool; for fixing into steel the minimum recommended distance is 15 mm using any type of fixing tool.

The safety and maintenance aspects of these fixing tools cannot be overstressed. In particular, consideration must be given to the following points:

- Pins should not be driven into brittle or hard materials such as vitreous-faced bricks, cast iron or marble.
- Pins should not be driven into base materials where there is a danger of the pin passing through.
- Pins should not be fired into existing holes.
- During firing the tool should be held at right angles to the surface with the whole of the splinter guard flush with the surface of the base material. Some tools are designed so that they will not fire if the angle of the axis of the gun to the perpendicular exceeds 7°.
- Operatives under the age of 18 years should not be allowed to use cartridge-operated tools.
- All operators should have a test for colour blindness before being allowed to use these fixing tools.
- No one apart from the operator and an assistant should be in the immediate vicinity of firing to avoid accidents due to ricochet, splintering or re-emergence of pins.
- Operators should receive instructions as to the manufacturer's recommended method of loading, firing and the action to be taken in cases of misfiring.
- Protective items such as goggles and earmuffs should be worn as recommended by the manufacturers.

■ ■ ■ VIBRATORS

After concrete has been placed it should be consolidated either by hand tamping or by using special vibrators. The power for vibrators can be supplied by a small petrol engine, an electric motor or in some cases by compressed air. Three basic forms of vibrator are used in building works: poker vibrators, vibration tampers and clamp vibrators. **Poker vibrators** are immersed into the wet concrete, and because of their high rate of vibration they induce the concrete to consolidate. The effective radius of a poker vibrator is about 1.000 m: therefore the poker or pokers should be inserted at approximately 600 mm centres to achieve an overall consolidation of the concrete.

Vibration tampers are small vibrating engines that are fixed to the top of a tamping board for consolidating concrete pavings and slabs. **Clamp vibrators** are similar devices but are attached to the external sides of formwork to vibrate the whole of the form. Care must be taken when using this type of vibrator to ensure that the formwork has sufficient in-built strength to resist the load of the concrete and to withstand the vibrations.

In concrete members that are thin and heavily reinforced, careful vibration will cause the concrete to follow uniformly around the reinforcement, and this increased fluidity due to vibration will occur with mixes that in normal circumstances would be considered too dry for reinforced concrete. Owing to the greater consolidation achieved by vibration up to 10% more material may be required than with hand-tamped concrete.

Separation of the aggregates can be caused by over-vibrating a mix: therefore vibration should be stopped when the excess water rises to the surface. Vibration of concrete saves time and labour in the placing and consolidating of concrete, but does not always result in a saving in overall costs because of the high formwork costs, extra materials costs and the cost of providing the necessary plant.

■■■ POWER FLOATS

Power floats are hand-operated rotary machines powered by a petrol engine or an electric motor that drives the revolving blades or a revolving disc. The objective of power floats is to produce a smooth, level surface finish to concrete beds and slabs suitable to receive the floor finish without the need for a cement/sand screed. The surface finish that can be obtained is comparable to that achieved by operatives using hand trowels but takes only one-sixth of the time, giving a considerable saving in both time and money. Most power floats can be fitted with either a revolving disc or blade head, and these are generally interchangeable.

The surfacing disc is used for surface planing after the concrete has been vibrated and will erase any transverse tamping line marks left by a vibrator beam as well as filling in any small cavities in the concrete surface. The revolving blades are used after the disc-planing operation to provide the finishing and polishing, which can usually be achieved with two passings. The time at which disc planing can be started is difficult to specify: it depends on factors such as the workability of the concrete, temperature, relative humidity and the weight of the machine to be used. Experience is usually the best judge but, as a guide, if imprints of not more

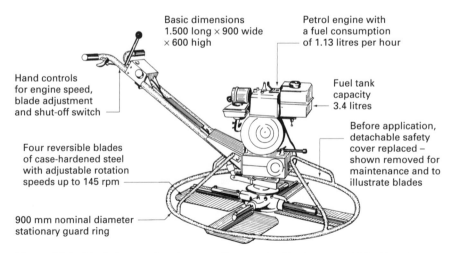

Basic dimensions
1.500 long × 900 wide
× 600 high

Petrol engine with
a fuel consumption
of 1.13 litres per hour

Hand controls
for engine speed,
blade adjustment
and shut-off switch

Fuel tank
capacity
3.4 litres

Before application,
detachable safety
cover replaced –
shown removed for
maintenance and to
illustrate blades

Four reversible blades
of case-hardened steel
with adjustable rotation
speeds up to 145 rpm

900 mm nominal diameter
stationary guard ring

Figure 2.2.4   Typical power float (*courtesy*: Construction Equipment and Machinery)

than 2 to 4 mm deep are made in the concrete when walked upon it is generally suitable for disc planing. If a suitable surface can be produced by a traditional concrete placing method the disc planing operation prior to rotary blade finishing is often omitted. Generally, blade finishing can be commenced once the surface water has evaporated; a typical power float is shown in Fig. 2.2.4. Power floats can also be used for finishing concrete floors with a granolithic or similar topping.

■■■ PUMPS

Pumps are among the most important items of small plant for the building contractor, because they must be reliable in all conditions, easy to maintain, easily transported and efficient. The basic function of a pump is to move liquids vertically or horizontally, or in a combination of the two directions. Before selecting a pump a builder must consider what task the pump is to perform, and this could be any of the following:

- keeping excavations free from water;
- lowering the water table to a reasonable depth;
- moving large quantities of water such as the dewatering of a cofferdam;
- supplying water for general purposes.

Having defined the task the pump is required to carry out, the next step is to choose a suitable pump, taking into account the following factors:

- Volume of water to be moved.
- Rate at which water is to be pumped.
- Height of pumping, which is the vertical distance from the level of the water to the pump and is usually referred to as **suction lift** or head. It should be the shortest distance practicable to obtain economic pumping.
- Height and distance to outfall or discharge point, usually called the **delivery head**.
- Loss due to friction in the length of hose or pipe, which increases as the diameter decreases. In many pumps the suction and delivery hoses are marginally larger in diameter than the pump inlet to reduce these frictional losses.
- Power source for pump, which can be a petrol engine, diesel engine or electric motor. Pumps powered by compressed air are available, but these are unusual on general building contracts.

Pumps in common use for general building works can be classified as follows:

- centrifugal;
- displacement;
- submersible.

CENTRIFUGAL PUMPS

These are classed as normal or self-priming. They consist of a rotary impeller that revolves at high speed, forcing the water to the sides of the impeller chamber and thus creating a vortex that sucks air out of the suction hose. Atmospheric pressure acting on the surface of the water to be pumped causes the water to rise into the pump, initiating the pumping operation. **Normal centrifugal pumps** are easy to maintain, but they require priming with water at the commencement of each pumping operation. Where continuous pumping is required, such as in a basement excavation, a **self-priming pump** should be specified. These pumps have a reserve supply of water in the impeller chamber so that if the pump runs dry the reserve water supply will remain in the chamber to reactivate the pumping sequence if the water level rises in the area being pumped.

DISPLACEMENT PUMPS

These are either reciprocating or diaphragm pumps. **Reciprocating pumps** work by the action of a piston or ram moving within a cylinder. The action of the piston draws water into the cylinder with one stroke and forces it out with the return stroke, resulting in a pulsating delivery. Pumps of this type can have more than one cylinder, forming what is called a duplex (two-cylinder) or triplex (three-cylinder) pump. Some reciprocating pumps draw water into the cylinder in front of the piston and discharge at the rear of the piston: these are called double-acting pumps, as opposed to the single-acting pumps where the water moves in one direction only with the movement of the piston. Although highly efficient and capable of increased capacity with increased engine speed, these pumps have the disadvantage of being unable to handle water containing solids.

Displacement pumps of the **diaphragm** type can, however, handle liquids containing 10–15% of solids, which makes them very popular. They work on the principle of raising and lowering a flexible diaphragm of rubber or rubberised canvas within a cylinder by means of a pump rod connected via a rocker bar to an engine crank. The upward movement of the diaphragm causes water to be sucked into the cylinder through a valve; the downward movement of the diaphragm closes the inlet valve and forces the water out through another valve into the delivery hose. Diaphragm or lift and force pumps are available with two cylinders and two diaphragms, giving greater output and efficiency. Typical pump examples are shown in Fig. 2.2.5.

SUBMERSIBLE PUMPS

These are used for extracting water from deep wells and sumps (see Fig. 3.1.6), and are suspended in the water to be pumped. The power source is usually an electric motor to drive a centrifugal unit, which is housed in a casing with an annular space to allow the water to rise upwards into the delivery pipe or rising main. Alternatively an electric submersible pump with a diaphragm arrangement can be used where large quantities of water are not involved.

■■■ ROLLERS

Rollers are designed to consolidate filling materials and to compact surface finishes such as tarmacadam for paths and pavings. The roller equipment used by building contractors is basically a smaller version of the large rollers used by civil engineering contractors for roadworks. Rollers generally rely upon deadweight to carry out

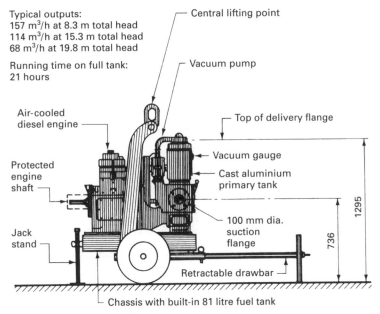

Typical outputs:
157 m³/h at 8.3 m total head
114 m³/h at 15.3 m total head
68 m³/h at 19.8 m total head

Running time on full tank:
21 hours

Central lifting point

Vacuum pump

Air-cooled
diesel engine

Top of delivery flange

Vacuum gauge

Cast aluminium
primary tank

Protected
engine
shaft

100 mm dia.
suction
flange

Jack
stand

1295

736

Retractable drawbar

Chassis with built-in 81 litre fuel tank

Typical self-priming centrifugal pump

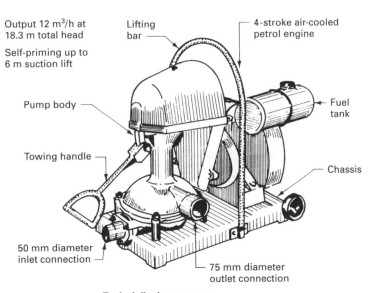

Output 12 m³/h at
18.3 m total head

Self-priming up to
6 m suction lift

Lifting
bar

4-stroke air-cooled
petrol engine

Pump body

Fuel
tank

Towing handle

Chassis

50 mm diameter
inlet connection

75 mm diameter
outlet connection

Typical diaphragm pump

Figure 2.2.5 Typical pumps (*courtesy*: Sykes Pumps and William R. Selwood)

Overall length 4.500, overall width 1.840, overall height 2.680, turning circle 5.500, overlap of rolls 100 mm giving total rolling width of 1.600

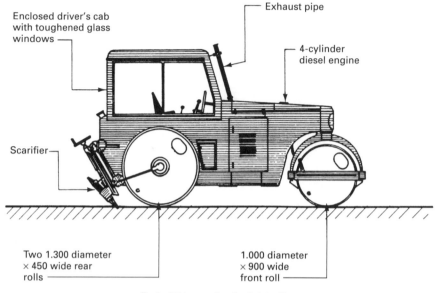

Exhaust pipe

Enclosed driver's cab with toughened glass windows

4-cylinder diesel engine

Scarifier

Two 1.300 diameter × 450 wide rear rolls

1.000 diameter × 900 wide front roll

Typical 6 tonne deadweight roller

Overall length 3.376, overall width 1.092, overall height 1.092, width of rollers 890, deadweight 1,250 kg

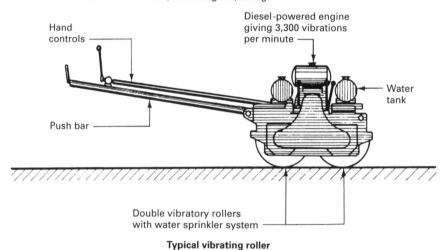

Diesel-powered engine giving 3,300 vibrations per minute

Hand controls

Water tank

Push bar

Double vibratory rollers with water sprinkler system

Typical vibrating roller

Figure 2.2.6   Typical rollers (*courtesy*: Marshall Sons and Duomat)

the consolidating operation or vibration as in the case of lightweight rollers. **Deadweight rollers** are usually diesel powered and driven by a seated operator within a cab. These machines can be obtained with weights ranging from 1 to 16 tonnes, which is distributed to the ground through two large-diameter rear wheels and a wider but small front steering wheel. Many of these rollers carry water tanks to add to the dead load and to supply small sparge or sprinkler pipes fixed over the wheels to dampen the surfaces, thus preventing the adhesion of tar or similar material when being rolled. These rollers are also available fitted with a scarifier to the rear of the vehicle for ripping up the surfaces of beds or roads (see Fig. 2.2.6).

Vibrating rollers, which depend mainly upon the vibrations produced by the petrol- or diesel-powered engine, can be hand guided or towed and are available with weights ranging from 500 kg to 5 tonnes (see Fig. 2.2.6). These machines will give the same degree of consolidation and compaction as their heavier deadweight counterparts, but being lighter and smaller they can be manoeuvred into buildings for consolidating small areas of hardcore or similar bed material. Single or double rollers are available with or without water sprinkler attachments, and with vibrations within the region of 3,000 vibrations per minute. Vibrating rollers are particularly effective for the compaction and consolidation of granular soils.

2.3

EARTH-MOVING AND EXCAVATION PLANT

The selection, management and maintenance of builders' plant are particularly important when considered in the context of earth-moving and excavation plant. Before deciding to use any form of plant for these activities the site conditions and volume of work entailed must be such that it will be an economic venture. The difference between plant that is classified as earth-moving equipment and excavating machines is very slight, because a piece of plant that is designed primarily to excavate will also be capable of moving the spoil to an attendant transporting vehicle, and, likewise, machines basically designed to move loose earth will also be capable of carrying out excavation works to some degree.

To browse through the catalogues of plant manufacturers and hirers to try and select a particular piece of plant is a bewildering exercise because of the wide variety of choice available for all classes of plant. Final choice is usually based upon experience, familiarity with a particular manufacturer's machines, availability or personal preference. There are many excellent works of reference devoted entirely to the analysis of the various machines to aid the would-be buyer or hirer: therefore in a text of this nature it is necessary only to consider the general classes of plant, pointing out their intended uses and amplifying this with typical examples of the various types without claiming that the example chosen is the best of its type but only representative.

■■■ BULLDOZERS AND ANGLEDOZERS

These machines are primarily a high-powered tractor with caterpillar or crawler tracks and fitted with a mould board or blade at the front for stripping and oversite excavations up to a depth of 400 mm (depending upon machine specification) by pushing the loosened material ahead of the machine. For backfilling operations the angledozer with its mould board set at an angle, in plan, to the machine's centreline can be used. Most mould boards can be set at an angle in either the vertical or the horizontal plane to act as an angledozer, and on some models the leading edge of the mould board can be fitted with teeth for excavating in hard ground. These machines can be very large, with mould boards of 1.200 to 4.000 m in width × 600 mm to 1.200 m in height and a depth of cut up to 400 mm. Most bulldozers and angledozers are mounted on crawler tracks, although small bulldozers with a wheeled base are available. The control of the mould board is hydraulic, as shown in Fig. 2.3.1. In common with other tracked machines one of the disadvantages of this arrangement is the need for a special transporting vehicle such as a low loader to move the equipment between sites.

Before any earth-moving work is started a drawing should be produced indicating the areas and volumes of cut and fill required to enable a programme to be prepared to reduce machine movements to a minimum. When large quantities of earth have to be moved on a cut and fill basis to form a predetermined level or gradient it is good practice to draw up a mass haul diagram indicating the volumes of earth to be moved, the direction of movement, and the need to import more spoil or alternatively remove the surplus spoil from site (see Fig. 2.3.2 for a typical example).

▨▨▨ SCRAPERS

This piece of plant consists of a power unit and a scraper bowl, and is used to excavate and transport soil where surface stripping, site levelling and cut and fill activities are planned. They are particularly appropriate where large volumes are encountered over a wide area, typical of civil engineering projects such as airfields and highways. These machines are capable of producing a very smooth and accurate formation level, and come in three basic types:

■ crawler-drawn scraper;
■ two-axle scraper;
■ three-axle scraper.

The design and basic operation of the scraper bowl is similar in all three types. It consists of a shaped bowl with a cutting edge that can be lowered to cut the top surface of the soil up to a depth of 300 mm. As the bowl moves forward the loosened earth is forced into the container, and when full the cutting edge is raised to seal the bowl. To ensure that a full load is obtained, many contractors use a bulldozer to act as a pusher over the last few metres of scrape. The bowl is emptied by raising the front apron and ejecting the collected spoil or, on some models, by raising the rear portion and spreading the collected spoil as the machine moves forwards.

The **crawler-drawn scraper** consists of a four-wheeled scraper bowl towed behind a crawler power unit. The speed of operation is governed by the speed of the towing vehicle, which does not normally exceed 8 km/h when hauling and 3 km/h when scraping. For this reason this type of scraper should be used only on small hauls of up to 300.000 m. The **two-axle** scraper which has a two-wheeled bowl pulled by a two-wheeled power unit, has advantages over its four-wheeled power unit or three-axle counterpart in that it is more manoeuvrable, offers less rolling

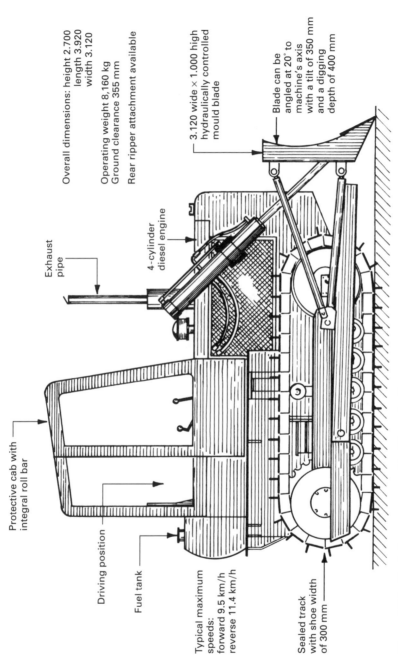

Overall dimensions: height 2.700
length 3.920
width 3.120

Operating weight 8,160 kg
Ground clearance 355 mm

Rear ripper attachment available

3.120 wide × 1.000 high
hydraulically controlled
mould blade

Blade can be
angled at 20° to
machine's axis
with a tilt of 350 mm
and a digging
depth of 400 mm

Exhaust
pipe

4-cylinder
diesel engine

Protective cab with
integral roll bar

Driving position

Fuel tank

Typical maximum
speeds:
forward 9.5 km/h
reverse 11.4 km/h

Sealed track
with shoe width
of 300 mm

Figure 2.3.1   Typical tractor-powered bulldozer details

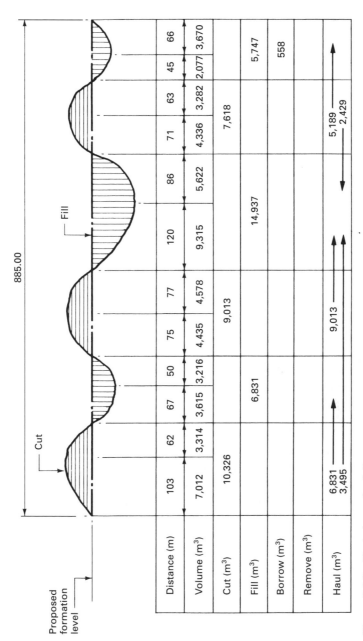

Distance (m)	103	62	67	50	75	77	120	86	71	63	45	66
Volume (m³)	7,012	3,314	3,615	3,216	4,435	4,578	9,315	5,622	4,336	3,282	2,077	3,670
Cut (m³)	10,326				9,013				7,618			
Fill (m³)			6,831				14,937				5,747	
Borrow (m³)											558	
Remove (m³)												
Haul (m³)	6,831 ⟶ 3,495				9,013 ⟶				5,189 ⟶ 2,429			

Figure 2.3.2 Typical mass haul diagram

resistance, and has better traction because the engine is mounted closer to the driving wheels. The **three-axle scraper**, however, can use its top speed more frequently, is generally easier to control, and the power unit can be used for other activities, which is not possible with most two-axle scraper power units. Typical examples are shown in Fig. 2.3.3. Scraper bowl heaped capacities of the machines described above range from 5 to 50 m^3.

To achieve maximum output and efficiency of scrapers the following should be considered:

- When working in hard ground the surface should be pre-broken by a ripper or scarifier, and assistance in cutting should be given by a pushing vehicle. Usually one bulldozer acting as a pusher can assist three scrapers if the cycle of scrape, haul, deposit and return is correctly balanced.
- Where possible, the cutting operation should take place downhill to take full advantage of the weight of the unit.
- Haul roads should be kept smooth to enable the machine to obtain maximum speeds.
- Recommended tyre pressures should be maintained, otherwise extra resistance to forward movement will be encountered.

▨■■ GRADERS

These are similar machines to bulldozers in that they have an adjustable mould blade either fitted at the front of the machine or slung under the centre of the machine's body. They are used for finishing to fine limits large areas of ground that have been scraped or bulldozed to the required formation level. These machines can be used only to grade the surface because their low motive power is generally insufficient to enable them to be used for oversite excavation work.

▨■■ TRACTOR SHOVEL

This machine, which is sometimes called a **loading shovel**, is basically a power unit in the form of a wheeled or tracked tractor with a hydraulically controlled bucket mounted in front of the vehicle. It is one of the most versatile pieces of plant available to the building contractor. Its primary function is to scoop up loose material in the bucket, raise the loaded spoil and manoeuvre into a position to discharge its load into an attendant lorry or dumper. The tractor shovel is driven towards the spoil heap with its bucket lowered almost to ground level, and uses its own momentum to force the bucket to bite into the spoil heap, thus filling the scoop or bucket.

Instead of the straight cutting edge to the lower lip of the bucket the shovel can be fitted with excavating teeth, enabling the machine to carry out excavating activities such as stripping topsoil or reduce-level digging in loose soils. Another popular version of the tractor shovel is fitted with a 4-in-1 bucket, which enables the machine to perform the functions of bulldozing, excavating and loading (see Fig. 2.3.4). Other alternatives to the conventional front-discharging machine are

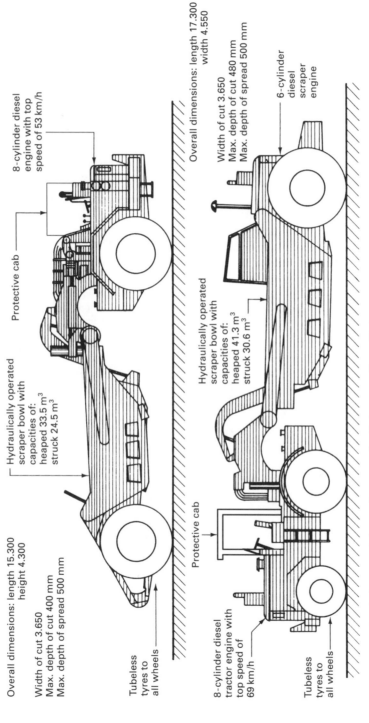

Overall dimensions: length 17.300
width 4.550

Width of cut 3.650
Max. depth of cut 480 mm
Max. depth of spread 500 mm

6-cylinder
diesel
scraper
engine

8-cylinder diesel
engine with top
speed of 53 km/h

Protective cab

Hydraulically operated
scraper bowl with
capacities of:
heaped 41.3 m³
struck 30.6 m³

Hydraulically operated
scraper bowl with
capacities of:
heaped 33.5 m³
struck 24.5 m³

Overall dimensions: length 15.300
height 4.300

Width of cut 3.650
Max. depth of cut 400 mm
Max. depth of spread 500 mm

Tubeless
tyres to
all wheels

Protective cab

8-cylinder diesel
tractor engine with
top speed of
69 km/h

Tubeless
tyres to
all wheels

Figure 2.3.3   Typical two- and three-axle scraper details (*courtesy:* Caterpillar)

Overall dimensions: length (bucket on ground) 4.300
width (bucket) 1.800

Maximum lifting height 4.675
Top speeds: forward 6.7 km/h, reverse 8.2 km/h
Width of tracks 300 mm

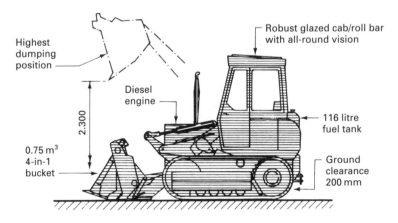

Highest
dumping
position

Robust glazed cab/roll bar
with all-round vision

Diesel
engine

2.300

116 litre
fuel tank

0.75 m³
4-in-1
bucket

Ground
clearance
200 mm

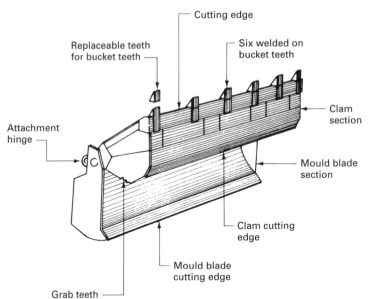

Cutting edge

Replaceable teeth
for bucket teeth

Six welded on
bucket teeth

Clam
section

Attachment
hinge

Mould blade
section

Clam cutting
edge

Mould blade
cutting edge

Grab teeth

Typical 4-in-1 bucket details

Figure 2.3.4   Typical tractor shovel

shovels that discharge at the rear by swinging the bucket over the top of the tractor, and machines equipped with shovels that have a side discharge facility enabling the spoil to be tipped into the attendant haul unit parked alongside, thus saving the time normally taken by the tractor in manoeuvring into a suitable position to discharge its load. Output of these machines is governed largely by the bucket capacity, which can be from 0.5 to 4 m³, and the type of soil encountered.

■■■ EXCAVATING MACHINES

Most excavating machines consist of a power unit, which is normally a diesel engine, and an excavating attachment designed to perform a specific task in a certain manner. These machines can be designed to carry out one specific activity with the excavating attachment hydraulically controlled, or the plant can consist of a basic power unit capable of easy conversion by changing the boom, bucket and rigging arrangement to carry out all the basic excavating functions. Such universal machines are usually chosen for this adaptability because the bucket sizes and outputs available of both versions are comparable.

■■■ SKIMMER

These machines are invariably based on the universal power unit, and consist of a bucket sliding along a horizontal jib. The bucket slides along the jib, digging away from the machine. Skimmers are used for oversite excavation up to a depth of 300 mm where great accuracy in level is required, and they can achieve an output of some 50 bucket loads per hour. To discharge the spoil the boom or jib is raised and the power unit is rotated until the raised bucket is over the attendant haulage vehicle, enabling the spoil to be discharged through the opening bottom direct into the haul unit (see Fig. 2.3.5).

■■■ FACE SHOVEL

This type of machine can be used as a loading shovel or for excavating into the face of an embankment or berm. Universal power unit or hydraulic machines are available with a wide choice of bucket capacities, achieving outputs in the region of 80 bucket loads per hour. The discharge operation is similar to that described above for the skimmer except that in the universal machine the discharge opening is at the rear of the bucket whereas in the hydraulic machines discharge is from the front of the bucket (see Figs 2.3.4 and 2.3.6). These machines are limited in the depth to which they can dig below machine level; this is generally within the range 300 mm to 2.000 m.

■■■ BACKACTER

This piece of plant is probably the most common form of excavating machinery used by building contractors for excavating basements, pits and trenches. Universal power unit and hydraulic versions are available, the latter often

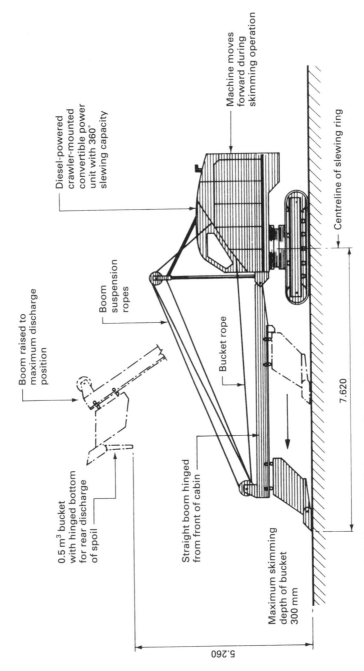

Diesel-powered
crawler-mounted
convertible power
unit with 360°
slewing capacity

Machine moves
forward during
skimming operation

Centreline of slewing ring

Boom raised to
maximum discharge
position

Boom
suspension
ropes

Bucket rope

0.5 m³ bucket
with hinged bottom
for rear discharge
of spoil

Straight boom hinged
from front of cabin

Maximum skimming
depth of bucket
300 mm

7.620

5.260

Figure 2.3.5   Typical skimmer details (*courtesy*: Ruston–Bucyrus)

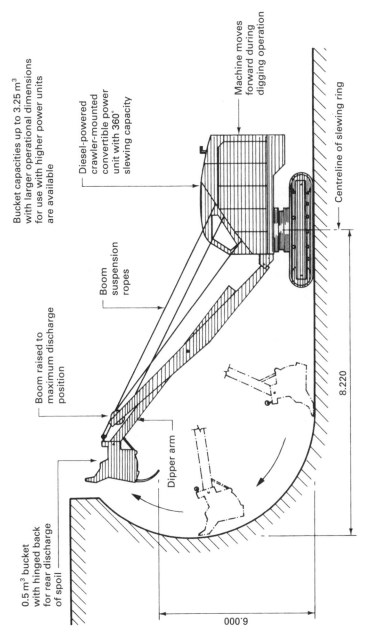

Bucket capacities up to 3.25 m³
with larger operational dimensions
for use with higher power units
are available

Diesel-powered
crawler-mounted
convertible power
unit with 360°
slewing capacity

Machine moves
forward during
digging operation

Centreline of slewing ring

Boom
suspension
ropes

Boom raised to
maximum discharge
position

Dipper arm

0.5 m³ bucket
with hinged back
for rear discharge
of spoil

8.220

6.000

Figure 2.3.6 Typical face shovel details (*courtesy:* Ruston–Bucyrus)

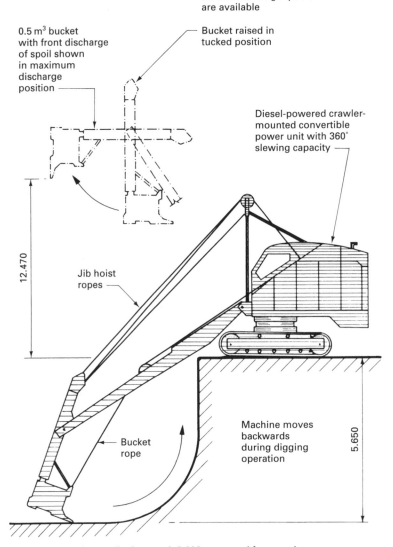

Bucket capacities up to 1.53 m³
with larger operational dimensions
for use with larger power machines
are available

0.5 m³ bucket
with front discharge
of spoil shown
in maximum
discharge
position

Bucket raised in
tucked position

Diesel-powered crawler-
mounted convertible
power unit with 360°
slewing capacity

12.470

Jib hoist
ropes

Bucket
rope

Machine moves
backwards
during digging
operation

5.650

Maximum digging reach 9.600 measured from cutting
edge of bucket to centreline of slewing ring

Figure 2.3.7 Typical backacter details (*courtesy*: Ruston–Bucyrus)

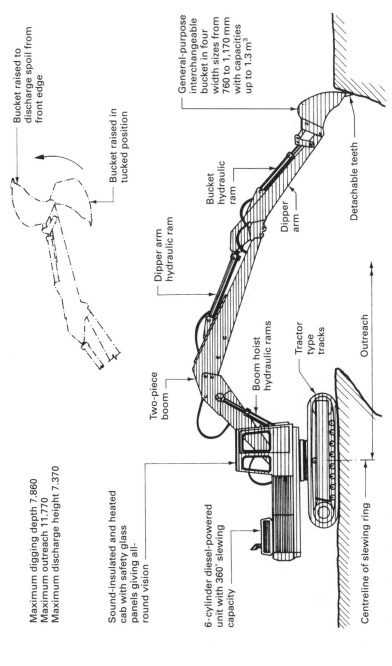

Bucket raised to discharge spoil from front edge

Bucket raised in tucked position

General-purpose interchangeable bucket in four width sizes from 760 to 1,170 mm with capacities up to 1.3 m³

Bucket hydraulic ram

Dipper arm

Detachable teeth

Dipper arm hydraulic ram

Two-piece boom

Boom hoist hydraulic rams

Tractor type tracks

Outreach

Maximum digging depth 7.860
Maximum outreach 11.770
Maximum discharge height 7.370

Sound-insulated and heated cab with safety glass panels giving all-round vision

6-cylinder diesel-powered unit with 360° slewing capacity

Centreline of slewing ring

Figure 2.3.8   Typical hydraulic backacter details (*courtesy:* Hymac)

sacrificing bucket capacity to achieve a greater reach from a set position.
Discharge in both types is by raising the bucket in a tucked position and
emptying the spoil through the open front end into the attendant haul unit
or alongside the trench. Outputs will vary from 30 to 60 bucket loads per hour,
depending upon how confined is the excavation area. Typical details are shown
in Figs 2.3.7 and 2.3.8.

▨▤▮ DRAGLINE

This type of excavator is essentially a crane with a long jib to which is attached a
drag bucket for excavating in loose and soft soils below the level of the machine.
This machine is for bulk excavation where fine limits are not of paramount
importance, because this is beyond the capabilities of the machine's design. The
accuracy to which a dragline can excavate depends upon the skill of the operator.
Discharge of the collected spoil is similar to that of a backacter, being through the
open front end of the bucket (see Fig. 2.3.9). A machine rigged as a dragline can be
fitted with a grab bucket as an alternative for excavating in very loose soils below
the level of the machine. Outputs of dragline excavators will vary according to
operating restrictions from 30 to 80 bucket loads per hour.

▨▤▮ MULTI-PURPOSE EXCAVATORS

These machines are based upon a tractor power unit and are very popular with
the small to medium-sized building contractor because of their versatility. The
tractor is usually a diesel-powered wheeled vehicle, although tracked versions are
available; both are fitted with a hydraulically controlled loading shovel at the front
and a hydraulically controlled backacting bucket or hoe at the rear of the vehicle
(see Fig. 2.3.10). It is essential that the weight of the machine is removed from the
axles during a backacting excavation operation. This is achieved by outrigger jacks
at the corners or by jacks at the rear of the power unit working in conjunction with
the inverted bucket at the front of the machine.

▨▤▮ TRENCHERS

These are machines designed to excavate trenches of constant width with
considerable accuracy and speed. Widths available range from 250 to 450 mm with
depths up to 4.000 m. Most trenchers work on a conveyor principle, having a series
of small cutting buckets attached to two endless chains, which are supported by a
boom that is lowered into the ground to the required depth. The spoil is transferred
to a cross-conveyor to deposit the spoil alongside the trench being dug; alternatively
it is deposited onto plough-shaped deflection plates that direct the spoil into
continuous heaps on both sides of the trench being excavated as the machine digs
along the proposed trench run. With a depth of dig of some 1.500 m outputs of
up to 2.000 m^3 per minute can be achieved, according to the nature of the subsoil.
Some trenchers are fitted with an angled mould blade to enable the machine to
carry out the backfilling operation (see Fig. 2.3.11 for a typical example).

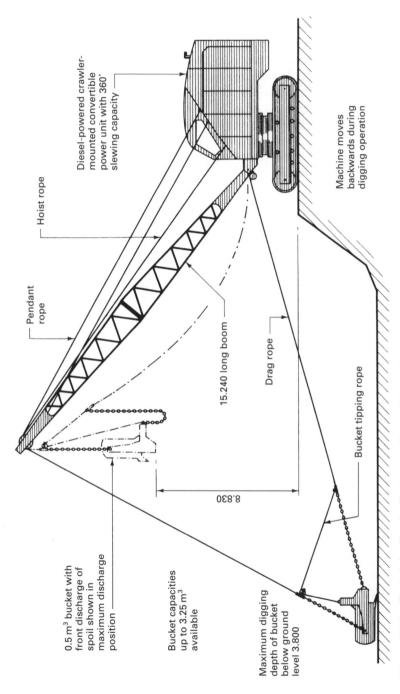

Diesel-powered crawler-mounted convertible power unit with 360° slewing capacity

Hoist rope

Pendant rope

15.240 long boom

Drag rope

Bucket tipping rope

Machine moves backwards during digging operation

8.830

0.5 m³ bucket with front discharge of spoil shown in maximum discharge position

Bucket capacities up to 3.25 m³ available

Maximum digging depth of bucket below ground level 3.800

Figure 2.3.9 Typical dragline details (*courtesy:* Ruston–Bucyrus)

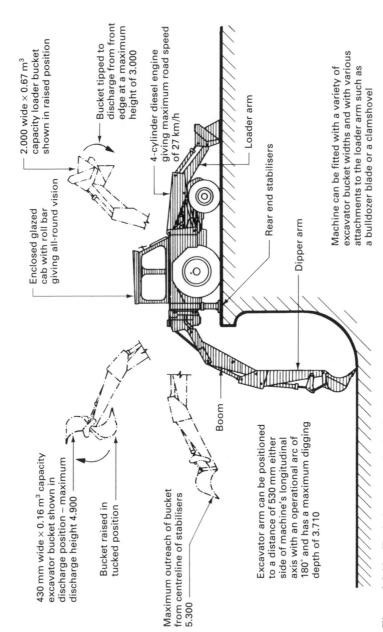

2.000 wide × 0.67 m³
capacity loader bucket
shown in raised position

Bucket tipped to
discharge from front
edge at a maximum
height of 3.000

Enclosed glazed
cab with roll bar
giving all-round vision

4-cylinder diesel engine
giving maximum road speed
of 27 km/h

Loader arm

Rear end stabilisers

Dipper arm

Machine can be fitted with a variety of
excavator bucket widths and with various
attachments to the loader arm such as
a bulldozer blade or a clamshovel

430 mm wide × 0.16 m³ capacity
excavator bucket shown in
discharge position – maximum
discharge height 4.900

Bucket raised in
tucked position

Maximum outreach of bucket
from centreline of stabilisers
5.300

Boom

Excavator arm can be positioned
to a distance of 530 mm either
side of machine's longitudinal
axis with an operational arc of
180° and has a maximum digging
depth of 3.710

Figure 2.3.10 Typical excavator/loader details (*courtesy*: JCB Sales)

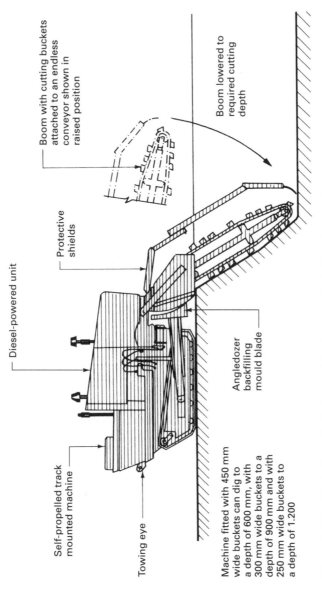

Boom with cutting buckets attached to an endless conveyor shown in raised position

Boom lowered to required cutting depth

Diesel-powered unit

Protective shields

Angledozer backfilling mould blade

Self-propelled track mounted machine

Towing eye

Machine fitted with 450 mm wide buckets can dig to a depth of 600 mm, with 300 mm wide buckets to a depth of 900 mm and with 250 mm wide buckets to a depth of 1.200

Figure 2.3.11 Typical trench-digging machine (*courtesy*: Davis Manufacturing)

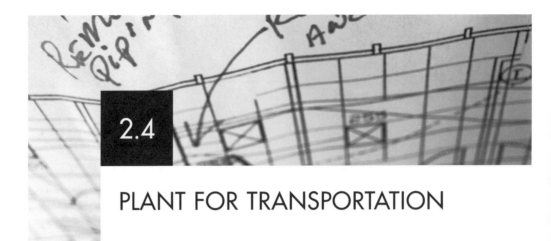

2.4

PLANT FOR TRANSPORTATION

Movement of materials and/or personnel around and between building sites can be very time-consuming and non-productive: therefore wherever economically possible most contractors will use some form of mechanical transportation. The movement required can be horizontal, vertical, or a combination of both directions. In the case of horizontal and vertical movement of large quantities of water the usual plant employed is a pump (already described in Chapter 2.2 on small powered plant). Similarly the transportation of large quantities of concrete can be carried out by using special pumping equipment, and this form of material transportation will be considered in Chapter 2.5 on concreting plant.

■■■ LORRIES AND TRUCKS

Transportation of operatives, machines and materials between sites is usually carried out by using suitably equipped or adapted lorries or trucks, ranging from the small pickup vehicle weighing less than 800 kg unladen to the very long low-loaders used to convey tracked plant such as cranes and bulldozers. The small pickup vehicle is usually based upon the manufacturer's private car range and has the advantages of lower vehicle excise duty than heavier lorries and the fact that the driver does not require a heavy goods vehicle licence for trucks of less than 7.5 tonnes gross vehicle weight. To satisfy EU regulations trucks will be fitted with tachographs for general use and specifically where travel distances exceed 50 km. These are particularly appropriate where the user is not regularly employed as the driver, e.g. a roofer using the vehicle to transport tiles to a contract. Most lorries designed and developed for building contractors' use are powered by a diesel engine, which is more economical and robust than the petrol engine although they are heavier and dearer. With mileages in excess of 20,000 (32,000 km) per year experienced by most contractors, diesel engines usually prove to be a worthwhile proposition. A vast range of lorries are produced by the leading motor manufacturers, and are available with refinements such as tipping, tailhoist and self-loading facilities using hydraulic lifting gear. Because lorries, trucks and vans are standard forms of transportation encountered in all aspects of daily living they will not be considered in detail in this text, but regulations made under the Road Traffic Act, Vehicles (Excise) Act and Customs and Excise Act must be noted.

The above legislation is very extensive and complex, dealing in detail with such requirements as: the minimum driving ages for various types of vehicle; limitations of hours of duty; maximum speed limits of certain classes of vehicle; vehicle lighting regulations; construction, weight and equipment of motor vehicles and trailers; testing requirements; and the rear-marking regulations for vehicles over 13.000 m long. Most of the statutory requirements noted above will be incorporated into the design and finish of the vehicle as purchased, but certain regulations regarding such matters as projection of loads, maximum loading of vehicles and notifications to be given to highway authorities, police forces and the Department for Transport are the direct concern of the building contractor. Precise information on these requirements can be found in the Road Traffic (Authorisation of Special Types) (General) Order 2002 together with any subsequent amendments. Note that excise regulations are the builder's and vehicle supplier's responsibility.

DUMPERS

These are one of the most versatile, labour-saving and misused pieces of plant available to the builder for the horizontal movement of materials ranging from bricks to aggregates, sanitary fittings to scaffolding, and fluids such as wet concrete. These diesel-powered vehicles require only one operative, the driver, and can traverse the rough terrain encountered on many building sites. Many sizes and varieties are produced, giving options such as two- or four-wheel drive, hydraulic- or gravity-operated container, side or high-level discharge, self-loading facilities and specially equipped dumpers for collecting and transporting crane skips. Specification for dumpers is usually given by quoting the container capacity in litres for heaped, struck and water levels (see Fig. 2.4.1 for typical examples).

FORKLIFT TRUCKS

Forklift trucks for handling mainly palleted materials quickly and efficiently around building sites over the rough terrain normally encountered have rapidly gained in popularity since their introduction in the early 1970s. This popularity was probably promoted by the shortage and cost of labour at that time, together with the need for the rapid movement of materials with a low breakage factor. Designs now available offer the choice of front- or rear-wheel drive and four-wheel drive with various mast heights and lifting capacities (typical details are shown in Fig. 2.4.2). Although these machines can carry certain unpalleted materials, this activity will require hand loading, which reduces considerably the economic advantages of machine-loading palleted materials.

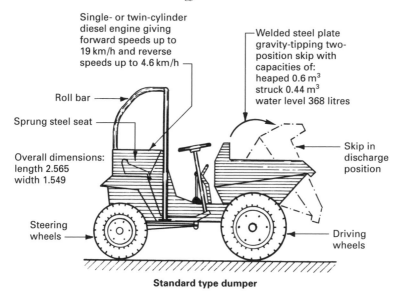

Single- or twin-cylinder
diesel engine giving
forward speeds up to
19 km/h and reverse
speeds up to 4.6 km/h

Roll bar

Sprung steel seat

Overall dimensions:
length 2.565
width 1.549

Steering
wheels

Welded steel plate
gravity-tipping two-
position skip with
capacities of:
heaped 0.6 m^3
struck 0.44 m^3
water level 368 litres

Skip in
discharge
position

Driving
wheels

Standard type dumper

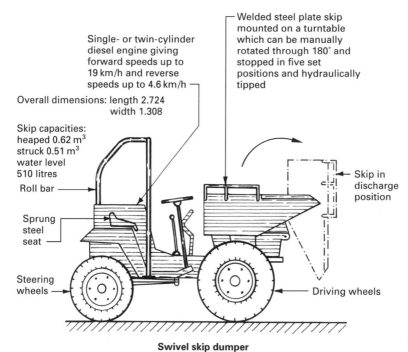

Single- or twin-cylinder
diesel engine giving
forward speeds up to
19 km/h and reverse
speeds up to 4.6 km/h

Overall dimensions: length 2.724
width 1.308

Skip capacities:
heaped 0.62 m^3
struck 0.51 m^3
water level
510 litres

Roll bar

Sprung
steel
seat

Steering
wheels

Welded steel plate skip
mounted on a turntable
which can be manually
rotated through 180° and
stopped in five set
positions and hydraulically
tipped

Skip in
discharge
position

Driving wheels

Swivel skip dumper

Figure 2.4.1 Typical diesel dumpers

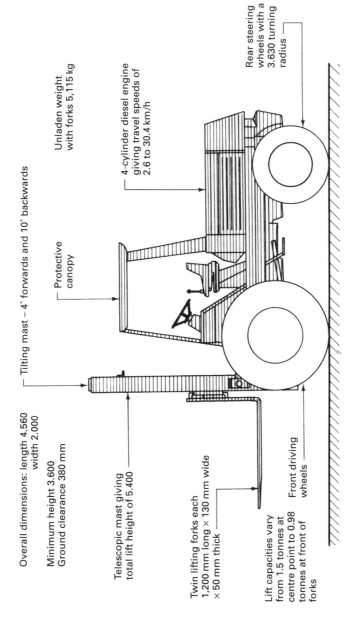

Overall dimensions: length 4.560
width 2.000

Minimum height 3.600
Ground clearance 380 mm

Tilting mast – 4° forwards and 10° backwards

Protective
canopy

Unladen weight
with forks 5,115 kg

4-cylinder diesel engine
giving travel speeds of
2.6 to 30.4 km/h

Rear steering
wheels with a
3.630 turning
radius

Telescopic mast giving
total lift height of 5.400

Twin lifting forks each
1,200 mm long × 130 mm wide
× 50 mm thick

Lift capacities vary
from 1.5 tonnes at
centre point to 0.98
tonnes at front of
forks

Front driving
wheels

Figure 2.4.2 Typical fork lift truck details (*courtesy*: Manitou)

▨ ▧ ■ ELEVATORS AND CONVEYORS

The distinction between an elevator and a conveyor is usually one of direction of movement, in that elevators are considered as those belts moving materials mainly in the vertical direction, whereas conveyors are a similar piece of plant moving materials mainly in the horizontal direction. Elevators are not common on building sites but, if used wisely, can be economical for such activities as raising bricks or roofing tiles to the fixing position. Most elevators consist of an endless belt with raised transverse strips at suitable spacings against which can be placed the materials to be raised, usually to a maximum height of 7.000 m. Conveyors or endless belts are used mainly for transporting aggregates and concrete and are generally considered economic only on large sites where there may be a large concrete-mixing complex.

HOISTS

Hoists are a means of transporting materials or passengers vertically by means of a moving level platform. Generally, hoists are designed specifically to lift materials or passengers, but recent designs have been orientated towards the combined materials/passenger hoist. Note that under no circumstances should passengers be transported on hoists designed specifically for lifting materials only.

Materials hoists come in basically two forms, the static and mobile models. The static version consists of a mast or tower with the lift platform either cantilevered from the small section mast or centrally suspended with guides on either side within an enclosing tower. Both forms need to be plumb and tied to the structure or scaffold at the intervals recommended by the manufacturer to ensure stability. Mobile hoists usually have a maximum height of 24.000 m and do not need tying to the structure unless extension pieces are fitted, when they are then treated as a cantilever hoist. All mobile hoists should be positioned on a firm level base and jacked to ensure stability (see Fig. 2.4.3). The operation of a materials hoist should be entrusted to a trained driver who has a clear view from the operating position. Site operatives should be instructed as to the correct loading procedures, such as placing barrows on the hoist platform at ground level with the handles facing the high-level exit so that walking onto the raised platform is reduced to a minimum.

Passenger hoists, like the materials hoist, can be driven by a petrol, diesel or electric motor, and can be of a cantilever or enclosed variety. The cantilever type consists of one or two passenger hoist cages operating on one or both sides of the cantilever tower; the alternative form consists of a passenger hoist cage operating within an enclosing tower. Tying-back requirements are similar to those needed for the materials hoist. Passenger and material hoists should conform to the recommendations of BS 7212: *Code of practice for safe use of construction hoists.* Typical hoist details are shown in Fig. 2.4.4.

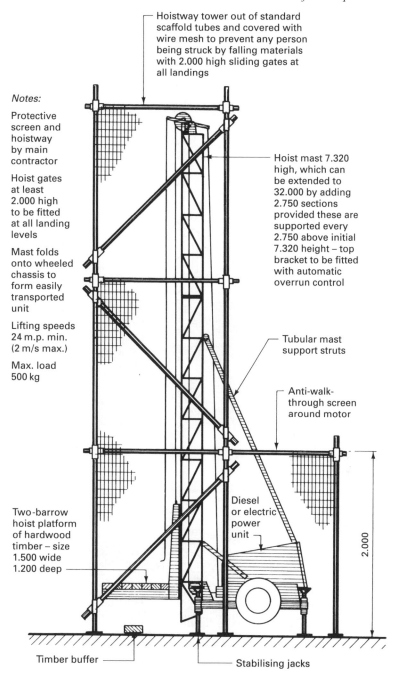

Hoistway tower out of standard
scaffold tubes and covered with
wire mesh to prevent any person
being struck by falling materials
with 2.000 high sliding gates at
all landings

Notes:

Protective
screen and
hoistway
by main
contractor

Hoist gates
at least
2.000 high
to be fitted
at all landing
levels

Mast folds
onto wheeled
chassis to
form easily
transported
unit

Lifting speeds
24 m.p. min.
(2 m/s max.)

Max. load
500 kg

Hoist mast 7.320
high, which can
be extended to
32.000 by adding
2.750 sections
provided these are
supported every
2.750 above initial
7.320 height – top
bracket to be fitted
with automatic
overrun control

Tubular mast
support struts

Anti-walk-
through screen
around motor

Two-barrow
hoist platform
of hardwood
timber – size
1.500 wide
1.200 deep

Diesel
or electric
power
unit

2.000

Timber buffer

Stabilising jacks

Figure 2.4.3   Typical materials hoist (*courtesy*: Wickham Engineering)

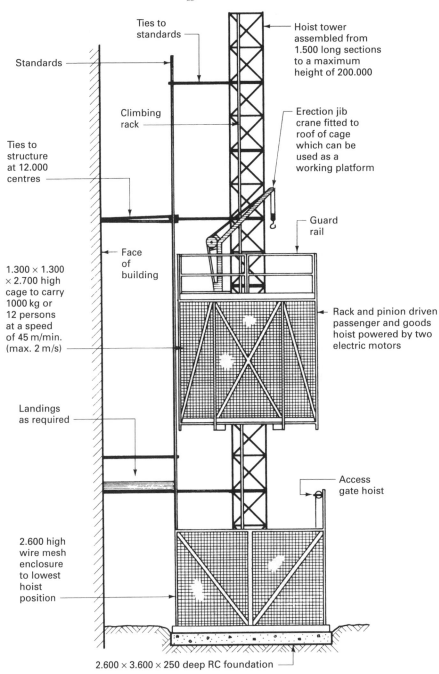

Ties to
standards

Standards

Climbing
rack

Ties to
structure
at 12.000
centres

1.300 × 1.300
× 2.700 high
cage to carry
1000 kg or
12 persons
at a speed
of 45 m/min.
(max. 2 m/s)

Face
of
building

Landings
as required

2.600 high
wire mesh
enclosure
to lowest
hoist
position

Hoist tower
assembled from
1.500 long sections
to a maximum
height of 200.000

Erection jib
crane fitted to
roof of cage
which can be
used as a
working platform

Guard
rail

Rack and pinion driven
passenger and goods
hoist powered by two
electric motors

Access
gate hoist

2.600 × 3.600 × 250 deep RC foundation

Figure 2.4.4   Typical passenger hoist (*courtesy*: Linden Alimark)

◼◼◼ THE LIFTING OPERATIONS AND LIFTING EQUIPMENT REGULATIONS 1998

These regulations, known by the acronym LOLER, were made under the Health and Safety at Work etc. Act 1974. The regulations provide an extensive checklist with regard to assessment of risk in the use of hoists. The following is a brief summary of the main points:

- Wherever access can be gained, and wherever anyone at ground level could be struck by the platform or counterweight, enclosures and gates should be at least 2.000 m high. Access gates must be kept closed at all times except for the necessary loading and unloading of the platform. The platform itself must be fitted with a device capable of supporting a full load in the event of the hoist ropes or hoisting gear failing. Furthermore, the hoist must be fitted with an automatic device to prevent the platform or cage over-running.
- Hoist operation is to be from one point only at all times if not controlled from the cage itself. The operator must have a clear view of the hoist throughout its entire travel. If this is not possible, auxiliary visibility devices or a signaller or banksman must be located to cover all landings.
- Winches and carriers must have independent devices to prevent free fall and an automatic braking system that is applied whenever the controls are not in the normal operating position. Also, multiple roping or cylinders (hydraulic hoists) and equipment with high factors of safety should be a standard facility.
- Safe working loads (SWL) in terms of materials and/or maximum number of passengers to be carried must be displayed on all platforms and associated equipment. Other acceptable terms that may be used include **rated capacity** or **working load limit**.
- Examination and inspection include the engagement of appropriately trained and competent employees, capable of ensuring that a hoist is safe to use. Pre-use checks should be conducted at the beginning of each working day, after alterations to height and after repairs. Thorough and detailed examination of all components will be necessary at periodic intervals determined by the degree of exposure and nature of the work. All examinations and inspections must be recorded and filed.
- Passenger-carrying hoists specifically require the cages to be constructed in such a manner that passengers cannot fall out, become trapped or struck by objects falling down the hoistway. Other requirements include the need for gates that will prevent the hoists being activated until they are closed and which can be opened only at landing levels. Over-run devices must also be fitted at the top and bottom of the hoistway.
- Facilities for the prevention of movement or tipping of materials during transportation by hoist are essential. Security of loads such as loose materials should be provided by suitable containers such as wheelbarrows that are scotched or otherwise restrained.

▩▩■ CRANES

A crane may be defined as a device or machine for lifting loads by means of a rope. The use of cranes has greatly increased in the construction industry, mainly because of the need to raise the large and heavy prefabricated components often used in modern structures. The range of cranes available is very wide, and therefore actual choice must be made on a basis of sound reasoning, overall economics, capabilities of cranes under consideration, prevailing site conditions and the anticipated utilisation of the equipment.

The simplest crane of all consists of a single-grooved wheel, over which the rope is passed, suspended from a scaffold or beam, and is called a **gin wheel**. The gin wheel is manually operated and always requires more effort than the weight of the load to raise it to the required height. It is suitable only for light loads such as, for example, a bucketful of mortar and is normally used only on very small contracts. To obtain some mechanical advantage the gin wheel can be replaced by a pulley block that contains more than one pulley or sheave; according to the number and pattern of sheaves used the lesser or greater is the saving in effort required to move any given load.

Another useful but simple crane that can be employed for small, low-rise structures is the **scaffold crane**, which consists of a short jib counterbalanced by the small petrol or electric power unit. The crane is fastened to a specially reinforced scaffold standard incorporated within the general scaffold framework, with extra bracing to overcome the additional stresses as necessary. The usual maximum lifting capacity of this form of crane is 200 kg.

Apart from these simple cranes for small loads most cranes come in the more recognisable form. Subdivision of crane types can be very wide and varied, but one simple method of classification is to consider cranes under three general headings:

- mobile cranes;
- static or stationary cranes;
- tower cranes.

MOBILE CRANES

Mobile cranes come in a wide variety of designs and capacities, generally with a 360° rotation or slewing circle, a low pivot and luffing jib, the main exception being the mast crane. Mobile cranes can be classed into five groups:

- self-propelled cranes;
- lorry-mounted cranes;
- track-mounted cranes;
- mast cranes;
- gantry cranes.

Self-propelled cranes

These are wheel-mounted mobile cranes that are generally of low lifting capacities of up to 10 tonnes. They can be distinguished from other mobile cranes by the fact that the driver has only one cab position for both driving and operating the crane. They are extremely mobile, but to be efficient they usually require a hard, level surface from which to work. Road speeds obtained are in the region of 30 km/h. The small-capacity machines have a fixed boom or jib length, whereas the high-capacity cranes can have a sectional lattice jib or a telescopic boom to obtain various radii and lifting capacities. In common with all cranes, the shorter the lifting radius the greater will be the lifting capacity (see Fig. 2.4.5 for a typical example).

Lorry-mounted cranes

These consist of a crane mounted on a specially designed lorry or truck. The operator drives the vehicle between sites from a conventional cab but has to operate the crane engine and controls from a separate crane-operating position. The capacity of lorry-mounted cranes ranges from 5 to 20 tonnes in the free-standing position, but this can be increased by using the jack outriggers built into the chassis. Two basic jib formats for this type of crane are available: the folding lattice jib and the telescopic jib. Most cranes fitted with folding jibs are designed for travelling on the highway with the basic jib supported by a vertical frame extended above the driving cab; extra jib lengths and fly jibs can be added upon arrival on site if required (see Fig. 2.4.6). Telescopic jib cranes are very popular because of the short time period required to prepare the crane for use upon arrival on site, making them ideally suitable for short-hire periods (see Fig. 2.4.7 for a typical example). Mobile lorry cranes can travel between sites at speeds of up to 48 km/h, which makes them very mobile, but to be fully efficient they need a firm and level surface from which to operate.

Track-mounted cranes

This form of mobile crane is usually based upon the standard power unit capable of being rigged as an excavator. These cranes can traverse around most sites without the need for a firm level surface, and have capacity ranges similar to those of the lorry-mounted cranes. The jib is of lattice construction with additional sections and fly jibs to obtain the various lengths and capacities required (see Fig. 2.4.8). The main disadvantage of this form of mobile crane is the general need for a special low-loading lorry to transport the crane between sites.

Mast cranes

These cranes are often confused with mobile tower cranes. The main differences are:

- The mast is mounted on the jib pivots and held in the vertical position by ties.
- Cranes are high-pivot machines with a luffing jib.
- Operation is usually from the chassis of the machine.

Mast cranes can be either lorry- or track-mounted machines: see Fig. 2.4.9 for a typical example. The main advantages of the high-pivot mast crane are that it is less

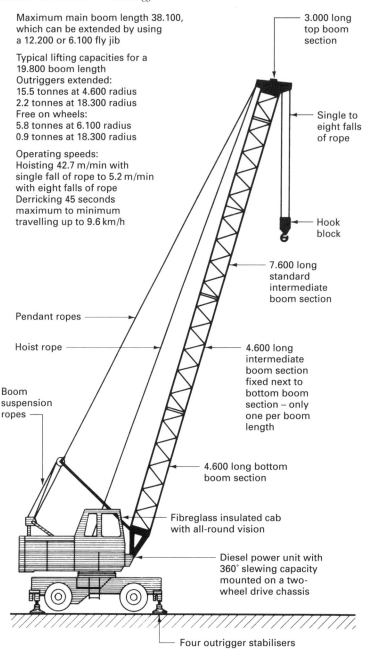

Maximum main boom length 38.100, which can be extended by using a 12.200 or 6.100 fly jib

Typical lifting capacities for a 19.800 boom length
Outriggers extended:
15.5 tonnes at 4.600 radius
2.2 tonnes at 18.300 radius
Free on wheels:
5.8 tonnes at 6.100 radius
0.9 tonnes at 18.300 radius

Operating speeds:
Hoisting 42.7 m/min with single fall of rope to 5.2 m/min with eight falls of rope
Derricking 45 seconds maximum to minimum travelling up to 9.6 km/h

3.000 long top boom section

Single to eight falls of rope

Hook block

7.600 long standard intermediate boom section

Pendant ropes

Hoist rope

4.600 long intermediate boom section fixed next to bottom boom section – only one per boom length

Boom suspension ropes

4.600 long bottom boom section

Fibreglass insulated cab with all-round vision

Diesel power unit with 360° slewing capacity mounted on a two-wheel drive chassis

Four outrigger stabilisers

Figure 2.4.5   Typical self-propelled crane (*courtesy*: Jones Cranes)

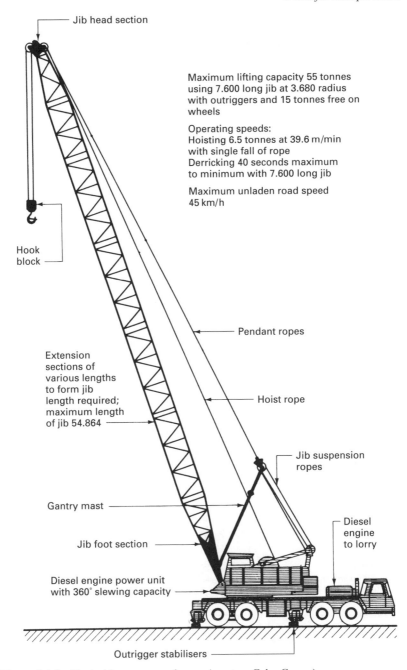

Jib head section

Maximum lifting capacity 55 tonnes
using 7.600 long jib at 3.680 radius
with outriggers and 15 tonnes free on
wheels

Operating speeds:
Hoisting 6.5 tonnes at 39.6 m/min
with single fall of rope
Derricking 40 seconds maximum
to minimum with 7.600 long jib

Maximum unladen road speed
45 km/h

Hook
block

Pendant ropes

Extension
sections of
various lengths
to form jib
length required;
maximum length
of jib 54.864

Hoist rope

Jib suspension
ropes

Gantry mast

Jib foot section

Diesel
engine
to lorry

Diesel engine power unit
with 360° slewing capacity

Outrigger stabilisers

Figure 2.4.6 Typical lorry-mounted crane (*courtesy*: Coles Cranes)

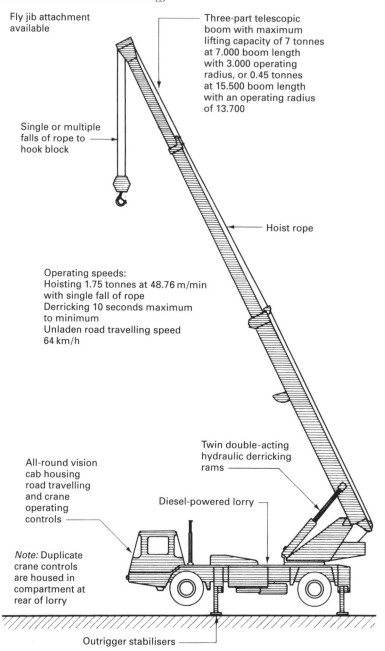

Fly jib attachment
available

Three-part telescopic
boom with maximum
lifting capacity of 7 tonnes
at 7.000 boom length
with 3.000 operating
radius, or 0.45 tonnes
at 15.500 boom length
with an operating radius
of 13.700

Single or multiple
falls of rope to
hook block

Hoist rope

Operating speeds:
Hoisting 1.75 tonnes at 48.76 m/min
with single fall of rope
Derricking 10 seconds maximum
to minimum
Unladen road travelling speed
64 km/h

Twin double-acting
hydraulic derricking
rams

All-round vision
cab housing
road travelling
and crane
operating
controls

Diesel-powered lorry

Note: Duplicate
crane controls
are housed in
compartment at
rear of lorry

Outrigger stabilisers

Figure 2.4.7   Lorry-mounted telescopic crane (*courtesy*: Coles Cranes)

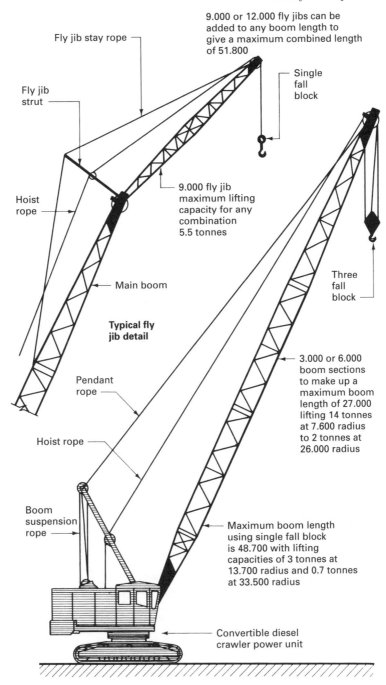

9.000 or 12.000 fly jibs can be added to any boom length to give a maximum combined length of 51.800

Fly jib stay rope

Single fall block

Fly jib strut

Hoist rope

9.000 fly jib maximum lifting capacity for any combination 5.5 tonnes

Three fall block

Main boom

Typical fly jib detail

3.000 or 6.000 boom sections to make up a maximum boom length of 27.000 lifting 14 tonnes at 7.600 radius to 2 tonnes at 26.000 radius

Pendant rope

Hoist rope

Boom suspension rope

Maximum boom length using single fall block is 48.700 with lifting capacities of 3 tonnes at 13.700 radius and 0.7 tonnes at 33.500 radius

Convertible diesel crawler power unit

Figure 2.4.8 Typical track-mounted crane (*courtesy*: Thomas Smith & Sons)

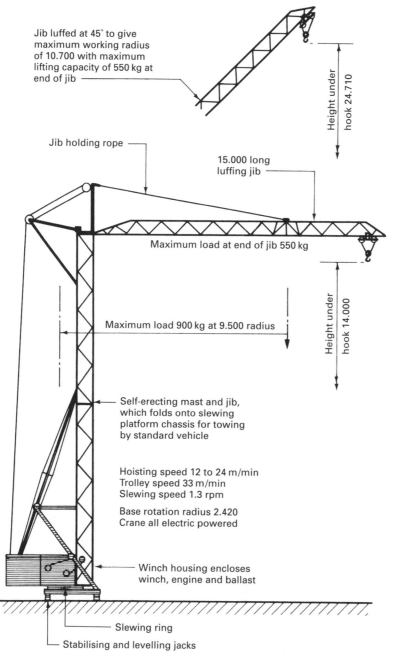

Jib luffed at 45° to give
maximum working radius
of 10.700 with maximum
lifting capacity of 550 kg at
end of jib

Height under hook 24.710

Jib holding rope

15.000 long
luffing jib

Maximum load at end of jib 550 kg

Height under hook 14.000

Maximum load 900 kg at 9.500 radius

Self-erecting mast and jib,
which folds onto slewing
platform chassis for towing
by standard vehicle

Hoisting speed 12 to 24 m/min
Trolley speed 33 m/min
Slewing speed 1.3 rpm

Base rotation radius 2.420
Crane all electric powered

Winch housing encloses
winch, engine and ballast

Slewing ring

Stabilising and levelling jacks

Figure 2.4.9   Typical mobile mast crane (*courtesy*: Manitou)

likely to foul the side of a building under construction, and it can approach closer to the structure than a low-pivot machine of equivalent capacity and reach. This can be of paramount importance on congested sites.

Gantry cranes

The gantry or portal crane is a rail-mounted crane consisting of a horizontal transverse beam that carries a combined driver's cab and hook-supporting saddle. The beam is supported by rail-mounted 'A' frames on powered bogies situated on both sides of the building under construction. This is a particularly safe form of crane as it requires no ballast, gives the driver an excellent all-round view, and allows the hook three-way movement in vertical, horizontal and transverse directions. Although limited in application, this special form of mobile crane can be very usefully and economically employed on repetitive and partially prefabricated blocks of medium-rise dwellings.

STATIC OR STATIONARY CRANES

These cranes are fixed at their working position and are used primarily for lifting heavy loads such as structural steelwork. Although more common on civil engineering contracts they can be successfully and economically employed by building contractors.

Guyed derrick

This is a simple and inexpensive form of static crane powered by a diesel engine or electric motor. It consists of a lattice mast with a pedestal bearing stabilised by five or more anchored guy ropes. The jib is of the low-pivot type and is slightly shorter in length than the mast height so that it can rotate through the whole 360° without fouling the guy ropes if raised in the near-vertical position (see Fig. 2.4.10 for typical details).

Scotch derrick

This consists of a slewing mast and a luffing jib that is usually longer than the jib used on a similar-capacity guyed derrick. Stabilisation of a scotch derrick is obtained by using lattice members called guys and stays. Two guys are fixed to the top of the slewing mast at an angle of 45° to the horizontal and at an angle of 90° in plan; the lower ends of the guys are connected to the ends of horizontal stays fixed to the base of the mast, forming an angle of 90° in plan. A horizontal brace is fixed between the ends of the guys and stays, forming a complete triangulation of the stabilising members together with the mast. Scotch derricks are capable of slewing only 270°, being restricted in further rotational movement by the sloping lattice guys. Resistance to overturning can be provided by kentledge applied to the struts and brace, or these members can be bolted to temporary concrete bases. The power can be supplied by a diesel engine, but the common power source is an electric motor (for typical details see Fig. 2.4.10).

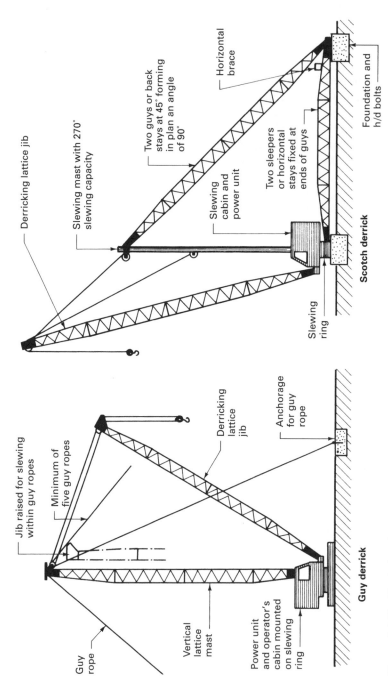

Figure 2.4.10　Typical static derrick cranes

Monotower cranes

These are basically an elevated scotch derrick crane consisting of a fabricated and well-braced tower surmounted by a derrick crane, the mast of which extends to a pivot bearing well down the height of the tower. The tower can be up to 60.000 m in height with a jib length of some 40.000 m, capable of raising 2 tonnes at its maximum radius. To be economic this form of crane needs to be centrally sited to give maximum site coverage.

TOWER CRANES

Since their introduction in 1950 by the then Department of Scientific and Industrial Research the tower crane has been universally accepted by the building industry as a standard piece of plant required for construction of medium- to high-rise structures. These cranes are available in several forms with a horizontal jib carrying a saddle or a trolley, or alternatively with a luffing or derricking jib with a lifting hook at its extreme end. Horizontal jibs can bring the load closer to the tower, whereas luffing jibs can be raised to clear obstructions such as adjacent buildings, an advantage on confined sites. The basic types of tower crane available are:

- self-supporting static tower cranes;
- supported static tower cranes;
- travelling tower cranes;
- climbing cranes.

Self-supporting static tower cranes

These cranes generally have a greater lifting capacity than other types of crane. The mast of the self-supporting tower crane must be firmly anchored at ground level to a concrete base with holding-down bolts or alternatively to a special mast base section cast into a foundation. They are particularly suitable for confined sites, and should be positioned in front or to one side of the proposed building with a jib of sufficient length to give overall coverage of the new structure. Generally these cranes have a static tower, but types with a rotating or slewing tower and luffing jib are also available (see Fig. 2.4.11 for a typical self-supporting crane example).

Supported static tower cranes

These are similar in construction to self-supporting tower cranes but are used for lifting to a height in excess of that possible with self-supporting or travelling tower cranes. The tower or mast is fixed or tied to the structure using single or double steel stays to provide the required stability. This tying back will induce stresses in the supporting structure, which must therefore be of adequate strength. Supported tower cranes usually have horizontal jibs, because the rotation of a luffing jib mast renders it unsuitable for this application (see Fig. 2.4.12 for a typical example).

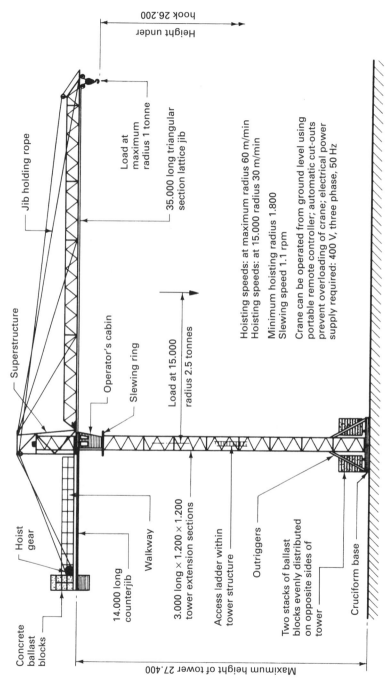

Figure 2.4.11 Typical self-supporting static tower crane (*courtesy:* Stothert & Pitt)

Height under
hook 26.200

Load at
maximum
radius 1 tonne

35.000 long triangular
section lattice jib

Jib holding rope

Superstructure

Operator's cabin

Slewing ring

Load at 15.000
radius 2.5 tonnes

Hoisting speeds: at maximum radius 60 m/min
Hoisting speeds: at 15.000 radius 30 m/min

Minimum hoisting radius 1.800
Slewing speed 1.1 rpm

Crane can be operated from ground level using
portable remote controller; automatic cut-outs
prevent overloading of crane; electrical power
supply required: 400 V, three phase, 50 Hz

Hoist
gear

Concrete
ballast
blocks

14.000 long
counterjib

Walkway

3.000 long × 1.200 × 1.200
tower extension sections

Access ladder within
tower structure

Outriggers

Two stacks of ballast
blocks evenly distributed
on opposite sides of
tower

Cruciform base

Maximum height of tower 27.400

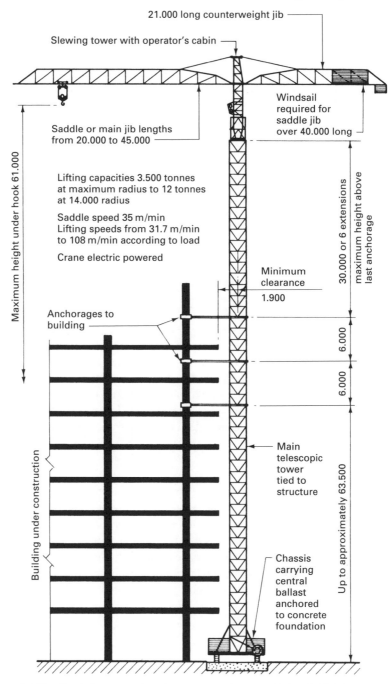

21.000 long counterweight jib

Slewing tower with operator's cabin

Maximum height under hook 61.000

Saddle or main jib lengths from 20.000 to 45.000

Windsail required for saddle jib over 40.000 long

Lifting capacities 3.500 tonnes at maximum radius to 12 tonnes at 14.000 radius

Saddle speed 35 m/min
Lifting speeds from 31.7 m/min to 108 m/min according to load

Crane electric powered

30.000 or 6 extensions maximum height above last anchorage

Minimum clearance 1.900

Anchorages to building

6.000

6.000

Building under construction

Main telescopic tower tied to structure

Up to approximately 63.500

Chassis carrying central ballast anchored to concrete foundation

Figure 2.4.12 Typical supported static tower crane (*courtesy*: Babcock Weitz)

Travelling tower cranes

To obtain better site coverage with a tower crane a rail-mounted or travelling crane could be used. The crane travels on heavy wheeled bogies mounted on a wide-gauge (4.200 m) rail track with gradients not exceeding 1 in 200 and curves not less than 11.000 m radius depending on mast height. It is essential that the base for the railway track sleepers is accurately prepared, well drained, regularly inspected and maintained if the stability of the crane is to be ensured. The motive power is electricity, the supply of which should be attached to a spring-loaded drum, which will draw in the cable as the crane reverses to reduce the risk of the cable becoming cut or trapped by the wheeled bogies. Travelling cranes can be supplied with similar lifting capacities and jib arrangements as given for static cranes (see Fig. 2.4.13 for a typical example).

Climbing cranes

These are designed for tall buildings, being located within and supported by the structure under construction. The mast, which extends down through several storeys, requires only a small (1.500 to 2.000 m square) opening in each floor. Support is given at floor levels by special steel collars, frames and wedges. The raising of the static mast is carried out using a winch that is an integral part of the system. Generally, this form of crane requires a smaller horizontal or luffing jib to cover the construction area than a static or similar tower crane. The jib is made from small, easy-to-handle sections, which are lowered down the face of the building, when the crane is no longer required, by means of a special winch attached to one section of the crane. The winch is finally lowered to ground level by hand when the crane has been dismantled (see Fig. 2.4.14 for typical crane details).

CRANE SKIPS AND SLINGS

Cranes are required to lift all kinds of materials, ranging from prefabricated components to loose and fluid materials. Various skips or containers have been designed to carry loose or fluid materials (see Fig. 2.4.15). Skips should be of sound construction, easy to attach to the crane hook, easily cleaned, easy to load and unload and of a suitable capacity. Prefabricated components are usually hoisted from predetermined lifting points by using wire or chain slings (see Fig. 2.4.15).

WIRE ROPES

Wire ropes consist of individual wires twisted together to form strands, which are then twisted together around a steel core to form a rope or cable. **Ordinary lay** ropes are formed by twisting the wires in the individual strands in the opposite direction to the group of strands, whereas in **Lang's lay** ropes the wires in the individual strands are twisted in the same direction as the groups of strands.

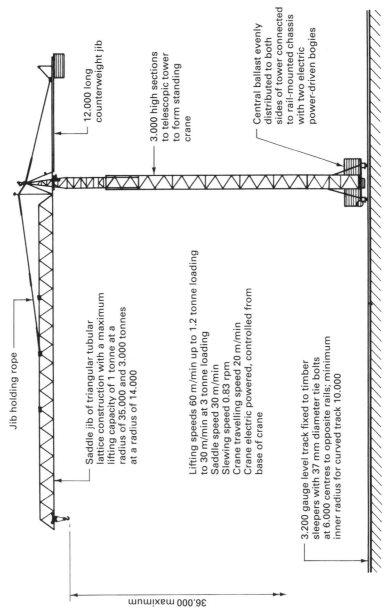

12.000 long counterweight jib

3.000 high sections to telescopic tower to form standing crane

Central ballast evenly distributed to both sides of tower connected to rail-mounted chassis with two electric power-driven bogies

Jib holding rope

Saddle jib of triangular tubular lattice construction with a maximum lifting capacity of 1 tonne at a radius of 35.000 and 3.000 tonnes at a radius of 14.000

Lifting speeds 60 m/min up to 1.2 tonne loading to 30 m/min at 3 tonne loading
Saddle speed 30 m/min
Slewing speed 0.83 rpm
Crane travelling speed 20 m/min
Crane electric powered, controlled from base of crane

3.200 gauge level track fixed to timber sleepers with 37 mm diameter tie bolts at 6.000 centres to opposite rails; minimum inner radius for curved track 10.000

36.000 maximum

Figure 2.4.13   Typical travelling tower crane (*courtesy:* Richier International)

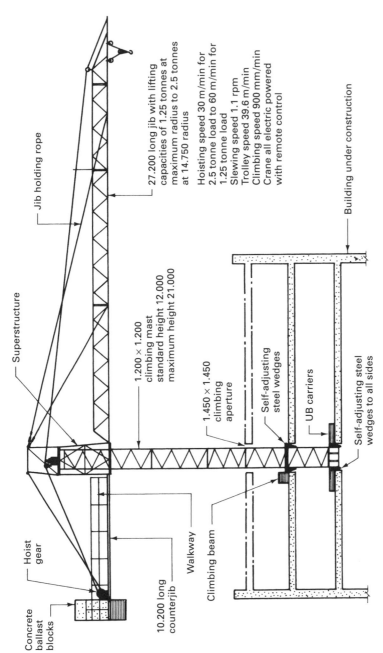

Figure 2.4.14   Typical climbing tower crane (*courtesy*: Stothert & Pitt Ltd)

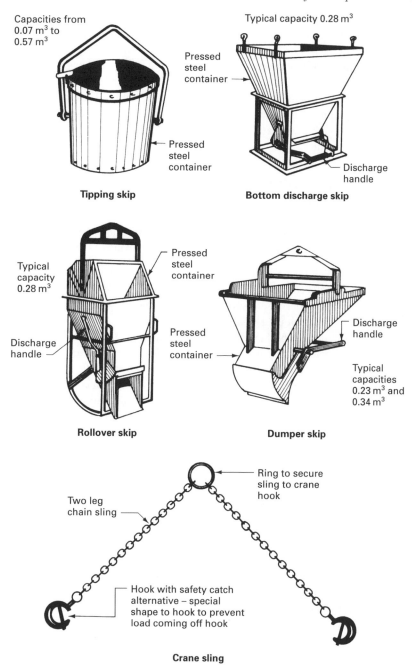

Figure 2.4.15 Typical crane skips and sling

Lang's lay ropes generally have better wearing properties thanks to the larger surface area of the external wires, but they have the tendency to spin if the ends of the rope are not fixed. For this reason ordinary lay ropes with a working life of up to two years are usually preferred for cranes. All wire ropes are lubricated during manufacture, but this does not preclude the need to clean and lubricate wire ropes when exposed to the elements. Under Regulation No. 9 of the Lifting Operations and Lifting Equipment Regulations 1998 wire ropes must be inspected before use and given a thorough examination every six months, the results being recorded on the appropriate form.

■■■ THE LIFTING OPERATIONS AND LIFTING EQUIPMENT REGULATIONS 1998 – SEE ALSO PAGE 93

These regulations detail the minimum requirements for lifting appliances, chains, ropes and lifting gear. The main points include the following:

- Examination of all forms of lifting gear to ensure sound construction, materials appropriate for the conditions of use, adequate strength for the intended task, retention in good order, and inspection at regular intervals depending on use and exposure as determined by a competent person.
- Adequate support, strength, stability, anchoring, fixing and erection of lifting appliances to include an appropriate factor of safety against failure.
- Travelling and slewing cranes require a 500 mm (preferably 600 mm) wide minimum clearance wherever practicable between the appliance and fixtures such as a building or access scaffold. If such a clearance cannot be provided, movement between the appliance and fixture should be prevented.
- Safe means of access is required for the crane operator and the signaller. If the access platform is above ground it must be of adequate size, close boarded or plated, provided with guard rails at least 950 mm above platform level with toe boards at least 150 mm high. Any gap between the toe board and intermediate guard rail or between guard rails must be no more than 470 mm. (See Chapter 1.5 in *Construction Technology*.)
- A cab or cabin is required for the crane operator; this must provide an unrestricted view for safe use of the applicance. The cabin must have adequate protection from the weather and harmful substances, with a facility for ventilation and heating. The cabin must also allow access to the machinery for maintenance work.
- Equipment that can be adapted for various operating radii and other configurations must be clearly marked with corresponding safe working loads for these variables.
- Brakes, controls and safety devices must be clearly marked to prevent accidental operation or misuse.
- Safe means of access are to be provided for examination, repair and servicing, particularly where there is a possibility of a person falling from any height.
- Strength and fixing requirements for equipment supporting pulley blocks and gin wheels must be ascertained.

■ Stability of lifting appliances on soft ground, uneven surfaces and slopes must be considered. Cranes must be either anchored to the ground or to a foundation, or suitably counterweighted or stabilised to prevent overturning.

■ Rail-mounted cranes must have a track laid and secured on a firm foundation to prevent risk of derailment. There must be provision for buffers, effective braking systems and adequate maintenance of both track and equipment.

■ Strength requirements are to be assessed for mounting cranes on bogies, trolley or wheeled carriages.

■ Cranes (particularly those used for lifting persons) require hoisting and lowering limiters in addition to capacity indicators and limiters.

■ Measures must be taken to prevent a freely suspended load from moving uncontrollably. Devices that could be fitted include: multiple ropes, pawl and ratchet, check valves for hydraulics and safety nets.

■ Cranes must be erected under planned conditions and the supervision of a competent person.

■ There are specific requirements for persons operating lifting equipment and for signalling duties. Operators of lifting appliances must be trained, experienced and over 18 years of age. If the operator cannot see the whole passage of a lift, an efficient signalling system must be used. A signaller must be over 18 years of age, and capable of giving clear and distinct communications by hand, mechanical or electrical means.

■ Testing, examination and inspections are required for all equipment. Chains, slings, ropes, hooks, shackles, eyebolts, pulleys, blocks, gin wheels, sheer legs and other small components are no less important than grabs and winches. All must be tested and thoroughly examined before being put into operation. Further thorough examination is required every 12 months (6 months if for lifting persons), and inspections are to be conducted weekly and documented.

■ All cranes must be clearly marked with their safe maximum working loads relevant to lifting radius and maximum operating radius, particularly when fitted with a derricking jib. Lifting equipment not designed for personnel must be clearly marked as such.

■ Jib cranes must be fitted with an automatic safe load indicator such as a warning light for the operator and a warning bell for persons nearby.

■ Except for testing purposes, the safe working load must not be exceeded.

■ When loads are approaching the safe maximum load, the initial lift should be short. A check should then be made to establish safety and stability before proceeding to complete the lift.

Apart from the legal requirements summarised above, commonsense precautions on site must be taken, such as the clear marking of high-voltage electric cables and leaving the crane in an 'out of service' position when unattended or if storm or high wind conditions prevail. Most tower cranes can operate in wind conditions of up to 60 km/h. The usual 'out of service' position for tower cranes is as follows:

1. Jibs to be left on free slew and pointed in the direction of the wind on the leeward side of the tower.
2. Fuel and power supplies switched off.

3. Load removed.
4. Hook raised to highest position.
5. Hook positioned close to tower.
6. Rail-mounted cranes should have their wheels chocked or clamped.

■ ■ ■ ERECTION OF CRANES

Before commencing to erect a crane, careful consideration must be given to its actual position on the site. As for all forms of plant, maximum utilisation is the ultimate aim: therefore a central position within reach of all storage areas, loading areas and activity areas is required. Generally, output will be in the region of 18 to 20 lifts per hour: therefore the working sequence of the crane needs to be carefully planned and coordinated if full advantage is to be made of the crane's capabilities.

The erection of mast and tower cranes varies with the different makes, but there are several basic methods. Mast cranes are usually transported in a collapsed and folded position and are quickly unfolded and erected on site, using built-in lifting and erection gear. Tower cranes, however, have to be assembled on site. In some cases the superstructure that carries the jib and counterjib is erected on the base frame. The top section of the tower or pintle is raised by internal climbing gear housed within the superstructure; further 3.000 m tower lengths can be added as the pintle is raised until the desired tower height has been reached. The jib and counterjib are attached at ground level to the superstructure, which is then raised to the top of the pintle; this whole arrangement then slews around the static tower.

Another method of assembly and erection adopted by some manufacturers is to raise the first tower section onto a concrete base, and then assemble the jib and counterjib and fix these to the first tower section with the aid of a mobile crane. Using the facilities of the jib, further tower sections can be fitted inside the first section and elevated hydraulically on a telescopic principle; this procedure is repeated until the desired height has been reached. A similar approach to the last method is to have the jib and top tower section fixed to a cantilever bracket arrangement so that it is offset from the main tower. Further sections can be added to the assembly until the required height is reached when the jib assembly can be transferred to the top of the tower.

A procedure using hydraulic jacks and climbing mast is shown in Fig. 2.4.16.

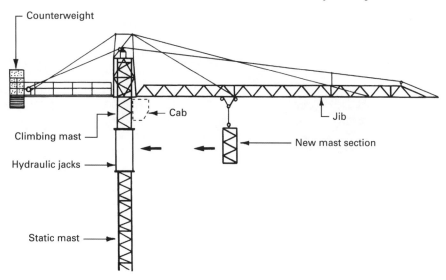

Figure 2.4.16 Crane mast assembly

CONCRETE MIXERS AND PUMPS

The mixing and transporting of concrete and mortar mixes are important activities on most building sites, from the very small to the very large contract. The choice of method for mixing and transporting the concrete or mortar must be made on the basis of the volume of mixed material required in any given time and also on the horizontal and vertical transportation distances involved. Consideration must also be given to the use of ready-mixed concrete, especially where large quantities are required and/or site space is limited.

■ ■ ■ CONCRETE MIXERS

Most concrete mixers used on building sites are derived from the minimum recommendations of BS 1305: *Specification for batch type concrete mixers*. Although obsolescent, this British Standard provides useful references for two basic forms: the drum type or free-fall concrete mixer, and the pan type or forced action concrete mixer. The drum mixers are subdivided into three distinct forms:

- **Tilting drum** (T) – in which the single-compartment drum has an inclinable axis with loading and discharge through the front opening. This form of mixer is primarily intended for small batch outputs ranging from 100T to 200T litres. Note that mixer output capacities are given in litres for sizes up to 1,000 litres and in cubic metres for outputs over 1,000 litres; a letter suffix designating the type is also included in the title. In common with all drum mixers, tilting mixers have fixed blades inside the revolving drum that lift the mixture and at a certain point in each revolution allow the mixture to drop towards the bottom of the drum to recommence the mixing cycle. The complete cycle time for mixing one batch from load to reload is usually specified as $2^{1}/_{2}$ minutes. Typical examples of tilting drum mixers are shown in Fig. 2.5.1.
- **Non-tilting drum** (NT) – in which the single-compartment drum has two openings, and rotates on a horizontal axis; output capacities range from 200NT

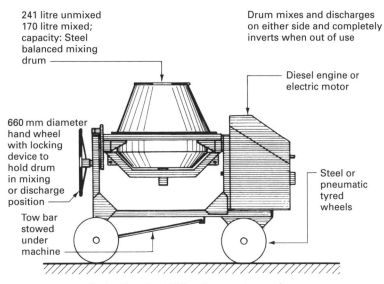

241 litre unmixed 170 litre mixed; capacity: Steel balanced mixing drum

Drum mixes and discharges on either side and completely inverts when out of use

Diesel engine or electric motor

660 mm diameter hand wheel with locking device to hold drum in mixing or discharge position

Steel or pneumatic tyred wheels

Tow bar stowed under machine

Typical (one bag) tilting drum concrete mixer

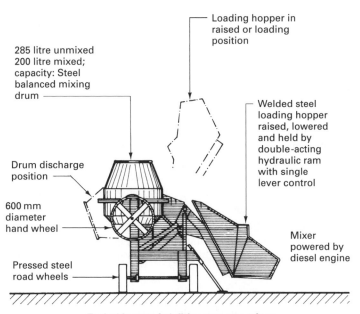

285 litre unmixed 200 litre mixed; capacity: Steel balanced mixing drum

Loading hopper in raised or loading position

Welded steel loading hopper raised, lowered and held by double-acting hydraulic ram with single lever control

Drum discharge position

600 mm diameter hand wheel

Mixer powered by diesel engine

Pressed steel road wheels

Typical hopper fed tilting concrete mixer

Figure 2.5.1 Typical tilting drum mixers (*courtesy*: Liner Concrete Machinery)

to 750NT. Loading is through the front opening and discharge through the rear opening by means of a discharge chute collecting the mixture from the top of the drum. The chute should form an angle of not less than 40° with the horizontal axis of the drum.

■ **Reversing drum** (R) – a more popular version of a mixer with a drum rotating on a horizontal axis than the non–tilting drum mixer described above. Capacities of this type of mixer range from 200R to 500R. Loading is through a front opening and discharge from a rear opening carried out by reversing the rotation of the drum (see Fig. 2.5.2).

Generally, mixers with an output capacity exceeding 200 litres are fitted with an automatic or manually operated water system, which will deliver a measured volume of water to the drum of the mixer. Table 1 of BS 1305 gives recommended minimum water tank capacities for the various mixer sizes.

Forced action mixers (P) are generally for larger capacity outputs than the drum mixers described above, and can be obtained within the range from 200P to 2000P. The mixing of the concrete is achieved by the relative movements between the mix, pan and blades or paddles. Usually the pan is stationary while the paddles or blades rotate, but rotating pan models are also available consisting of a revolving pan and a revolving mixer blade or star giving a shorter mixing time of 30 seconds with large outputs. In general, pan mixers are not easily transported, and for this reason are usually employed only on large sites where it would be an economic proposition to install this form of mixer.

▨ ■ ■ CEMENT STORAGE

Cement for the mixing of mortars or concrete can be supplied in 25 kg bags or in bulk for storage on site prior to use. Bagged cement requires a dry and damp–free store to prevent air setting taking place (see Chapters 1.2 and 2.4 of *Construction Technology*). If large quantities of cement are required an alternative method of storage is the silo, which will hold cement supplied in bulk under ideal conditions. A typical cement silo consists of an elevated welded steel cylindrical container supported on four crossed braced legs with a bottom discharge outlet to the container. Storage capacities range from 12 to 50 tonnes. Silos can be incorporated into an on-site static batching plant, or they can have their own weighing attachments. Some of the advantages of silo storage for large quantities of cement are:

■ Cost of bulk cement is cheaper per tonne than bagged cement.
■ Unloading is by direct pumping from delivery vehicle to silo.
■ Less site space is required for any given quantity to be stored on site.
■ First cement delivered is the first to be used because it is pumped into the top of the silo and extracted from the bottom.

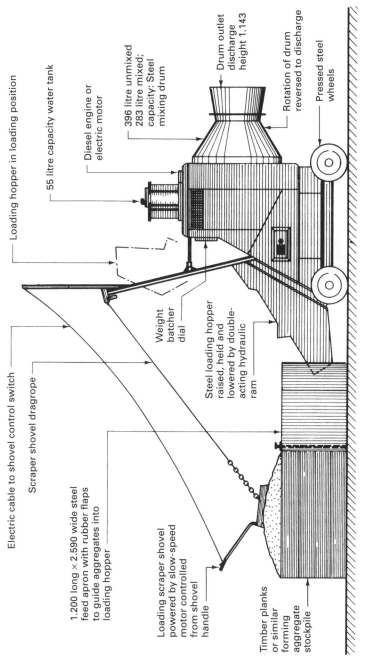

Electric cable to shovel control switch

Loading hopper in loading position

Scraper shovel dragrope

55 litre capacity water tank

Diesel engine or electric motor

396 litre unmixed 283 litre mixed; capacity: Steel mixing drum

Drum outlet discharge height 1.143

Rotation of drum reversed to discharge

Pressed steel wheels

1.200 long × 2.590 wide steel feed apron with rubber flaps to guide aggregates into loading hopper

Weight batcher dial

Steel loading hopper raised, held and lowered by double-acting hydraulic ram

Loading scraper shovel powered by slow-speed motor controlled from shovel handle

Timber planks or similar forming aggregate stockpile

Figure 2.5.2 Typical reversible drum mixer (*courtesy*: Liner Concrete Machinery)

■■■ READY-MIXED MORTAR

Many sites are congested, particularly in town and city centres where land is at a premium and density of building is high. Space for cement-mixing plant and materials is not always available. Furthermore, labour requirements for mixing and transporting mortar about the site have reduced the cost viability of site preparation. By comparison, quality-controlled factory preparation and delivery of mortar offers many cost-saving advantages, including a guarantee of consistency batch after batch. If required, the manufacturer can also declare the following:

- workability and workable life;*
- chloride content;
- air content;
- compressive strength;
- bond strength;
- water absorption;
- density;
- water vapour permeability;
- thermal conductivity;
- prescribed mix proportions (by volume or weight).**

* It is usual to incorporate a cement set retarder in the mix to extend the mortar working life. This is to allow for any delay in transport and the time taken to work through a large batch delivery.
** Mix selection is normally to BS 5628: *Code of practice for use of masonry* and BS EN 1052: *Methods of test for masonry.*

■■■ READY-MIXED CONCRETE

The popularity of ready-mixed concrete has increased tremendously since 1968, when the British Ready-Mixed Concrete Association laid down minimum standards for plant, equipment, personnel and quality control for all BRMCA-approved depots. The ready-mixed concrete industry consumes a large proportion of the total cement output of the United Kingdom in supplying many millions of cubic metres of concrete per annum to all parts of the country.

Ready-mixed concrete is supplied to sites in specially designed truck mixers, which are basically a mobile mixing drum mounted on a lorry chassis. Truck mixers can be employed in one of three ways:

- Loaded at the depot with dry batched materials plus the correct quantity of water, the truck mixer is used to complete the mixing process at the depot before leaving for the site. During transportation to the site the mix is kept agitated by the revolving drum; on arrival the contents are remixed before being discharged.
- Fully or partially mixed concrete is loaded into the truck mixer at the depot. During transportation to the site the mix is agitated by the drum revolving at 1 to 2 revolutions per minute. On arrival the mix is finally mixed by increasing the drum's revolutions to between 10 and 15 revolutions per minute for a few minutes before being discharged.

■ When the time taken to deliver the mix to the site may be unacceptable the mixing can take place on site by loading the truck mixer at the depot with dry batched materials and adding the water upon arrival on site before completing the mixing operation and subsequent discharge.

All forms of truck mixer carry a supply of water that is normally used to wash out the drum after discharging the concrete and before returning to the depot. See Fig. 2.5.3 for typical truck mixer details.

Truck mixers are heavy vehicles, weighing up to 24 tonnes when fully laden, with a turning circle of some 15.000 m; they require both a firm surface and turning space on site. The site allowance time for unloading is usually 30 minutes, allowing for the discharge of a full load in 10 minutes and leaving 20 minutes of free time to permit for a reasonable degree of flexibility in planning and programming to both the supplier and the user. Truck mixer capacities vary with the different models but 4, 5 and 6 m³ are common sizes. Consideration must be given by the contractor as to the best unloading position because most truck mixers are limited to a maximum discharge height of 1.500 m and using a discharging chute to a semicircular coverage around the rear of the vehicle within a radius of 3.000 m.

To obtain maximum advantage from the facilities offered by ready-mixed concrete suppliers, building contractors must place a clear order of the exact requirements, which should follow the recommendations given in BS EN 206-1 and BS's 8500-1 and 2. The supply instructions should contain the following:

■ type of cement;
■ types and maximum sizes of aggregate;
■ test and strength requirements;
■ testing methods;
■ slump or workability requirements;
■ volume of each separate mix specified;
■ delivery programme;
■ any special requirements such as a pumpable mix.

Note: Much of the above can be rationalised by specifying the concrete grade, e.g. C30, and the mix category, e.g. 'Designed'. For more detail, see BS EN 206-1: *Concrete. Specification, performance, production and conformity* and the complementary British Standards to BS EN 206-1, i.e. BS 8500-1: *Concrete. Method of specifying and guidance for the specifier*, and BS 8500-2: *Concrete. Specification for constituent materials and concrete*. See also, the summary in Chapter 2.4 of *Construction Technology*.

■ ■ ■ CONCRETE PUMPS

The advantages of moving large volumes of concrete by using a pump and pipeline are as follows:

■ Concrete is transported from point of supply to placing position in one continuous operation.

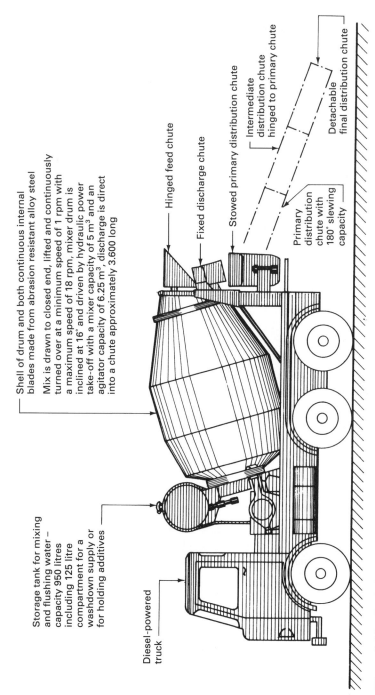

Shell of drum and both continuous internal blades made from abrasion resistant alloy steel

Mix is drawn to closed end, lifted and continuously turned over at a minimum speed of 1 rpm with a maximum speed of 18 rpm, mixer drum is inclined at 16° and driven by hydraulic power take-off with a mixer capacity of 5 m³ and an agitator capacity of 6.25 m³, discharge is direct into a chute approximately 3.600 long

Hinged feed chute

Fixed discharge chute

Stowed primary distribution chute

Intermediate distribution chute hinged to primary chute

Detachable final distribution chute

Primary distribution chute with 180° slewing capacity

Storage tank for mixing and flushing water – capacity 950 litres including 125 litre compartment for a washdown supply or for holding additives

Diesel-powered truck

Figure 2.5.3   Typical ready-mixed concrete truck details

■ Faster pours can be achieved with less labour. Typical placing figures are up to 100 m³ per hour using a two-person crew consisting of the pump operator and an operator at the discharge end.

■ No segregation of mix is experienced with pumping, and a more consistent placing and compaction is obtained, requiring less vibration.

■ Generally, site plant and space requirements are reduced.

■ Only method available for conveying wet concrete both vertically and horizontally in one operation.

■ No shock loading of formwork is experienced.

■ Generally the net cost of placing concrete is reduced.

Against the above advantages must be set the following limitations:

■ Concrete supply must be consistent and regular, which can usually be achieved by well-planned and organised deliveries of ready-mixed concrete. Note that under ideal conditions the discharge rate of each truck mixer can be in the order of 10 minutes.

■ Concrete mix must be properly designed and controlled, because not all concrete mixes are pumpable. The concrete is pumped under high pressure, which can cause bleeding and segregation of the mix: therefore the mix must be properly designed to avoid these problems, as well as having good cohesive, plasticity and self-lubricating properties to enable it to be pumped through the system without excessive pressure and without causing blockages.

■ More formwork will be required to receive the high output of the pump to make its use an economic proposition.

Most pumps used today are of the twin-cylinder hydraulically driven design, as either a trailer pump or a lorry-mounted pump using a small-bore (100 mm diameter) pipeline capable of pumping concrete 85.000 m vertically and 200.000 m horizontally, although these figures will vary with the actual pump used (see Fig. 2.5.4 for a typical example). The delivery pipes are usually of rigid seamless steel in 3.000 m lengths except where flexibility is required, as on booms and at the delivery end. Large-radius bends of up to 1.000 m radius giving $22^1/_2°$, $45°$ and $90°$ turning are available to give flexible layout patterns. Generally, small-diameter pipes of 75 and 100 mm are used for vertical pumping, whereas larger diameters of up to 150 mm are used for horizontal pumping. If a concrete mix with large aggregates is to be pumped, the pipe diameter should be at least three or four times the maximum aggregate size.

The time required on site to set up a pump is approximately 30 to 45 minutes. The pump operator will require a supply of water and grout for the initial coating of the pipeline: this usually requires about two or three bags of cement. A hardstanding should be provided for the pump, with adequate access and turning space for the attendant ready-mixed concrete vehicles. The output of a concrete pump will be affected by the distance the concrete is to be pumped: therefore the pump should be positioned so that it is as close to the discharge point as is practicable. Pours should be planned so that they progress backwards towards the pump, removing the redundant pipe lengths as the work proceeds.

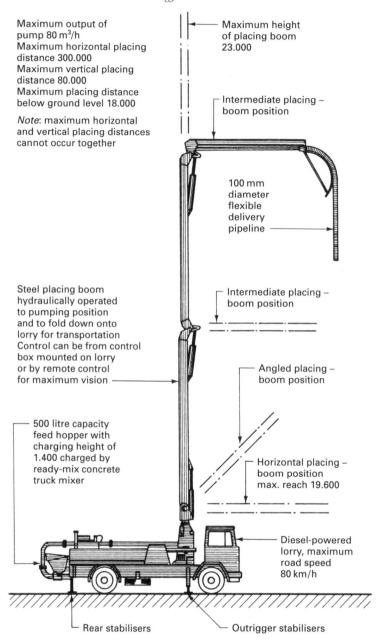

Maximum output of
pump 80 m³/h
Maximum horizontal placing
distance 300.000
Maximum vertical placing
distance 80.000
Maximum placing distance
below ground level 18.000

Note: maximum horizontal
and vertical placing distances
cannot occur together

Maximum height
of placing boom
23.000

Intermediate placing –
boom position

100 mm
diameter
flexible
delivery
pipeline

Steel placing boom
hydraulically operated
to pumping position
and to fold down onto
lorry for transportation
Control can be from control
box mounted on lorry
or by remote control
for maximum vision

Intermediate placing –
boom position

Angled placing –
boom position

500 litre capacity
feed hopper with
charging height of
1.400 charged by
ready-mix concrete
truck mixer

Horizontal placing –
boom position
max. reach 19.600

Diesel-powered
lorry, maximum
road speed
80 km/h

Rear stabilisers

Outrigger stabilisers

Figure 2.5.4   Typical lorry-mounted concrete pump (*courtesy*: Schwing)

Generally, if the volume of concrete to be placed is sufficient to warrant hiring a pump and operators it will result in an easier, quicker and usually cheaper operation than placing the concrete by the traditional method of crane and skip with typical outputs of 15 to 20 m^3 per hour as opposed to the 60 to 100 m^3 per hour output of the concrete pump. Concrete pumping and placing demands a certain amount of skill and experience, and for this reason most pumps in use are hired out and operated by specialist contractors.

■■■ SAFETY NOTE

Mortar and concrete contain cement. Mortar may also contain lime. Both ingredients can burn, and skin contact with fresh mortar or concrete may result in skin ulceration and/or dermatitis. Contact with the skin should be prevented by wearing suitable protective clothing – in particular, reinforced plastic gloves, waterproof overalls and footwear and protection for the eyes. Face protection may also be appropriate for certain applications. Where skin contact occurs, immediate washing with soap and water is recommended. Eye contact requires immediate washing of the affected area with clean water. If cement is swallowed, the mouth should be washed out, followed by drinking plenty of clean water. Cement products are packaged with a safety label, similar to that shown in Fig. 2.5.5.

References:
Control of Substances Hazardous to Health Regulations 1999
Personal Protective Equipment at Work Regulations 1992

HEALTH AND SAFETY WARNING

- Contains chromium (VI).
  May produce an allergic reaction.
- Risk of serious damage to eyes.
- Contact with wet cement, wet concrete or wet mortar may cause irritation, dermatitis or burns.
- Contact between cement powder and body fluids (e.g. sweat and eye fluid) may also cause skin and respiratory irritations, dermatitis or burns.
- Avoid eye and skin contact by wearing suitable eye protection, clothing and gloves.
- Avoid breathing dust.
- On contact with eyes or skin, rinse immediately with plenty of clean water. Seek medical advice after eye contact.
- Keep out of the reach of children.

IRRITANT

Figure 2.5.5   Safety sign on cement materials

2.6

SCAFFOLDING

A scaffold is a temporary frame, usually constructed from steel or aluminium alloy tubes clipped or coupled together to provide a means of access to high-level working areas as well as providing a safe platform from which to work. The two basic forms of scaffolding, namely the putlog scaffold with its single row of uprights or standards set outside the perimeter of the building and partly supported by the structure, and independent scaffolds which have two rows of standards, have been covered in Chapter 1.5 of *Construction Technology*.

It is therefore necessary to consider in this text only the special scaffolds such as slung, suspended, truss-out and gantry scaffolds, as well as the easy-to-erect system scaffolds. It cannot be overemphasised that all scaffolds must comply fully with the minimum requirements set out in the Work at Height Regulations 2005 and BS EN 12811-1: 2003 *Temporary works equipment. Scaffolds. Performance requirements and general design.*

◼◼◼ SLUNG SCAFFOLDS

These are scaffolds that are suspended by means of wire ropes or chains and are not provided with a means of being raised or lowered by a lifting appliance. Their main use is for gaining access to high ceilings or the underside of high roofs. A secure anchorage must be provided for the suspension ropes, and this can usually be achieved by using the structural members of the roof over the proposed working area. Any member selected to provide the anchorage point must be inspected to assess its adequacy. At least six evenly spaced suspension wire ropes or chains should be used, and these must be adequately secured at both ends. The working platform is constructed in a similar manner to conventional scaffolds, consisting of ledgers, transoms and timber scaffold boards with the necessary guard rails and toe boards. Working platforms in excess of 2.400 m × 2.400 m plan size should be checked to ensure that the supporting tubular components are not being overstressed.

▨▦▩ TRUSS-OUT SCAFFOLDS

These are a form of independent tied scaffold that rely entirely on the building for support, and are used where it is impossible or undesirable to erect a conventional scaffold from ground level. The supporting scaffolding structure that projects from the face of the building is known as the truss-out. Anchorage is provided by adjustable struts fixed internally between the floor and ceiling, from which the cantilever tubes project. Except for securing rakers, only right-angle couplers should be used. The general format for the remainder of the scaffold is as used for conventional independent scaffolds (see Fig. 2.6.1).

▨▦▩ SUSPENDED SCAFFOLDS

These consist of a working platform suspended from supports such as outriggers that cantilever over the upper edge of a building, and in this form are a temporary means of access to the face of a building for the purposes of cleaning and light maintenance work. Many new tall structures have suspension tracks incorporated in the fascia or upper edge beam, or a cradle suspension track is fixed to the upper surface of the flat roof on which is supported a manual or power trolley with retractable davit arms for supporting the suspended working platform or cradle. All forms of suspended cradles must conform with the minimum requirements set out in the Work at Height Regulations 2005 with regard to platform boards, guard rails and toe boards. Cradles may be single units or grouped together to form a continuous working platform; if grouped together they are connected to one another at their abutment ends with hinges. Figure 2.6.2 shows typical suspended scaffold details.

▨▦▩ MOBILE TOWER SCAFFOLDS

These are used mainly by painters and maintenance staff to gain access to ceilings where it is advantageous to have a working platform that can be readily moved to a new position. The scaffold is basically a square tower constructed from scaffold tubes mounted on wheels fitted with brakes. Platform access is gained by short opposing inclined ladders or one inclined ladder within the tower base area. See Chapter 1.5 of *Construction Technology* for more detail.

▨▦▩ BIRDCAGE SCAFFOLDS

These are used to provide a complete working platform at high level over a large area, and consist basically of a two-directional arrangement of standards, ledgers and transoms to support a close-boarded working platform at the required height. To ensure adequate stability standards should be placed at not more than 2.400 m centres in both directions, and the whole arrangement must be adequately braced.

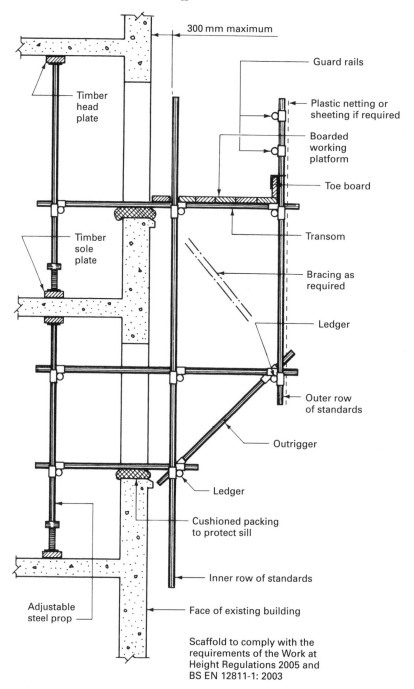

300 mm maximum

Guard rails

Plastic netting or
sheeting if required

Timber
head
plate

Boarded
working
platform

Toe board

Transom

Timber
sole
plate

Bracing as
required

Ledger

Outer row
of standards

Outrigger

Ledger

Cushioned packing
to protect sill

Inner row of standards

Adjustable
steel prop

Face of existing building

Scaffold to comply with the
requirements of the Work at
Height Regulations 2005 and
BS EN 12811-1: 2003

Figure 2.6.1   Typical truss-out scaffold details

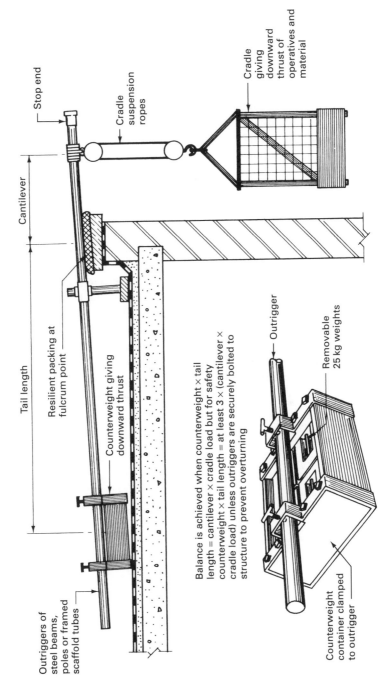

Figure 2.6.2 Typical suspended scaffold details

▨■■ GANTRIES

These are forms of scaffolding used primarily as elevated loading and unloading platforms over a public footpath where the structure under construction or repair is immediately adjacent to the footpath. As for hoardings, local authority permission is necessary, and their specific requirements such as pedestrian gangways, lighting and dimensional restrictions must be fully met. It may also be necessary to comply with police requirements as to when loading and unloading can take place. The gantry platform can also serve as a storage and accommodation area as well as providing the staging from which a conventional independent scaffold to provide access to the face of the building can be erected. Gantry scaffolds can be constructed from standard structural steel components as shown in Fig. 2.6.3 or from a system scaffold as shown in Fig. 2.6.4.

▨■■ SYSTEM SCAFFOLDS

These scaffolds are based upon the traditional independent steel tube scaffold, but instead of being connected together with a series of loose couplers and clips they usually have integral interlocking connections. They are easy to erect, adaptable, and generally can be assembled and dismantled by semi-skilled operatives. The design of these systems is such that the correct position of handrails, lift heights and all other aspects of the Work at Height Regulations 2005 are automatically met. Another advantage found in most of these system scaffolds is the elimination of internal cross-bracing, giving a clear walk-through space at all levels; facade-bracing, however, may still be required. Figure 2.6.5 shows details of a typical system scaffold and, like the illustrations chosen for items of builders' mechanical plant, is intended only to be representative of the many scaffolding systems available.

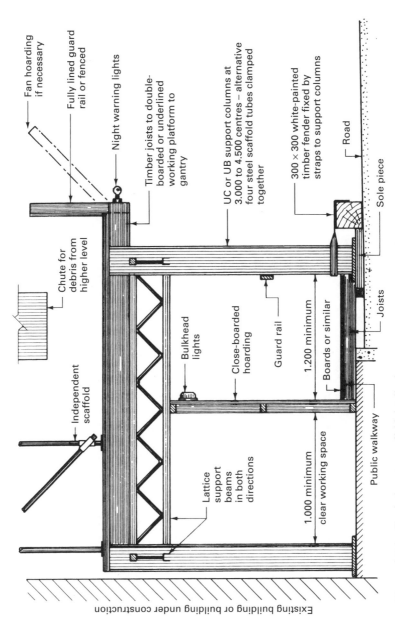

Fan hoarding if necessary

Fully lined guard rail or fenced

Night warning lights

Timber joists to double-boarded or underlined working platform to gantry

UC or UB support columns at 3.000 to 4.500 centres – alternative four steel scaffold tubes clamped together

300 × 300 white-painted timber fender fixed by straps to support columns

Road

Sole piece

Chute for debris from higher level

Bulkhead lights

Close-boarded hoarding

Guard rail

1.200 minimum

Boards or similar

Joists

Independent scaffold

Lattice support beams in both directions

1.000 minimum clear working space

Public walkway

Existing building or building under construction

Figure 2.6.3 Typical gantry scaffold details: 1

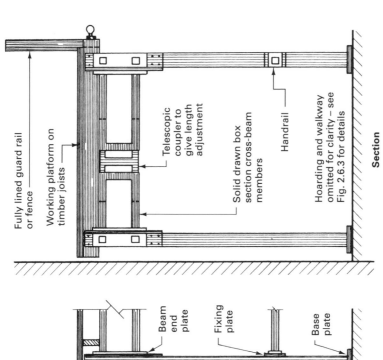

Fully lined guard rail or fence

Working platform on timber joists

Telescopic coupler to give length adjustment

Solid drawn box section cross-beam members

Handrail

Hoarding and walkway omitted for clarity – see Fig. 2.6.3 for details

Section

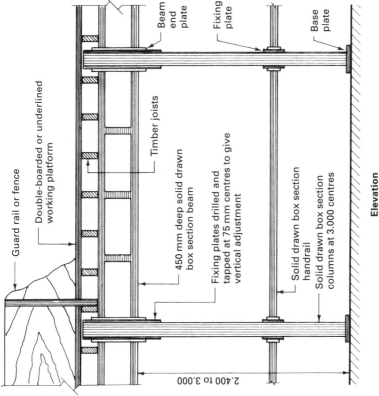

Guard rail or fence

Double-boarded or underlined working platform

Timber joists

Beam end plate

Fixing plate

Base plate

450 mm deep solid drawn box section beam

Fixing plates drilled and tapped at 75 mm centres to give vertical adjustment

Solid drawn box section handrail

Solid drawn box section columns at 3.000 centres

2.400 to 3.000

Elevation

Figure 2.6.4 Typical gantry scaffold details: 2

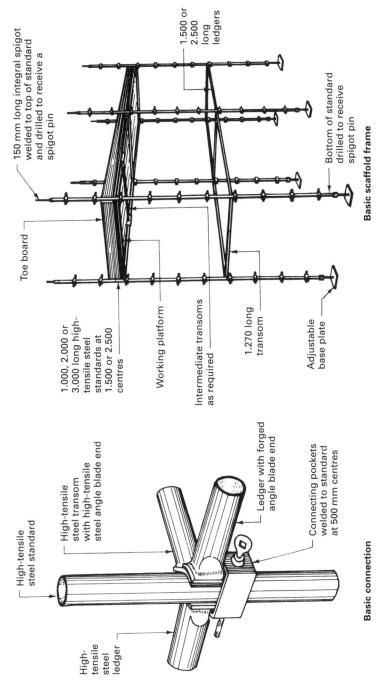

150 mm long integral spigot welded to top of standard and drilled to receive a spigot pin

1.500 or 2.500 long ledgers

Basic scaffold frame

Bottom of standard drilled to receive spigot pin

1.000, 2.000 or 3.000 long high-tensile steel standards at 1.500 or 2.500 centres

Toe board

Working platform

Intermediate transoms as required

1.270 long transom

Adjustable base plate

High-tensile steel standard

High-tensile steel transom with high-tensile steel angle blade end

Ledger with forged angle blade end

Connecting pockets welded to standard at 500 mm centres

High-tensile steel ledger

Basic connection

Figure 2.6.5   Typical system scaffold (*courtesy:* SGB Anglok Scaffolding)

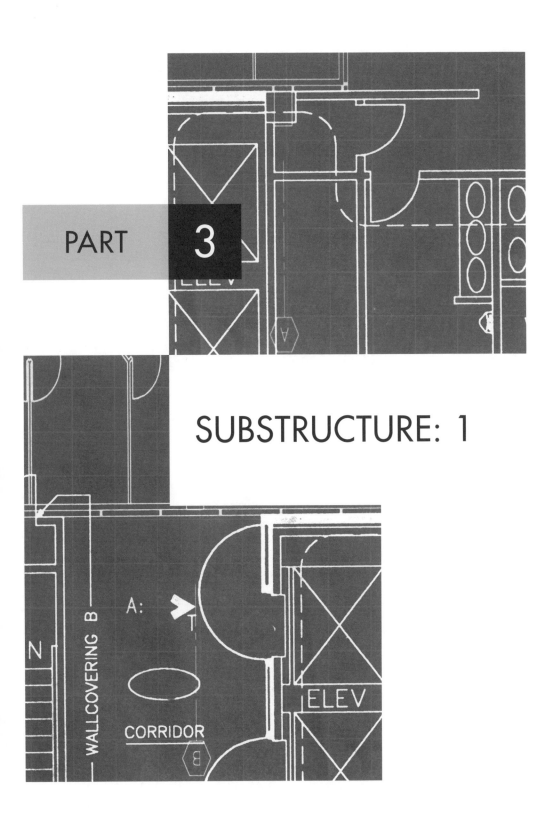

PART **3**

SUBSTRUCTURE: 1

GROUNDWATER CONTROL

Groundwater can be defined as water that is held temporarily in the soil above the level of the water table. Below the water table level is the subsoil water, which is the result of the natural absorption by subsoils of the groundwater. Both types of water can be effectively controlled by a variety of methods, which have been designed either to exclude the water from a particular area or merely to lower the water table to give reasonably dry working conditions, especially for excavation activities.

The extent to which water affects the stability and bearing capacity of a subsoil will depend upon the physical characteristics of the soil and in particular upon the particle size, which ranges from the very fine particles of clay soils to the larger particles or boulders of some granular soils. The effect of the water on these particles is that of a lubricant, enabling them to move when subjected to a force such as a foundation loading or simply causing them to flow by movement of the groundwater. The number and disposition of the particles together with the amount of water present will determine the amount of movement that can take place. The finer particles will be displaced more easily than the larger particles, which could create voids, thus encouraging settlement of the larger particles.

The voids caused by excavation works encourage water to flow, because the opposition to the groundwater movement provided by the soil has been removed. In cases where the flow of water is likely, an artificial opposition must be installed or the likelihood of water movement must be restricted by geotechnical processes. These processes can be broadly classified into one of two groups:

■ permanent exclusion of groundwater;
■ temporary exclusion of groundwater by lowering the water table.

■■■ PERMANENT EXCLUSION OF GROUNDWATER

Sheet piling

This is suitable for all types of soil except boulder beds, and is used to form a barrier or cut-off wall to the flow of groundwater. The sheet piling can be of a permanent nature, being designed to act as a retaining wall, or it can be a temporary enclosure to excavation works in the form of a cofferdam (for details see Chapter 3.3). Vibration and noise due to the driving process may render this method unacceptable, and the capital costs can be high unless they can be apportioned over several contracts on a use and re-use basis.

Diaphragm walls

These are suitable for all types of soil, and are usually of *in-situ* reinforced concrete installed using the bentonite slurry method (see Chapter 3.2 for details). This form of diaphragm wall has the advantages of low installation noise and vibration; it can be used in restricted spaces and can be installed close to existing foundations. Generally, unless the diaphragm wall forms part of the permanent structure, this method is uneconomic.

Slurry trench cut-off

These are non-structural thin cast *in-situ* unreinforced diaphragm walls suitable for subsoils of silts, sands and gravels. They can be used on sites where there is sufficient space to enclose the excavation area with a cut-off wall of this nature sited so that there is sufficient earth remaining between the wall and the excavation to give the screen or diaphragm wall support. Provided adequate support is given, these walls are rapidly installed and are cheaper than the structural version.

Thin-grouted membrane

This is an alternative method to the slurry trench cut-off wall when used in silt and sand subsoils, and is also suitable for installation in very permeable soils and made-up ground where bentonite methods are unsuitable. Like the previous example, ample earth support is required for this non-structural cut-off wall. The common method of formation is to drive into the ground a series of touching universal beam or column sections, sheet pile sections or, alternatively, small steel box sections to the required depth. A grout injection pipe is fixed to the web or face of the section, and this is connected by means of a flexible pipe to a grout pump at ground level. As the sections are withdrawn, the void created is filled with cement grout to form the thin membrane (see Fig. 3.1.1).

Contiguous piling

This is an alternative method to the reinforced concrete diaphragm wall, consisting of a series of interlocking reinforced concrete-bored piles. The formation of the

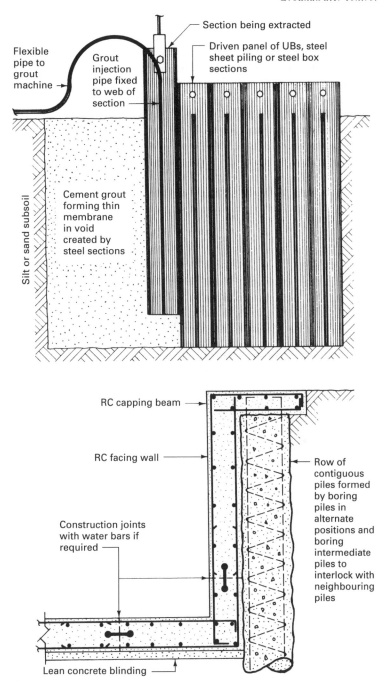

Flexible pipe to grout machine

Grout injection pipe fixed to web of section

Section being extracted

Driven panel of UBs, steel sheet piling or steel box sections

Silt or sand subsoil

Cement grout forming thin membrane in void created by steel sections

RC capping beam

RC facing wall

Construction joints with water bars if required

Row of contiguous piles formed by boring piles in alternate positions and boring intermediate piles to interlock with neighbouring piles

Lean concrete blinding

Figure 3.1.1 Thin grouted membrane and contiguous piling

bored piles can be carried out as described in Chapter 4.2, ensuring that the piles interlock for their entire length. This will require special cutting tools to form the key in the alternate piles for the interlocking intermediate piles. The pile diameter selected will be determined by the strength required after completion of the excavations to one side of the wall. The usual range of diameters used is between 300 and 600 mm. Contiguous piling can be faced with a reinforced rendering or covered with a mesh reinforcement sprayed with concrete to give a smooth finish. This latter process is called shotcrete or gunite. An alternative method is to cast in front of the contiguous piling a reinforced wall terminating in a capping beam to the piles (see Fig. 3.1.1).

Cement grouts

In common with all grouting methods, cement grouts are used to form a 'curtain' in soils that have high permeability, making temporary exclusion pumping methods uneconomic. Cement grouts are used in fissured and jointed rock strata, and are injected into the ground through a series of grouting holes bored into the ground in lines, with secondary intermediate borehole lines if necessary. The grout can be a mixture of neat cement and water, cement and sand up to a ratio of 1:4, or PFA (pulverised fuel ash) and cement in the ratio of 1:1 with 2 parts of water by weight. The usual practice is to start with a thin grout and gradually reduce the water:cement ratio as the process proceeds, to increase the viscosity of the mixture. To be effective this form of treatment needs to be extensive.

Clay/cement grouting

This is suitable for sands and gravels where the soil particles are too small for cement grout treatment. The grout is introduced by means of a sleeve grout pipe that limits its spread, and as for the cement grouting the equipment is simple and can be used in a confined space. The clay/cement grout is basically bentonite with additives such as Portland cement or soluble silicates to form the permanent barrier. One disadvantage of this method is that at least 4.000 m of natural cover is required to provide support for the non-structural barrier.

Chemical grouting

This is suitable for use in medium-to-coarse sands and gravels to stabilise the soil, and can also be used for underpinning works below the water-table level. The chemicals are usually mixed prior to their injection into the ground through injection pipes inserted at 600 mm centres. The chemicals form a permanent gel or sol in the earth, which increases the strength of the soil and also reduces its permeability. This method, in which a liquid base diluted with water is mixed with a catalyst to control the gel-setting time before being injected into the ground, is called the **one-shot method**. An alternative **two-shot method** can be used: it is carried out by injecting the first chemical (usually sodium silicate) into the ground and immediately afterwards injecting the second chemical (calcium chloride) to form a silica gel. The reaction of the two chemicals is immediate, whereas in the one-shot method the reaction of the chemicals can be delayed to allow for full penetration of the subsoil, which will in turn allow a wider spacing of the boreholes. One main disadvantage of chemical grouting is the need for at least 2.000 m of natural cover.

Resin grouts

These are suitable for silty fine sands or for use in conjunction with clay/cement grouts for treating fine strata, but like the chemical grouts described above they can be costly unless used on large works. Resin grouts are similar in application to the chemical grouts, but have a low viscosity, which enables them to penetrate the fine sands that are unsuitable for chemical grouting applications.

Bituminous grouts

These are suitable for injection into fine sands to decrease the permeability of the soil, but they will not increase the strength of the soil, and are therefore unsuitable for underpinning work.

Grout injection

Grouts of all kinds are usually injected into the subsoil by pumping in the mixture at high pressure through tubes placed at the appropriate centres according to the solution being used and/or the soil type. Soil investigation techniques will reveal the information required to enable the engineer to decide upon the pattern and spacing of the grout holes, which can be drilled with pneumatic tools or tipped drills. The pressure needed to ensure a satisfactory penetration of the subsoil will depend upon the soil conditions and results required, but is usually within the range of 1 N/mm^2 for fine soils to 7 N/mm^2 for cement grouting in fissured and jointed rock strata.

Freezing

This is a suitable method for all types of subsoil with a moisture content in excess of 8% of the voids. The basic principle is to insert freezing tubes into the ground and circulate a freezing solution around the tubes to form ice in the voids, thus creating a wall of ice to act as the impermeable barrier. This method will give the soil temporary extra mechanical strength, but there is a slight risk of ground heave, particularly when operating in clays and silts. The circulating solution can be a brine of magnesium chloride or calcium chloride at a temperature of between −15 and −25 °C, which would take between 10 to 17 days to produce an ice wall 1.000 m thick according to the type of subsoil. For works of short duration where quick freezing is required the more expensive liquid nitrogen can be used as the circulating medium. A typical freezing arrangement is shown in Fig. 3.1.2. Freezing methods of soil stabilisation are especially suitable for excavating deep shafts and driving tunnels.

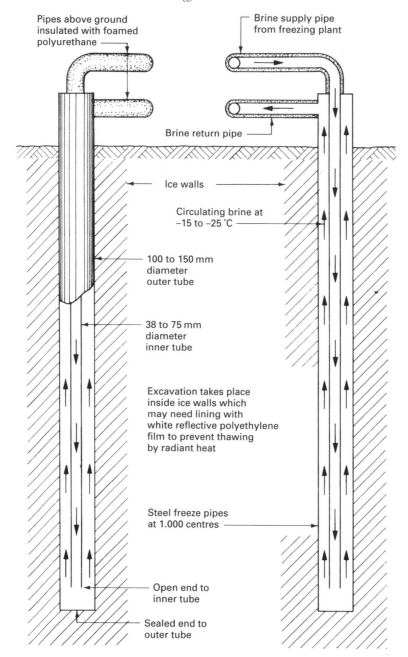

Pipes above ground
insulated with foamed
polyurethane

Brine supply pipe
from freezing plant

Brine return pipe

Ice walls

Circulating brine at
−15 to −25 °C

100 to 150 mm
diameter
outer tube

38 to 75 mm
diameter
inner tube

Excavation takes place
inside ice walls which
may need lining with
white reflective polyethylene
film to prevent thawing
by radiant heat

Steel freeze pipes
at 1.000 centres

Open end to
inner tube

Sealed end to
outer tube

Figure 3.1.2   Exclusion of groundwater by freezing

▨▨■ TEMPORARY EXCLUSION OF GROUNDWATER

Sump pumping

This is suitable for most subsoils and, in particular, gravels and coarse sands when working in open shallow excavations. The sump or water collection pit should be excavated below the formation level of the excavation and preferably sited in a corner position to reduce to a minimum the soil movement due to settlement, which is a possibility with this method. Open sump pumping is usually limited to a maximum depth of 7.500 m because of the limitations of suction lift of most pumps. An alternative method to the open sump pumping is the jetted sump, which will achieve the same objective and will also prevent the soil movement. In this method a metal tube is jetted into the ground and the void created is filled with a sand medium, a disposable hose and a strainer as shown in Fig. 3.1.3.

Wellpoint systems

These are popular methods for water lowering in non-cohesive soils up to a depth of between 5.000 and 6.000 m. To dewater an area beyond this depth requires a multi-stage installation (see Fig. 3.1.4). The basic principle is to water-jet into the ground a number of small diameter wells, which are connected to a header pipe attached to a vacuum pump (see Fig. 3.1.5). Wellpoint systems can be installed with the header pipe acting as a ring main enclosing the area to be excavated. The header pipe should be connected to two pumps, the first for actual pumping operations and the second as a standby pump, because it is essential to keep the system fully operational to avoid collapse of the excavation should a pump failure occur. The alternative system is the progressive line arrangement, where the header pipe is placed alongside a trench or similar excavation to one side or both sides according to the width of the excavation. A pump is connected to a predetermined length of header pipe, and further well points are jetted in ahead of the excavation works. As the work including backfilling is completed the redundant well points are removed and the header pipe is moved forwards.

Shallow-bored wells

These are suitable for sandy gravels and water-bearing rocks, and the action is similar in principle to wellpoint pumping but is more appropriate for installations that have to be pumped for several months because running costs are generally lower. This method is subject to the same lift restrictions as wellpoint systems and can be arranged as a multi-stage system if the depth of lowering exceeds 5.000 m.

Deep-bored wells

These can be used as an alternative to a multi-stage wellpoint installation where the groundwater needs to be lowered to a depth greater than 9.000 m. The wells are formed by sinking a 300 to 600 mm diameter steel lining tube into the ground to the

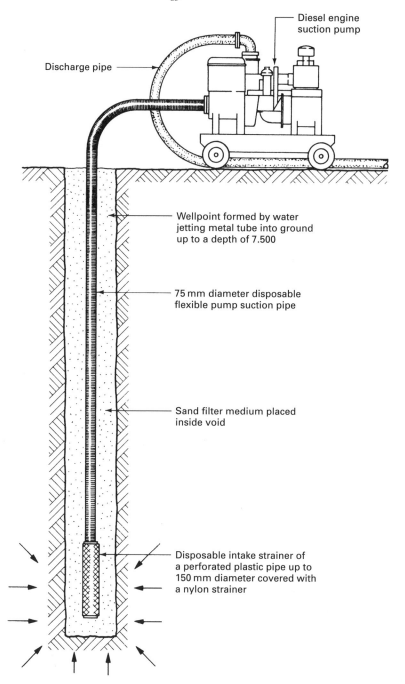

Diesel engine
suction pump

Discharge pipe

Wellpoint formed by water
jetting metal tube into ground
up to a depth of 7.500

75 mm diameter disposable
flexible pump suction pipe

Sand filter medium placed
inside void

Disposable intake strainer of
a perforated plastic pipe up to
150 mm diameter covered with
a nylon strainer

Figure 3.1.3 Typical jetted sump detail

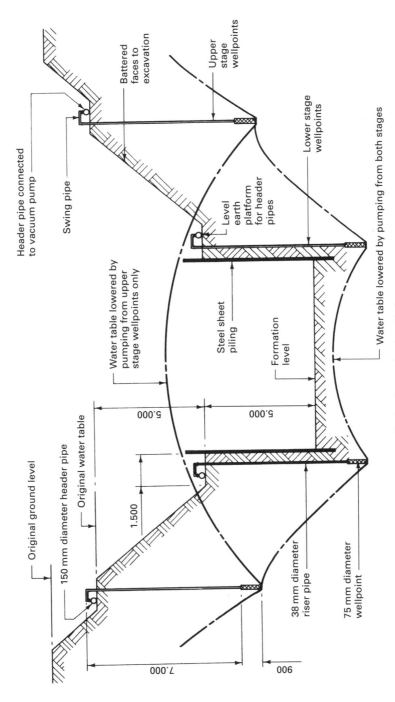

Figure 3.1.4 Typical example of a multi-stage wellpoint dewatering installation

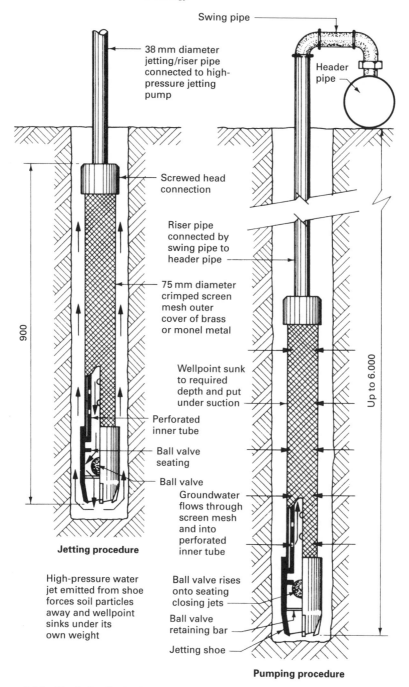

Swing pipe

38 mm diameter
jetting/riser pipe
connected to high-
pressure jetting
pump

Header
pipe

Screwed head
connection

Riser pipe
connected by
swing pipe to
header pipe

75 mm diameter
crimped screen
mesh outer
cover of brass
or monel metal

Wellpoint sunk
to required
depth and put
under suction

900

Perforated
inner tube

Ball valve
seating

Ball valve

Groundwater
flows through
screen mesh
and into
perforated
inner tube

Up to 6.000

Jetting procedure

High-pressure water
jet emitted from shoe
forces soil particles
away and wellpoint
sinks under its
own weight

Ball valve rises
onto seating
closing jets

Ball valve
retaining bar

Jetting shoe

Pumping procedure

Figure 3.1.5   Typical wellpoint installation details

required depth and at spacings to suit the subsoil being dewatered. This borehole allows a perforated well liner to be installed with an electro-submersible pump to extract the water. The annular space is filled with a suitable medium such as sand and gravel as the outer steel lining tube is removed (see Fig. 3.1.6).

Horizontal groundwater control

The pumping methods described above all work on a completely vertical system. An alternative is the horizontal system of dewatering, which consists of installing into the ground a 100 mm diameter PVC perforated suction pipe covered with a nylon filter sleeve to prevent the infiltration of fine particles. The pipe is installed using a special machine that excavates a narrow trench, lays the pipe, and backfills the excavation in one operation at speeds up to 180 m per hour with a maximum depth of 5.000 m. Under average conditions a single pump can handle approximately 230.000 m of pipe run; for distances in excess of the pumping length an overlap of consecutive pipe lengths of up to 4.000 m is required (see Fig. 3.1.7).

Electro-osmosis

This is an uncommon and costly method, which can be used for dewatering cohesive soils such as silts and clays where other pumping methods would not be adequate. It works on the principle that soil particles carry a negative charge, which attracts the positively charged ends of the water molecules, creating a balanced state; if this balance is disturbed the water will flow. The disturbance of this natural balance is created by inserting into the ground two electrodes and passing an electric charge between them. The positive electrode can be of steel rods or sheet piling, which will act as the anode, and a wellpoint is installed to act as the cathode or negative electrode. When an electric current is passed between the anode and cathode it causes the positively charged water molecules to flow to the wellpoint (cathode), where it is collected and pumped away to a discharge point. The power consumption for this method can vary from 1 kW/m^3 for large excavations up to 12 kW/m^3 of soil dewatered for small excavations, which will generally make this method uneconomic on running costs alone.

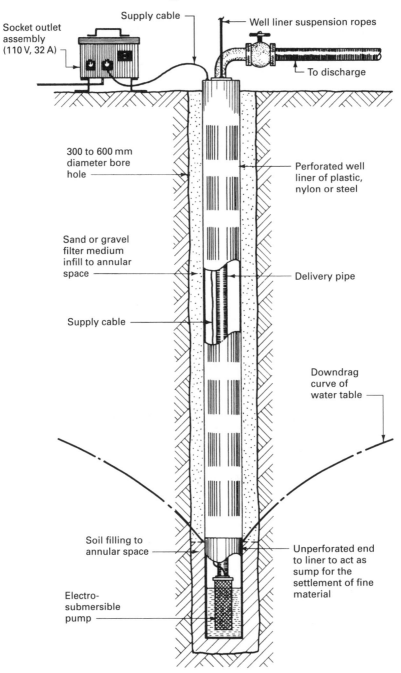

Figure 3.1.6   Typical deep-bored well details

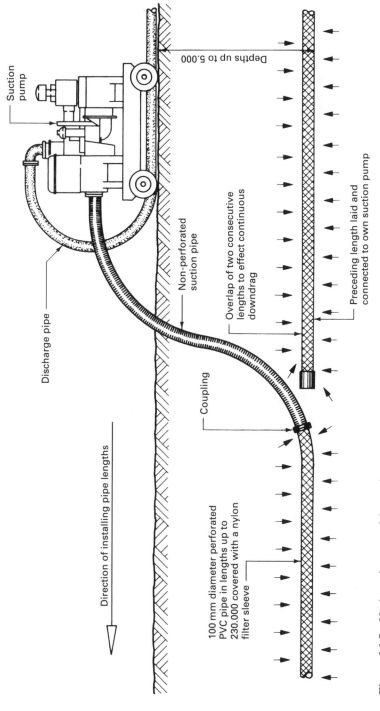

Suction pump

Depths up to 5.000

Non-perforated suction pipe

Overlap of two consecutive lengths to effect continuous downdrag

Preceding length laid and connected to own suction pump

Discharge pipe

Coupling

Direction of installing pipe lengths

100 mm diameter perforated PVC pipe in lengths up to 230.000 covered with a nylon filter sleeve

Figure 3.1.7 Horizontal system of dewatering

3.2

DEEP TRENCH EXCAVATIONS

Any form of excavation on a building site is a potential hazard, and although statistics show that, of the 46,000 or so reportable accidents occurring each year on building sites, excavations do not constitute the major hazard, they can often prove to be serious; indeed, approximately 1 in 10 accidents occurring in excavations are fatal.

■■■ THE CONSTRUCTION (DESIGN AND MANAGEMENT) REGULATIONS 1994, THE CONSTRUCTION (HEALTH, SAFETY AND WELFARE) REGULATIONS 1996, THE CONFINED SPACES REGULATIONS 1997 AND THE WORK AT HEIGHT REGULATIONS 2005

These statutory instruments are effected by the Factory Acts 1961 and the Health and Safety at Work etc. Act 1974. The Design and Management Regulations were introduced with particular regard for a minimum degree of safety for operatives working in the process of construction. The Health, Safety and Welfare Regulations have specific references to excavations, shafts, tunnels, demolitions and work on or adjacent to water, the Confined Spaces Regulations include, among other references, chambers, pits and trenches, and the Work at Height Regulations apply to work in any place, including at or below ground level. All of these regulations apply to building operations and works of engineering construction, because the risks encountered in the two industries are similar and therefore a common code of practice is desirable.

The Health, Safety and Welfare Regulations 12 and 13 establish the requirements for the supply and use of adequate support to excavations. These must be installed by competent, qualified or trained operatives, or be supervised and inspected by a competent person. Construction work involving excavation must be inspected every shift, with reports filed detailing each inspection and the approval of safe conditions.

All contractors and subcontractors are responsible for the safety of their own employees, and every person employed must cooperate in observing the various requirements of the regulations. If an employee discovers any defect or unsafe condition in a working area it is that person's duty to report the facts to the employer, the foreman, or a person appointed by the employer as safety supervisor.

■■■ DEEP TRENCH EXCAVATIONS

Deep trenches may be considered as those over 3.000 m deep; they are usually required for the installation of services, because deep foundations are not very often encountered owing to the more economic alternatives such as piling and raft techniques. Trench excavations should not be opened too far in advance of the proposed work, and any backfilling should be undertaken as soon as practicable after the completion of the work. These two precautions will lessen the risk of falls, flooding and damage to completed work as well as releasing the timbering for reuse at the earliest possible date. Great care must be taken in areas where underground services are present: these should be uncovered with care, protected and supported as necessary. The presence of services in an excavation area may restrict the use of mechanical plant to the point where its use becomes uneconomic. Hand trimming should be used for bottoming the trench, side trimming, end trimming and for forming the gradient just prior to the pipe, cable or drain laying.

In general, all deep excavations should be close boarded or sheeted as a precautionary measure, the main exception being hard and stable rock subsoils. Any excavation in a rock stratum should be carefully examined to ascertain its stability. Fissures or splits in the rock layers that slope towards the cut face may lead to crumbling or rock falls, particularly when exposed to the atmosphere for long periods. In this situation it would be prudent to timber the faces of the excavation according to the extent and disposition of the fissures.

In firm subsoils it might be possible to complete the excavation dig before placing the timbering in position. The general method of support for the excavation sides follows that used in shallow and medium–depth trench excavations studied in Chapter 2.1 of *Construction Technology*, except that larger sections are used to resist the increased pressures encountered as the depth of excavation increases (see Fig. 3.2.1).

If the subsoil is weak, waterlogged or running sand it will be necessary to drive timber runners, trench sheeting or interlocking sheet piles of steel, timber or precast concrete ahead of the excavation dig. This can be accomplished by driving to a depth exceeding the final excavation depth or by the drive-and-dig system, ensuring that timbering is always in advance of the excavating operation. Long runners or sheet piles will require a driving frame to hold and guide the members while being driven. To avoid the use of large piling frames and heavy driving equipment the tucking and pile framing techniques may be used. **Tucking framing** will give an approximately parallel working space width, but will

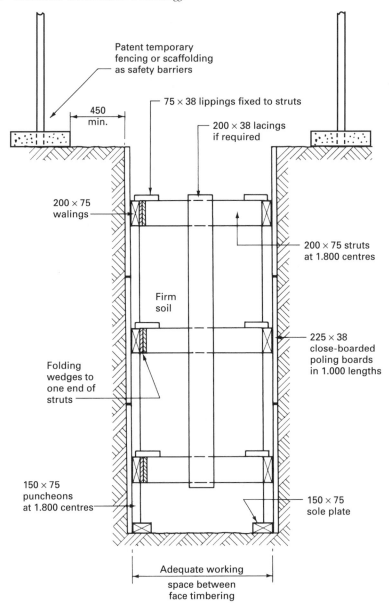

Patent temporary
fencing or scaffolding
as safety barriers

450
min.

75 × 38 lippings fixed to struts

200 × 38 lacings
if required

200 × 75
walings

200 × 75 struts
at 1.800 centres

Firm
soil

225 × 38
close-boarded
poling boards
in 1.000 lengths

Folding
wedges to
one end of
struts

150 × 75
puncheons
at 1.800 centres

150 × 75
sole plate

Adequate working
space between
face timbering

Figure 3.2.1   Traditional deep trench timbering

necessitate driving the short runners at an angle, whereas **pile framing** will give a diminishing working space but is easier to install. Both methods, having the bottom and top of consecutive members secured with a single strut, will give a saving on the total number of struts required over the more traditional methods (see Fig. 3.2.2).

The sizes of suitable timber walings can be calculated using Rankine's formula:

$$p = wh \left(\frac{1 - \sin \theta}{1 + \sin \theta} \right) \times 9.81$$

where p = approximate pressure at base of excavation (N/m²)
  w = mass of soil (typically 1,500–2,000 kg/m³)
  h = height of excavation (m)
  θ = soil angle of repose or shearing resistance (usually 30°)

The resultant thrust will act through the centre of gravity of the pressure diagram ($h/3$): see Fig. 3.2.3. The overturning moment (P) is expressed as a variation on Rankine's formula:

$$P = \left(\frac{wh^2}{2} \right) \left(\frac{1 - \sin \theta}{1 + \sin \theta} \right) \times 9.81$$

Therefore, for a 3.000 m high excavation in soil of 2,000 kg/m³ mass with $\theta = 30°$:

$$P = \frac{2,000 \times 3^2}{2} \times \frac{1 - 0.5}{1 + 0.5} \times 9.81 = 29.43 \text{ kN}$$

Bending moment = $wl^2/8 = (29.43 \times 1 \times 1) \div 8 = 3.68$ kN m

For timber beam design, the moment of resistance (M) or bending moment is given by

$$M = \frac{fbd^2}{6}$$

where f = permissible fibre stress of the timber, according to species, moisture content and stress grade
  b = breadth of timber section (usually assumed)
  d = depth of timber section to be calculated

Assuming breadth of waling (b) to be 75 mm and the fibre stress (f) of timber to be 7.4 N/mm², substituting in the design formula:

$$d^2 = \frac{6 \times M \times 10^6}{f \times b} = \frac{6 \times 3.68 \times 10^6}{7.4 \times 75} = 39783, \therefore \text{ d} = 200 \text{ mm}$$

Dimensions of waling timbers = 200 × 75 mm

Note: Bending moment of $wl^2/10$ for runners and $wl^2/8$ for short poling boards.

Remember that timbering is a general term used to cover all forms of temporary support to the sides of an excavation to:

Provide edge safety barriers as required

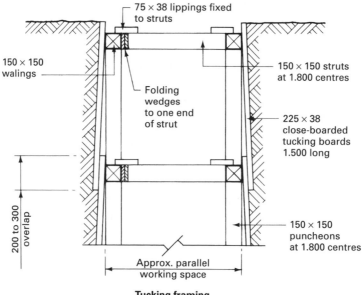

Tucking framing

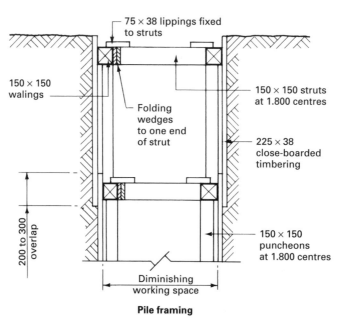

Pile framing

Figure 3.2.2   Timbering: tucking and pile framing

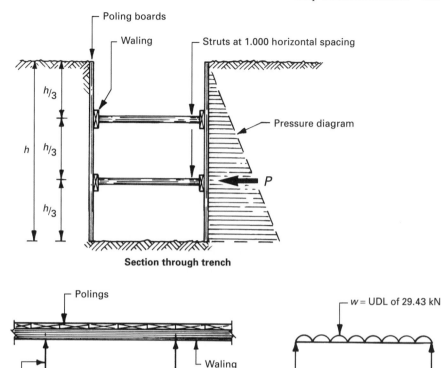

Figure 3.2.3   Design principles for earth support members

- prevent collapse of the excavation sides that would endanger the operatives working in the immediate vicinity;
- keep the excavation open during the required period.

The term is used both when timber is actually used and when a different material is employed to fulfil the same function.

■■■ DEEP BASEMENT EXCAVATIONS

The methods that can be used to support the sides of deep basement excavations can be considered under four headings:

- perimeter trench;
- raking struts;
- cofferdams;
- diaphragm walls.

PERIMETER TRENCH

This method is employed where weak subsoils are encountered, and is carried out by excavating a perimeter trench around the proposed basement excavation. The width and depth of the trench must be sufficient to accommodate the timbering and basement retaining wall, and provide adequate working space. The trench can be timbered by any of the methods described above for deep trenches in weak subsoils. The bottom of the trench should be graded and covered with a 50 to 75 mm blinding layer of weak concrete or coarse sand to protect the base of the excavation from drying and shrinking, and to form a definite level from which to set out and construct the basement wall. The base of the wall should be cast first, with a kicker formed for the stem and starter bars left projecting for the stem and the base slab. The stem or wall should be cast in suitable lifts, and as it cures the struts are transferred to the new wall. When the construction of the perimeter wall has been completed the mound of soil, or dumpling, in the centre of the basement area can be excavated and the base slab cast (see Fig. 3.2.4). Although this method is intended primarily for weak subsoils it can also be used in firm soils, where it may be possible to excavate the perimeter trench completely before placing the timbering in position.

RAKING STRUTS

This method is used where it is possible to excavate the basement area back to the perimeter line without the need for timbering: therefore firm subsoils must be present. The perimeter is trimmed and the timbering placed in position and strutted by using raking struts converging on a common grillage platform similar to raking shoring; alternatively each raker is taken down to a separate sole plate and the whole arrangement adequately braced. An alternative method is to excavate back to the perimeter line on the subsoil's natural angle of repose and then cast the base slab to protect the excavation bottom from undue drying out and subsequent shrinkage. The perimeter trimming can now be carried out, timbered and strutted using the base slab as the abutment (see Fig. 3.2.5). The retaining wall is cast in stages as with the perimeter trench method, and the strutting is transferred to the stem as the work proceeds.

COFFERDAMS

The term 'cofferdam' comes from the French word *coffre* meaning a box, which is an apt description as a cofferdam consists of a watertight perimeter enclosure, usually of interlocking steel sheet piles, used in conjunction with waterlogged sites or actually in water. The enclosing cofferdam is constructed, adequately braced, and the earth is excavated from within the enclosure. Any seepage of water through the cofferdam can normally be handled by small pumps. The sheet piles can be braced and strutted by using a system of raking struts, horizontal struts or tie rods and ground anchors (see Fig. 3.2.6). Cofferdams are considered in greater detail, along with caissons, in the next chapter.

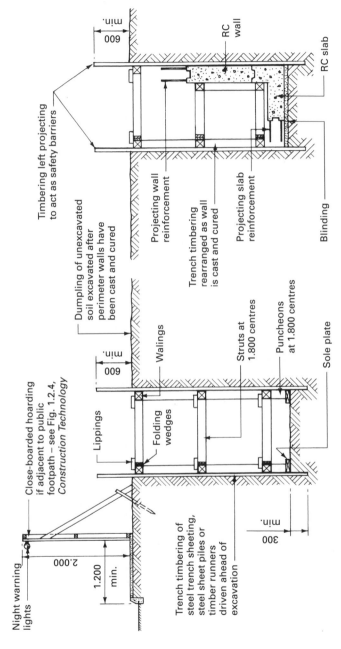

Figure 3.2.4   Basement timbering: perimeter trench method

Provide edge safety barriers as required

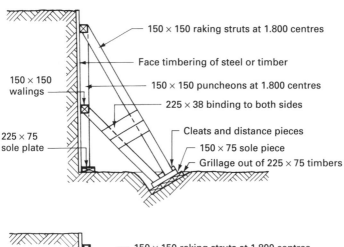

150 × 150 raking struts at 1.800 centres

Face timbering of steel or timber

150 × 150 walings

150 × 150 puncheons at 1.800 centres

225 × 38 binding to both sides

Cleats and distance pieces

225 × 75 sole plate

150 × 75 sole piece

Grillage out of 225 × 75 timbers

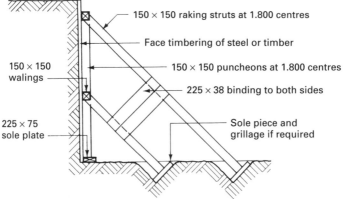

150 × 150 raking struts at 1.800 centres

Face timbering of steel or timber

150 × 150 walings

150 × 150 puncheons at 1.800 centres

225 × 38 binding to both sides

225 × 75 sole plate

Sole piece and grillage if required

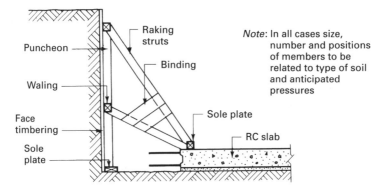

Puncheon

Raking struts

Binding

Waling

Face timbering

Sole plate

Sole plate

RC slab

Note: In all cases size, number and positions of members to be related to type of soil and anticipated pressures

Figure 3.2.5   Basement timbering: raking struts

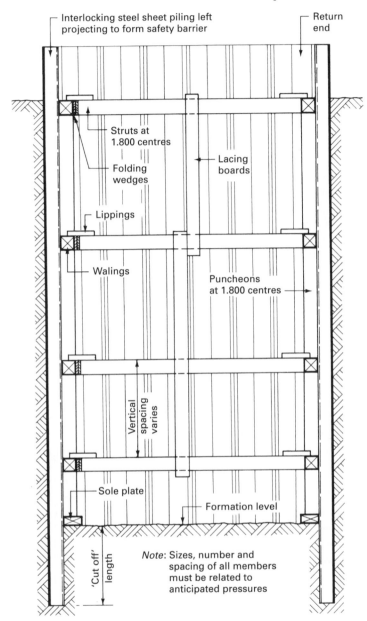

┌ Interlocking steel sheet piling left
  projecting to form safety barrier

┌ Return
  end

┌ Struts at
  1.800 centres

┌ Folding
  wedges

┌ Lacing
  boards

┌ Lippings

┌ Walings

Puncheons
at 1.800 centres →

Vertical
spacing
varies

┌ Sole plate

┌ Formation level

'Cut off'
length

Note: Sizes, number and
spacing of all members
must be related to
anticipated pressures

Figure 3.2.6   Cofferdams: basic principles

DIAPHRAGM WALLS

A diaphragm can be defined as a dividing membrane, and in the context of building a diaphragm wall can be used as a retaining wall to form the perimeter wall of a basement structure, to act as a cut-off wall for river or similar embankments, and to retain large masses of soil such as a side wall of a road underpass.

In-situ concrete diaphragm walls are being used to a large extent in modern construction work and can give the following advantages:

- The final wall can be designed and constructed as the required structural wall.
- Diaphragm walls can be constructed before the bulk excavation takes place, thus eliminating the need for temporary works such as timbering.
- Methods that can be employed to construct the wall are relatively quiet and have little or no vibration.
- Work can be carried out immediately adjacent to an existing structure.
- They may be designed to resist vertical and/or horizontal forces.
- Walls are watertight when constructed.
- Virtually any plan shape is possible.
- Overall they are an economic method for the construction of basement or retaining walls.

There are two methods by which a cast *in-situ* diaphragm wall may be constructed:

- touching or interlocking bored piles;
- excavation of a trench using the bentonite slurry method.

The formation of bored piles is fully described later in Chapter 4.2 on piling. The piles can be bored so that their interfaces are just touching; or by boring for piles in alternate positions and using a special auger the intermediate pile positions can be bored so that the interfaces interlock. The main advantage of using this method is that the wall can be formed within a restricted headroom of not less than 2.000 m. The disadvantages are the general need for a reinforced tie beam over the heads of the piles and the necessity for a facing to take out the irregularities of the surface. This facing usually takes the form of a cement rendering lightly reinforced with a steel-welded fabric mesh.

The general method used to construct diaphragm walls is the bentonite slurry system. Bentonite is manufactured from a montmorillonite clay that is commonly called fuller's earth because of its historical use by fullers in the textile industry to absorb grease from newly woven cloth. When mixed with the correct amount of water bentonite shows thixotropic properties, giving a liquid behaviour when agitated and a gel structure when undisturbed.

The basic procedure is to replace the excavated spoil with the bentonite slurry as the work proceeds. The slurry forms a soft gel or 'filter cake' at the interface of the excavation sides with slight penetration into the subsoil. Hydrostatic pressure caused by the bentonite slurry thrusting on the filter cake cushion is sufficient to hold back the subsoil and any groundwater that may be present. This alleviates the need for timbering and/or pumping, and can be successfully employed up to 36.000 m deep.

Diaphragm walls constructed by this method are executed in alternate panels from 4.500 m to 7.000 m long with widths ranging from 500 to 900 mm using a specially designed hydraulic grab attached to a standard crane, or by using a continuous cutting and recirculating machine. Before the general excavation commences, a guide trench about 1.000 m deep is excavated and lined with lightly reinforced walls. These walls act as a guide line for the excavating machinery and provide a reservoir for the slurry, enabling pavings and underground services to be broken out ahead of the excavation.

To form an interlocking and watertight joint at each end of the panel, circular stop end pipes are placed in the bentonite-filled excavation before the concrete is placed. The continuous operation of concreting the panel is carried out using a tremie pipe and a concrete mix designed to have good flow properties without the tendency to segregate. This will require a concrete with a high slump of about 200 mm but with high-strength properties ranging from 20 to 40 N/mm^2. Generally the rate of pour is in the region of 15 to 20 m^3 per hour, and as the concrete is introduced into the excavated panel it will displace the bentonite slurry, which is less dense than the concrete, which can be stored for reuse or transferred to the next panel being excavated. The ideal situation is to have the two operations acting simultaneously and in complete unison (see Fig. 3.2.7).

Before the concrete is placed, reinforcement cages of high yield or mild steel bars are fabricated on site in one or two lengths. Single length cages of up to 20.000 m are possible; where cages in excess of this are required they are usually spot-welded together when the first cage is projecting about 1.000 m above the slurry level. The usual recommended minimum cover is 100 mm, which is maintained by having spacing blocks or rings attached to the outer bars of the cage. Upon completion of the concreting the bentonite slurry must be removed from the site either by tanker or by diluting so that it can be discharged into the surface water sewers by agreement with the local authority.

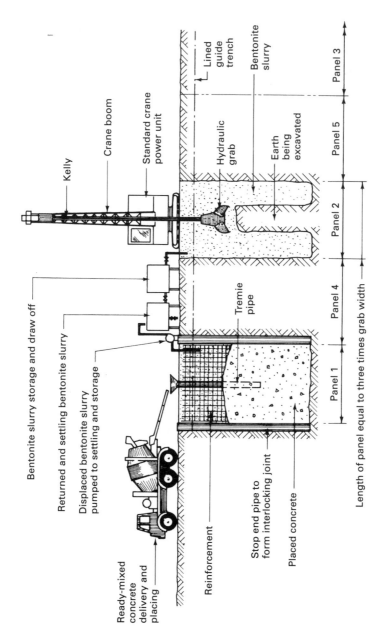

Kelly

Crane boom

Standard crane
power unit

Bentonite slurry storage and draw off

Returned and settling bentonite slurry

Displaced bentonite slurry
pumped to settling and storage

Ready-mixed
concrete
delivery and
placing

Reinforcement

Stop end pipe to
form interlocking joint

Placed concrete

Lined
guide
trench

Bentonite
slurry

Hydraulic
grab

Earth
being
excavated

Tremie
pipe

Panel 3

Panel 5

Panel 2

Panel 4

Panel 1

Length of panel equal to three times grab width

Figure 3.2.7   Diaphragm wall construction: bentonite slurry method

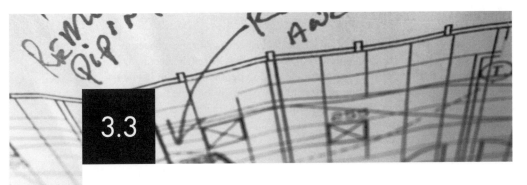

3.3

COFFERDAMS AND CAISSONS

A cofferdam may be defined as a temporary box structure constructed in earth or water to exclude soil and/or water from a construction area. It is usually formed to enable the formation of foundations to be carried out in safe working conditions. It is common practice to use interlocking steel trench sheeting or steel sheet piling to form the cofferdam, but any material that will fulfil the same function can be used, including timber piles, precast concrete piles, earth-filled crib walls and banks of soil and rock. It must be clearly understood that, to be safe, economic and effective, cofferdams must be the subject of structural design. Regulations 12 and 13 of the Construction (Health, Safety and Welfare) Regulations 1996 specifically require cofferdams and caissons to be properly designed, constructed and maintained. Of equal importance are Regulations 5 and 14. Regulation 5 requires a safe place to work and safe means of access to and egress from that place of work. Regulation 14 includes provision for water transport and rescue equipment for operatives working where water levels are a perceived danger. These regulations are complemented by the Confined Spaces Regulations 1997, which include specific reference to chambers and pits, i.e. implications for cofferdams and caissons. Under Regulation 5 of this Statutory Instrument there must be preparation of suitable contingency plans in the event of an emergency rescue and availability of resuscitation equipment. Under Regulation 4 of the Work at Height Regulations 2005, specific reference is made to adequate supervision, planning and safety procedures. This includes planning for emergencies and rescue.

■■■■ SHEET PILE COFFERDAMS

Cofferdams constructed from steel sheet piles or steel trench sheeting can be considered under two headings:

- single-skin cofferdams;
- double-skin cofferdams.

SINGLE-SKIN COFFERDAMS

These consist of a suitably supported single enclosing row of trench sheeting or sheet piles forming an almost completely watertight box. Trench sheeting could be considered for light loadings up to an excavation depth of 3.000 m below the existing soil or water level, whereas sheet piles are usually suitable for excavation depths of up to 15.000 m. The small amount of seepage that will occur through the interlocking joints must not be in excess of that which can be comfortably controlled by a pump; alternatively the joints can be sealed by caulking with suitable mastics, bitumastic compounds or silicone sealants.

Single-skin cofferdams constructed to act as cantilevers are possible in all soils, but the maximum amount of excavation height will be low relative to the required penetration of the toe of the pile, and this is particularly true in cohesive soils. Most cofferdams are therefore either braced and strutted or anchored using tie rods or ground anchors. Standard structural steel sections or structural timber members can be used to form the support system, but generally timber is economically suitable only for low loadings. The total amount of timber required to brace a cofferdam adequately would be in the region of 0.25 to 0.3 m^3 per tonne of steel sheet piling used, whereas the total weight of steel bracing would be in the region of 30–60% of the total weight of sheet piling used to form the cofferdam. Typical cofferdam support arrangements are shown in Figs 3.3.1 and 3.3.2. Single-skin cofferdams that are circular in plan can also be constructed using ring beams of concrete or steel to act as bracing without the need for strutting. Diameters up to 36.000 m are economically possible using this method.

Cofferdams constructed in water, particularly those being erected in tidal waters, should be fitted with sluice gates to act as a precaution against unanticipated weaknesses in the arrangement and, in the case of tidal waters, to enable the water levels on both sides of the dam to be equalised during construction and before final closure. Special piles with an integral sluice gate forming a 200 mm wide × 400 mm deep opening are available. Alternatively a suitable gate can be formed by cutting a pair of piles and fitting them with a top-operated screw gear so that they can be raised to form an opening of any desired depth.

DOUBLE-SKIN COFFERDAMS

These are self-supporting gravity structures constructed by using two parallel rows of piles with a filling material placed in the void created. Gravity-type cofferdams can also be formed by using straight-web sheet pile sections arranged as a cellular construction (see Figs 3.3.3 and 3.3.4).

The stability of these forms of cofferdam depends upon the design and arrangement of the sheet piling and upon the nature of the filling material. The inner wall of a double-skin cofferdam is designed as a retaining wall which is suitably driven into the substrata whereas the outer wall acts primarily as an anchor wall. The two parallel rows of piles are tied together with one or two rows of tie rods acting against external steel walings. Inner walls should have a series of low-level weepholes to relieve the filling material of high water pressures and thus

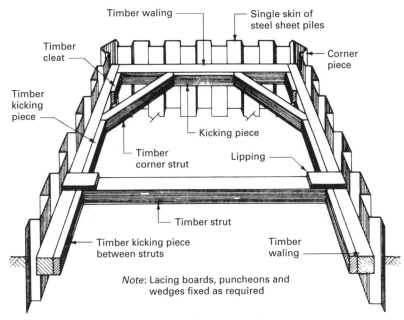

Timber waling — Single skin of steel sheet piles

Timber cleat —

Corner piece

Timber kicking piece —

Kicking piece

Timber corner strut

Lipping —

Timber strut

Timber kicking piece between struts

Timber waling

Note: Lacing boards, puncheons and wedges fixed as required

Cofferdam with timber strutting

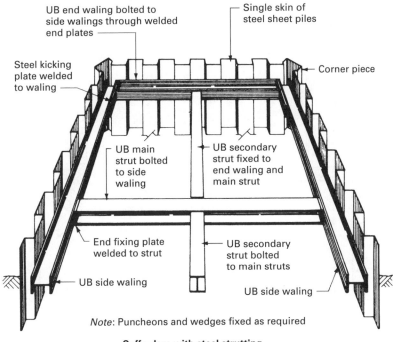

UB end waling bolted to side walings through welded end plates —

Single skin of steel sheet piles

Steel kicking plate welded to waling —

Corner piece

UB main strut bolted to side waling

UB secondary strut fixed to end waling and main strut

End fixing plate welded to strut

UB secondary strut bolted to main struts

UB side waling

UB side waling

Note: Puncheons and wedges fixed as required

Cofferdam with steel strutting

Figure 3.3.1   Typical cofferdam strutting arrangements

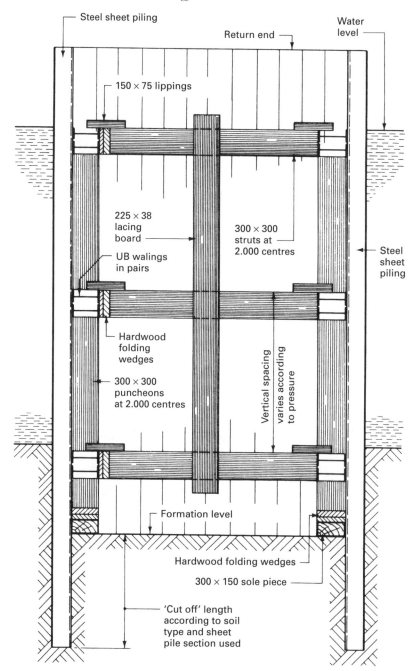

Steel sheet piling

Return end

Water level

150 × 75 lippings

225 × 38 lacing board

300 × 300 struts at 2.000 centres

UB walings in pairs

Steel sheet piling

Hardwood folding wedges

300 × 300 puncheons at 2.000 centres

Vertical spacing varies according to pressure

Formation level

Hardwood folding wedges

300 × 150 sole piece

'Cut off' length according to soil type and sheet pile section used

Figure 3.3.2   Typical cofferdam section using timber and steel strutting

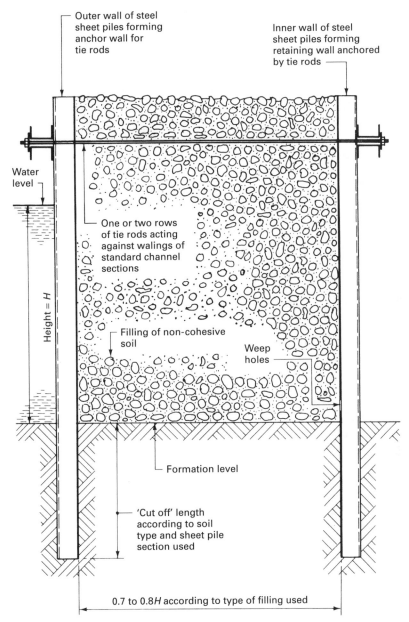

Outer wall of steel sheet piles forming anchor wall for tie rods

Inner wall of steel sheet piles forming retaining wall anchored by tie rods

Water level

One or two rows of tie rods acting against walings of standard channel sections

Height = H

Filling of non-cohesive soil

Weep holes

Formation level

'Cut off' length according to soil type and sheet pile section used

0.7 to 0.8H according to type of filling used

Figure 3.3.3   Typical double-skin cofferdam details

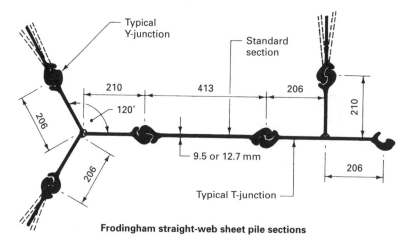

Frodingham straight-web sheet pile sections

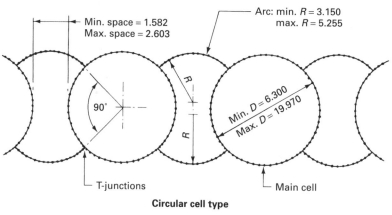

Circular cell type

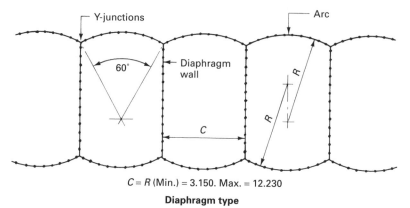

Diaphragm type

Figure 3.3.4 Cellular cofferdam arrangements

increase its shear resistance. For this reason the filling material selected should be capable of being well drained. Therefore materials such as sand, hardcore and broken stone are suitable, whereas cohesive soils such as clay are unsuitable. The width-to-height ratio shown in Fig. 3.3.3 of $0.7:0.8H$ can in some cases be reduced by giving external support to the inner wall by means of an earth embankment or berm.

Cellular cofferdams are entirely self-supporting, and do not require any other form of support such as that provided by struts, braces and tie rods. The straight web pile with its high web strength and specially designed interlocking joint is capable of resisting the high circumferential tensile forces set up by the non-cohesive filling materials. The interlocking joint also has sufficient angular deviation to enable the two common arrangements of circular cell and diaphragm cellular cofferdams to be formed (see Fig. 3.3.4). Like the double-skin cofferdam, the walls of cellular cofferdams should have weepholes to provide adequate drainage of the filling material. The circular cellular cofferdam has one major advantage over its diaphragm counterpart in that each cell can be filled independently, whereas care must be exercised when filling adjacent cells in a diaphragm type to prevent an unbalanced pressure being created on the cross-walls or diaphragms. In general, cellular cofferdams are used to exclude water from construction areas in rivers and other waters where large structures such as docks are to be built.

▨ ▨ ■ STEEL SHEET PILING

Steel sheet piling is the most common form of sheet piling; it can be used in temporary works such as timbering to excavations in soft and/or waterlogged soils and in the construction of cofferdams. This material can also be used to form permanent retaining walls, especially those used for river bank strengthening, and in the construction of jetties. Three common forms of steel sheet pile are the Larssen, Frodingham and straight-web piles, all of which have an interlocking joint to form a water seal, which may need caulking where high water pressures are encountered. Straight-web sheet piles are used to form cellular cofferdams as described above and illustrated in Fig. 3.3.4. Larssen and Frodingham sheet piles are suitable for all uses except for the cellular cofferdam, and can be obtained in lengths up to 18.000 m according to the particular section chosen – typical sections are shown in Fig. 3.3.5.

To erect and install a series of sheet piles and keep them vertical in all directions usually requires a guide frame or trestle constructed from large-section timbers. The piles are pitched or lifted by means of a crane, using the lifting holes sited near the top of each length, and positioned between the guide walings of the trestle (see Figs 3.3.6 and 3.3.7). When sheet piles are being driven there is a tendency for them to creep or lean in the direction of driving. Correct driving methods will help to eliminate this tendency, and the generally accepted method is to install the piles in panels as follows:

1. A pair of piles are pitched and driven until approximately one-third of their length remains above ground level to act as anchor piles to stop the remainder of the piles in the panel from leaning or creeping while being driven. It is essential that this first pair of piles are driven accurately and plumb in all directions.

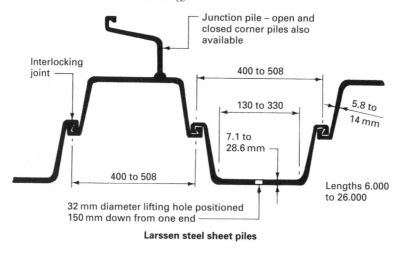

Junction pile – open and
closed corner piles also
available

Interlocking
joint

400 to 508

130 to 330

5.8 to
14 mm

7.1 to
28.6 mm

400 to 508

Lengths 6.000
to 26.000

32 mm diameter lifting hole positioned
150 mm down from one end

Larssen steel sheet piles

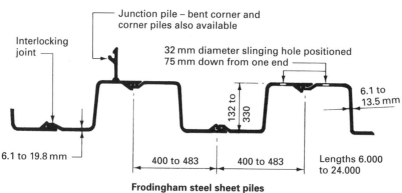

Junction pile – bent corner and
corner piles also available

Interlocking
joint

32 mm diameter slinging hole positioned
75 mm down from one end

6.1 to
13.5 mm

132 to
330

6.1 to 19.8 mm

400 to 483

400 to 483

Lengths 6.000
to 24.000

Frodingham steel sheet piles

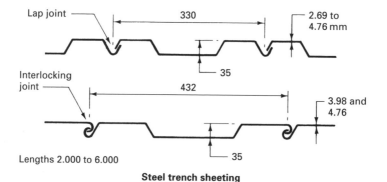

Lap joint

330

2.69 to
4.76 mm

35

Interlocking
joint

432

3.98 and
4.76

Lengths 2.000 to 6.000

35

Steel trench sheeting

Figure 3.3.5   Typical steel sheet pile and trench sheeting sections

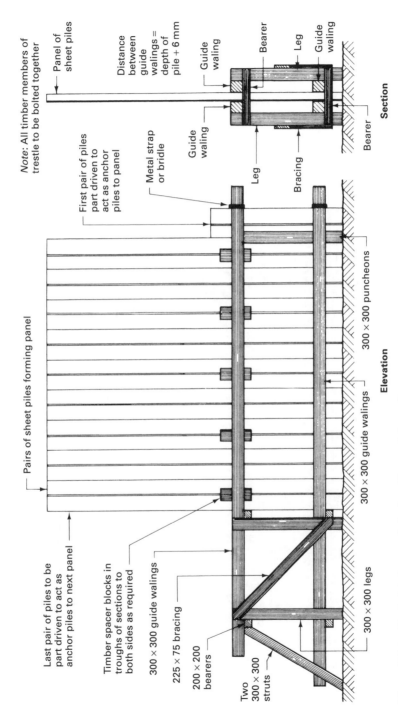

Figure 3.3.6 Typical timber trestle for installing steel sheet piles

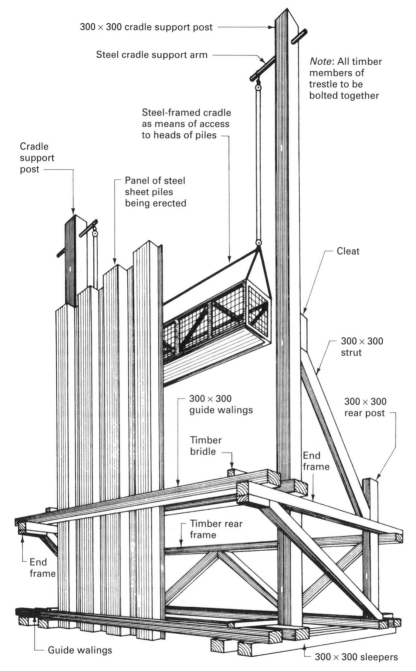

300 × 300 cradle support post

Steel cradle support arm

Note: All timber members of trestle to be bolted together

Steel-framed cradle as means of access to heads of piles

Cradle support post

Panel of steel sheet piles being erected

Cleat

300 × 300 strut

300 × 300 guide walings

300 × 300 rear post

Timber bridle

End frame

Timber rear frame

End frame

Guide walings

300 × 300 sleepers

Figure 3.3.7   Typical timber trestle for large steel sheet piles

2. Pitch a series of piles in pairs adjacent to the anchor piles to form a panel of 10 to 12 pairs of piles.
3. Partially drive the last pair of piles to the same depth as the anchor piles.
4. Drive the remaining piles in pairs including the anchor piles to their final set.
5. The last pair of piles remain projecting, for about one-third of their length, above the ground level to act as guide piles to the next panel.

To facilitate accurate and easy driving there should be about a 6 mm clearance between the pile faces and the guide walings, and timber spacer blocks should be used in the troughs of the piles (see Fig. 3.3.6).

Steel sheet piles may be driven to the required set using percussion hammers or hydraulic drivers. **Percussion hammers** activated by steam, compressed air or diesel power can be used, and these are usually equipped with leg grips bolted to the hammer and fitted with inserts to grip the face of the pile to ensure that the hammer is held in line with the axis of the pile. Wide, flat driving caps are also required to prevent damage to the head of the pile by impact from the hammer. **Hydraulic drivers** can be used to push the piles into suitable subsoils such as clays, silts and fine granular soils. These driving systems are vibrationless and almost silent, making them ideal for installing sheet piling in close proximity to other buildings. The driving head usually consists of a power pack crosshead containing eight hydraulic rams fitted with special pile connectors, each ram having a short stroke of 750 mm. The driving operation entails lowering and connecting the hydraulic driver to the heads of a panel of piles, activating two rams to drive a pair of piles for the full length of the stroke, and repeating this process until all the rams have been activated. The rams are retracted and the crosshead is lowered onto the top of the piles to recommence the whole driving cycle. This process is repeated until the required penetration of the piles has been reached.

The tendency for sheet piles to lean while being driven may occur even under ideal conditions with careful supervision, and should this occur immediate steps must be taken to correct the fault, or it may get out of control. The following correction methods can be considered:

■ Attach a winch rope to the pair of piles being driven and exert a corrective pulling force as driving continues.
■ Attach a winch rope to the previously driven adjacent pair of piles and exert a corrective pulling force as driving continues.
■ Reposition the hammer towards the previously driven pair of piles to give an eccentric blow.
■ Combinations of any of the above methods.

If the above correction techniques are not suitable or effective it is possible to form tapered piles by making use of the tolerances within the interlocking joint by welding steel straps across the face of the pair of piles to hold them in tapered form. This technique is suitable only if the total amount of taper required does not exceed 50 mm in the length of the pile. To insert a special tapered pile within a panel it will of course be necessary to extract a pair of piles from the panel. Typical examples of correction techniques are shown in Fig. 3.3.8.

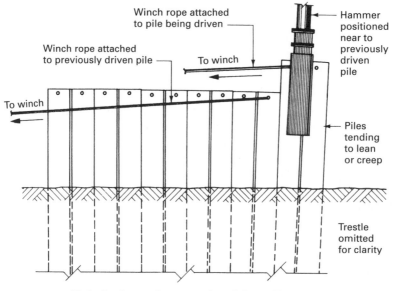

Methods of correcting creep of steel sheet piling

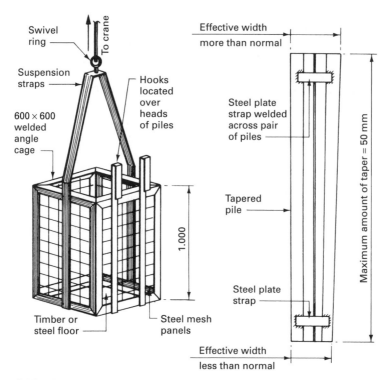

Figure 3.3.8   Creep correction techniques and access cage

WATER JETTING

In soft and silty subsoils the installation of steel sheet piles can be assisted by using high-pressure water jetting to the sides and toes of the piles. The jets should be positioned in the troughs of the piles and to both sides of the section. In very soft subsoils this method can sometimes be so effective that the assistance of a driving hammer is not required. To ensure that the pile finally penetrates into an undisturbed layer of subsoil the last few metres or so of driving should be carried out without jetting.

ACCESS

Access in the form of a working platform to the heads of sheet piles is often necessary during installation to locate adjacent piles, to position and attach the driving hammer, and to carry out inspections of the work in progress. Suitable means of access are:

- independent scaffolding;
- suspended cradle (see Fig. 3.3.7);
- mobile platform mounted on a hydraulic arm;
- working platform in the form of a cage suspended from a mobile crane and hooked onto the pile heads (see Fig. 3.3.8).

EXTRACTION OF PILES

Piles that are used in temporary works should have well-greased interlocking joints to enable them to be extracted with ease. Inverted double-acting hammers can be used to extract sheet piles, but these have generally been superseded by specially designed sheet pile extractors. These usually consist of a compressed-air- or steam-activated ram giving between 120 and 200 upward blows per minute, which causes the jaws at the lower end to grip the pile and force it out of the ground.

■■■ CAISSONS

These are box-like structures that can be sunk through ground or water to install foundations or similar structures below the water line or table. They differ from cofferdams in that they usually become part of the finished foundation or structure, and should be considered as an alternative to the temporary cofferdam if the working depth below the water level exceeds 18.000 m. The design and installation of the various types of caisson are usually the tasks of a specialist organisation, but building contractors should have a fundamental knowledge of the different types and their uses.

There are four basic types of caisson in general use:

- box caissons;
- open caissons;
- monolithic caissons;
- pneumatic or compressed-air caissons.

BOX CAISSONS

These are prefabricated precast concrete boxes that are open at the top and closed at the bottom. They are usually constructed on land, and are designed to be launched and floated to the desired position, where they are sunk onto a previously prepared dredged or rock blanket foundation (see Fig. 3.3.9). If the bed stratum is unsuitable for the above preparations it may be necessary to lay a concrete raft, by using traditional cofferdam techniques, onto which the caisson can be sunk. During installation it is essential that precautions are taken to overcome the problems of flotation by flooding the void with water or adding kentledge to the caisson walls. The sides of the caisson will extend above the water line after it has been finally positioned, providing a suitable shell for such structures as bridge piers, breakwaters and jetties. The void is filled with *in-situ* concrete placed by pump, tremie pipe or crane and skip. Box caissons are suitable for situations where the bed conditions are such that it is not necessary to sink the caisson below the prepared bed level.

OPEN CAISSONS

Sometimes referred to as **cylinder caissons** because of their usual plan shape, these are of precast concrete and open at both the top and bottom ends, with a cutting edge to the bottom rim. They are suitable for installation in soft subsoils where the excavation can be carried out by conventional grabs, enabling the caisson to sink under its own weight as the excavation proceeds. These caissons can be completely or partially pre-formed; in the latter case further sections can be added or cast on as the structure sinks to the required depth. When the desired depth has been reached a concrete plug in the form of a slab is placed in the bottom by tremie pipe to prevent further ingress of water. The cell void can now be pumped dry and filled with crushed rocks or similar material if necessary to overcome flotation during further construction works.

Open caissons can also be installed in land if the subsoil conditions are suitable. The shoe or cutting edge is formed so that it is wider than the wall above to create an annular space some 75 to 100 mm wide into which a bentonite slurry can be pumped to act as a lubricant and thus reduce the skin friction to a minimum. Excavation is carried out by traditional means within the caisson void, the caisson sinking under its own weight. The excavation operation is usually carried out simultaneously with the construction of the caisson walls above ground level (see Fig. 3.3.10).

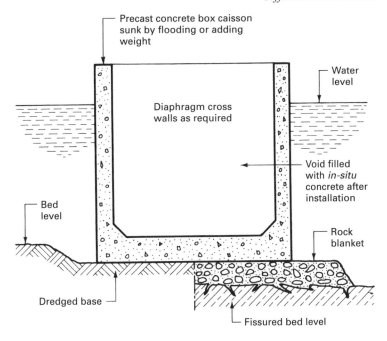

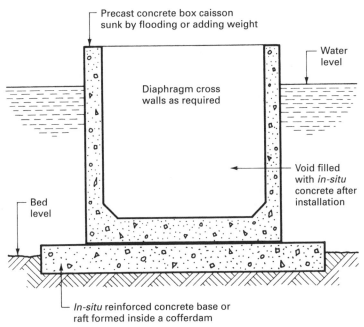

Figure 3.3.9 Typical box caisson details

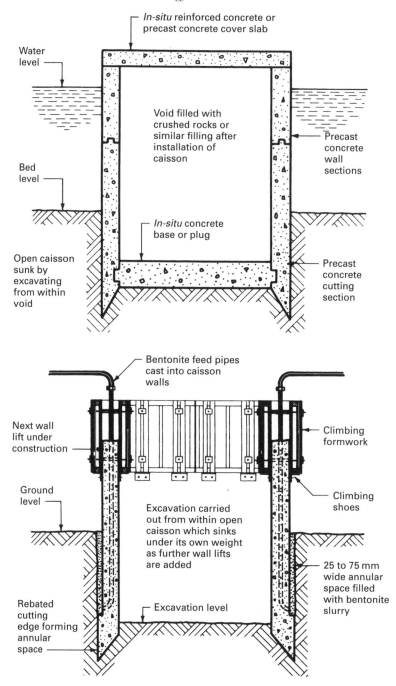

Figure 3.3.10   Typical open caisson details

MONOLITHIC CAISSONS

These are usually rectangular in plan and are divided into a number of voids or wells through which the excavation is carried out. They are similar to open caissons but have greater self-weight and wall thickness, making them suitable for structures such as quays, which may have to resist considerable impact forces in their final condition.

PNEUMATIC OR COMPRESSED-AIR CAISSONS

These are similar to open caissons except that there is an airtight working chamber some 3.000 m high at the cutting edge. They are used where difficult subsoils exist and where hand excavation in dry working conditions is necessary. The working chamber must be pressurised sufficiently to control the inflow of water and/or soil and at the same time provide safe working conditions for the operatives. The maximum safe working pressure is usually specified as 310 kN/m^2, which will limit the working depth of this type of caisson to about 28.000 m. When the required depth has been reached the floor of the working chamber can be sealed over with a 600 mm thick layer of well-vibrated concrete. This is followed by further well-vibrated layers of concrete until only a small space remains, which is pressure-grouted to finally seal the working chamber. The access shafts are finally sealed with concrete some three to four days after sealing off the working chamber: for typical details see Fig. 3.3.11.

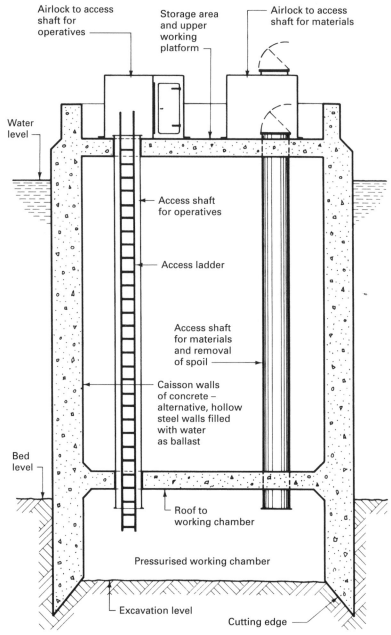

Figure 3.3.11 Typical pneumatic or compressed–air caisson details

3.4

TUNNELLING AND CULVERTS

A tunnel may be defined as an artificial underground passage. Tunnels have been used by man as a means of communication or for transportation for several thousand years. Prehistoric man is known to have connected his natural cave habitats by tunnels hewn in the rock. A Babylonian king *circa* 2180–2160 BC connected his royal palace to the Temple of Jupiter on the opposite bank of the Euphrates by a brick-arched tunnel under the river. Other examples of early tunnels are those hewn in the rock in the tomb of Mineptah at Thebes in Egypt, and the early Greek tunnel, constructed about 687 BC, used for conveying water on the Island of Sámos.

The majority of these early tunnels were constructed in rock subsoils and therefore required no permanent or temporary support. Today, tunnelling in almost any subsoil is possible. Permanent tunnels for underground railways and roads can be lined with metal and/or concrete, but such undertakings are the province of the civil engineer. The general building contractor would normally be involved only with temporary tunnelling for the purposes of gaining access to existing services or installing new services, constructing small permanent tunnels for pedestrian subways under road or railway embankments, or forming permanent tunnels for services.

When the depth of a projected excavation is about 6.000 m the alternative of working in a heading or tunnel should be considered, taking the following factors into account:

- **Nature of subsoils**   The amount of timbering that will be required in the tunnel as opposed to that required in deep trench excavations.
- **Depth of excavation**   Over 9.000 m deep it is usually cheaper to tunnel or use one of the alternative methods such as thrust boring. The cover of ground over a tunnel to avoid disturbance of underground services, roads, pavings and tree roots is generally recommended to be 3.000 m minimum.

- **Existing services**   In urban areas buried services can be a problem with open deep-trench excavations; this can generally be avoided by tunnelling techniques.
- **Carriageways**   It may be deemed necessary to tunnel under busy roads to avoid disturbance of the flow of traffic.
- **Means of access**   The proposed tunnel may be entered by means of an open trench if the tunnel excavation is into an embankment or access may be gained by way of a shaft.
- **Construction Regulations**   The Construction (Health, Safety and Welfare) Regulations 1996 establish the need for a safe place of work for the protection of operatives working in excavations, shafts and tunnels, covering such aspects as temporary timbering, supervision of works, and means of egress in case of flooding. Regulations 23 to 27 deal with site-wide issues such as ventilation of excavations and provision of adequate lighting.
- **Confined Spaces Regulations 1997**   Regulation 1 includes reference to any 'chamber, tank, vat, silo, pit, trench, pipe, sewer, flue, well or other similar place' as enclosed situations with foreseeable risks. Tunnels and culverts can apply to many of these categories: therefore Regulations 3 (Duties), 4 (Work in confined spaces) and 5 (Emergency arrangements) are applicable.
- **Work at Height Regulations 2005**   Work at height means work in any place. This includes a workplace at or below ground, i.e. tunnels and culverts. Access to or egress from that place of work by suitable means is a specific requirement.

▨■■ SHAFTS

These are by definition vertical passages, but in the context of building operations they can also be used to form the excavation for a deep isolated base foundation. In common with all excavations, the extent and nature of the temporary support or timbering required will depend upon:

- subsoil conditions encountered;
- anticipated ground and hydrostatic pressures;
- materials used to provide temporary support;
- plan size and depth of excavation.

In loose subsoils a system of sheet piling could be used by driving the piles ahead of the excavation operation. This form of temporary support is called a cofferdam and is explained in the preceding chapter.

Alternatives to the cofferdam techniques for shaft timbering are tucking framing and pile framing (see Chapter 3.2). These methods consist of driving short timber runners, 1.000 to 2.000 m long, ahead of the excavation operation and then excavating and strutting within the perimeter of the runners. The process is repeated until the required depth has been reached. It is essential with all drive and dig methods that the depth to which the runner has been driven is in excess of the excavation depth at all times. The installation of tucking framing and pile framing in shafts is as described for deep trench timbering (see Fig. 3.2.2). Both tucking framing and pile framing have the advantages over sheet piling of not requiring large guide trestles and heavy driving equipment.

In firm subsoils the shaft excavation would be carried out in stages of 1.000 to 2.000 m deep according to the ability of the subsoil to remain stable for short periods. Each excavated stage is timbered before proceeding with the next digging operation. The sides of the excavated shaft can be supported by a system of adequately braced and strutted poling boards (see Figs 3.4.1 and 3.4.2). Sometimes a series of crossbeams are used at the head of the shaft timbering to reduce the risk of the whole arrangement sliding down the shaft as excavation work proceeds at the lower levels.

Shafts are usually excavated square in plan with side dimensions of 1.200 to 3.000 m, depending on:

- total depth required;
- method of timbering;
- sizes of support lining to save unnecessary cutting to width;
- number of operatives using or working within the shaft;
- size of skip or container to be used for removing spoil;
- type of machinery used for bulk excavation.

If the shaft is for the construction of an isolated base then an access or ladder bay should be constructed. This bay would be immediately adjacent to the shaft and of similar dimensions, making in plan a rectangular shaft. The most vulnerable point in any shaft timbering is the corners, where high pressures are encountered, and these positions should be specially strengthened by using corner posts or runners of larger cross-section (see Fig. 3.4.1).

▨▨▧■ TUNNELS

The operational sequence of excavating and timbering a tunnel or heading by traditional methods is as follows:

1. In firm soils excavate first 1.000 m long stage or bay; if weak subsoil is encountered it may be necessary to drive head boards and lining boards as the first operation.
2. Head boards 1.000 m long are placed against the upper surface.
3. Sole plate and stretcher are positioned; these are partly bedded into the ground to prevent lateral movement and are levelled through from stretcher to stretcher.
4. Cut and position head tree.
5. Cut side trees or struts to fit tightly between sole plate and head tree and wedge into position.
6. Secure frames using wrought iron dogs, spikes and cleats as required.
7. Excavate next stage or bay by starting at the top and taking out just enough soil to allow the next set of head boards to be positioned.
8. If loose subsoils are encountered it will be necessary to line the sides with driven or placed horizontal poling boards as the work proceeds (see Figs 3.4.2 and 3.4.3).

After the construction work has been carried out within the tunnel it can be backfilled with hand-compacted material, extracting the timbering as the work proceeds. This method is time consuming and costly; the general procedure is to backfill the tunnel with pumped concrete and leave the temporary support work in position.

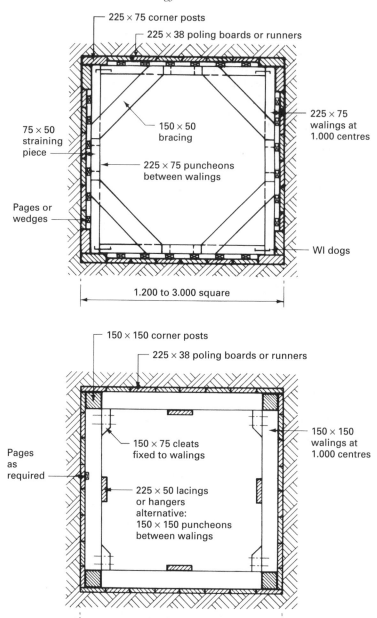

Figure 3.4.1   Shaft timbering: typical plans

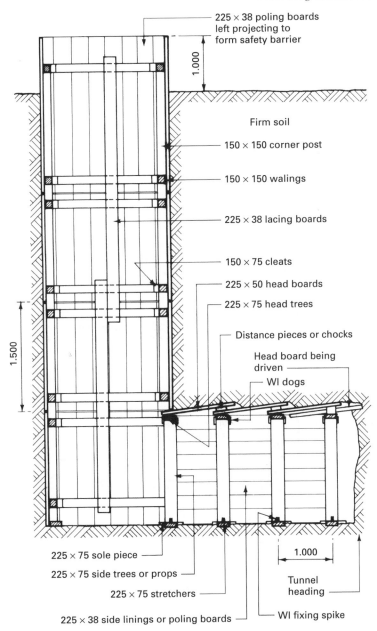

225 × 38 poling boards
left projecting to
form safety barrier

1.000

Firm soil

150 × 150 corner post

150 × 150 walings

225 × 38 lacing boards

150 × 75 cleats

225 × 50 head boards

225 × 75 head trees

Distance pieces or chocks

Head board being
driven

WI dogs

1.500

225 × 75 sole piece

225 × 75 side trees or props

225 × 75 stretchers

225 × 38 side linings or poling boards

1.000

Tunnel
heading

WI fixing spike

Figure 3.4.2   Typical shaft and tunnel timbering

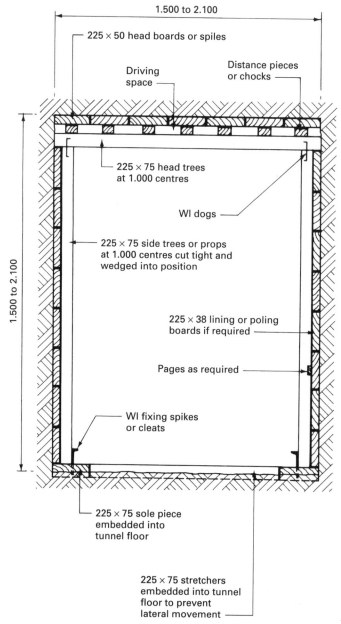

Figure 3.4.3   Typical section through tunnel timbering

▨▨■ ALTERNATIVE METHODS

Where the purpose of the excavation is the installation of pipework, alternative methods to tunnelling should be considered.

SMALL-DIAMETER PIPES

Two methods are available for the installation of small pipes up to 200 mm diameter:

■ **Thrust boring**   A bullet-shaped solid metal head is fixed to the leading end of the pipe to be installed and is pushed or jacked into the ground, displacing the earth.
■ **Auger boring**   This is carried out with a horizontal auger boring tool operating from a working pit having at least 2.400 m long × 1.500 m wide clear dimensions between any temporary supporting members. The boring operation can be carried out without casings, but where the objective is the installation of services, concrete or steel casings are usually employed. The auger removes the spoil by working within the bore of the casing, which is being continuously rammed or jacked into position. It is possible to use this method for diameters of up to 1.000 m.

PIPE JACKING

This method can be used for the installation of pipes from 150 to 3,600 mm diameter but is mainly employed on the larger diameters of over 1.000 m. Basically the procedure is to force the pipes into the subsoil by means of a series of hydraulic jacks, and excavate, as the driving proceeds, from within the pipes by hand or machine according to site conditions. The leading pipe is usually fitted with a steel shield or hood to aid the driving process. This is a very safe method, because the excavation work is carried out from within the casing or liner and the danger of collapsing excavations is eliminated; there is also no disruption of the surface or underground services, and it is a practical method for most types of subsoil.

The most common method is to work from a jacking or working pit, which is formed in a similar manner to traditional shafts except that a framed thrust pad is needed from which to operate the hydraulic jacks. The working pit must be large enough for the jacks to be extended and to allow for new pipe sections to be lowered into the working bay at the bottom (see Fig. 3.4.4).

Pipe jacking can also be carried out from ground level, and is particularly suitable for driving pipes through an embankment to form a pedestrian subway. A series of 300 mm diameter lined augered boreholes are driven through the embankment to accommodate tie bars, which are anchored to a bulkhead frame on the opposite side of the embankment. The reactions from the ramming jacks are thus transferred through the tie bars to the bulkhead frame, and the driving action becomes one of pushing and pulling (see Fig. 3.4.5). In firm soils the rate of bore by this method is approximately 3.000 m per day.

Pipes can also be jacked, from ground level, into the earth at a gradient of up to 1:12 using a jacking block attached to a row of tension piles sited below the commencing level.

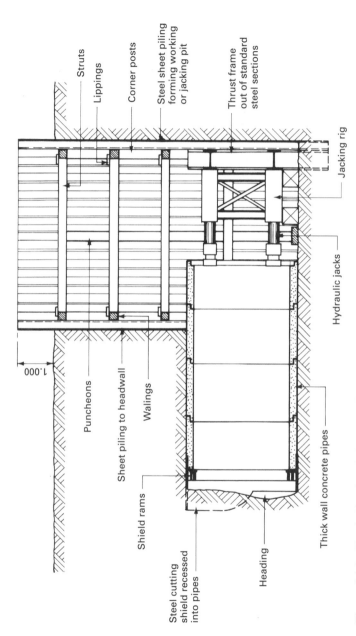

Figure 3.4.4 Pipe jacking from below ground level

PIPES

The pipes used in the above techniques are usually classified in diameter ranges thus:

- **Small pipes**   150 to 900 mm diameter: thrust or auger bored.
- **Medium pipes**   900 to 1,800 mm diameter: pipe-jacking techniques.
- **Large pipes**   1,800 to 3,600 mm diameter: pipe-jacking techniques.

Two materials are in common use for the pipes: concrete and steel. Spun concrete pipes are specially designed with thick walls, and have a rubber joint, making them especially suitable for sewers without the need for extra strengthening. Larger-diameter pipes for pedestrian subway constructions are usually made of cast concrete and can have special bolted connections making the joints watertight, which also renders them suitable for use as sewer pipes. Steel pipes have a wall thickness relative to their diameter, and usually have welded joints to give high tensile strength, the alternative being a flanged and bolted joint. They are obtainable with various coatings and linings to meet special requirements such as corrosive-bearing effluents.

CULVERTS

A culvert is a box, conduit or pipe of large cross-section usually used to convey pedestrians, water (foul or surface) or services in general under a road, canal or some other elevated construction. Figure 3.4.5 illustrates the concept, using precast concrete pipes under a railway embankment. In the true principle of culvert construction, a void for the pipe would be excavated and covered after the pipes were placed, thus differentiating the process from tunnelling. In practice this may not always be convenient or cost-effective: hence the preference for tunnelling in some situations.

The **cut and cover** technique involves considerable excavation, with sides usually temporarily restrained by interlocking sheet piles or steel trench sheeting. Groundwater controls may also be required, particularly if *in-situ* concrete is specified as the culvert material.

CONSTRUCTION

- *in-situ* concrete;
- *in-situ* concrete base with precast concrete crown;
- precast concrete.

See Fig. 3.4.6.

PRECAST CONCRETE BOX CULVERTS

These have proved the most popular specification for culverts, with manufacturers producing them in a wide range of sizes from about 1.000 to 4.000 m in width, 0.500 to 2.500 m in depth and in standard lengths of 1.000, 1.500 and 2.000 m to suit a variety of applications. These include subways, stream crossings, animal (badger) crossings, sewers and sea outfalls. They are not limited to horizontal use, and may also be used as vertical access shafts. Concrete is normally from ordinary Portland cement, but if aggressive soils are encountered, sulphate-resisting cement can be specified. Mild steel reinforcement with a minimum concrete cover of 30 mm is standard.

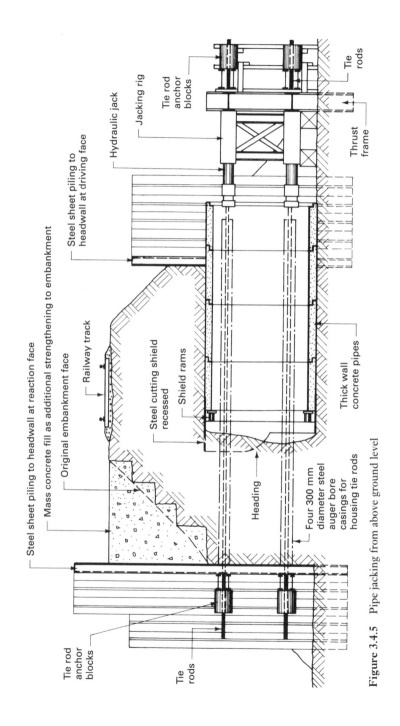

Steel sheet piling to headwall at reaction face

Mass concrete fill as additional strengthening to embankment

Steel sheet piling to headwall at driving face

Hydraulic jack

Jacking rig

Tie rod anchor blocks

Tie rods

Thrust frame

Original embankment face

Railway track

Steel cutting shield recessed

Shield rams

Thick wall concrete pipes

Heading

Four 300 mm diameter steel bore auger casings for housing tie rods

Tie rod anchor blocks

Tie rods

Figure 3.4.5   Pipe jacking from above ground level

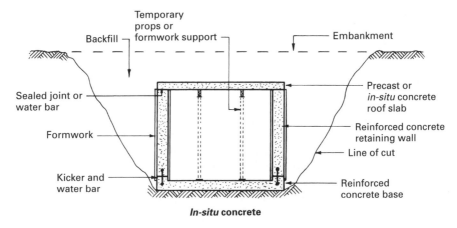

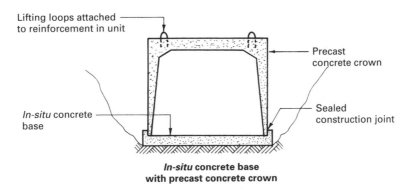

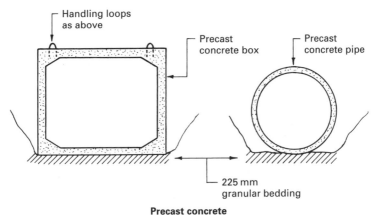

Figure 3.4.6   Concrete culverts: construction principles

Application

Site delivery and installation should be carefully planned to coordinate hire of suitable plant and operatives. Following excavation, a bed or foundation must be prepared from well-compacted selected granules of about 225 mm total thickness. This is levelled to suit the situation, or placed on a slight incline for drainage purposes. Projecting reinforcement loops or purpose-made threaded sockets provide crane lifting gear attachments for the box units. After placing and levelling the first unit, second and subsequent units are lowered and positioned for jointing. Pre-primed spigot and socket rebated joints are fitted with a mastic jointing strip or gasket-bonded to the socket as shown in Fig. 3.4.7. If preferred, and site conditions permit, the joints may be sealed with waterproof cement and sand mortar.

With the first unit restrained, second and subsequent units have their weight partially supported while tensioners draw the spigots into opposing sockets to compress the jointing material/gasket (see Fig. 3.4.7). Care must be taken to ensure that the joints remain clean, and that the bedding granules are excluded during this process.

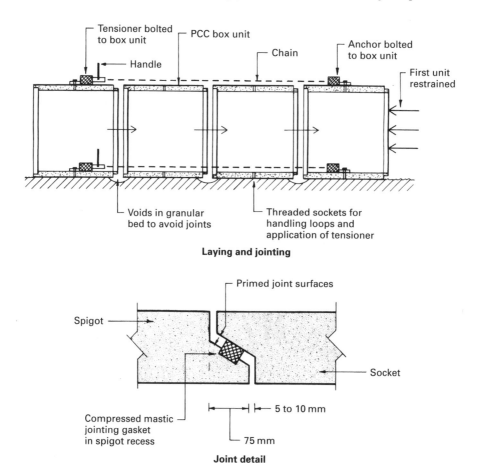

Laying and jointing

Joint detail

Figure 3.4.7   Precast concrete box unit culvert

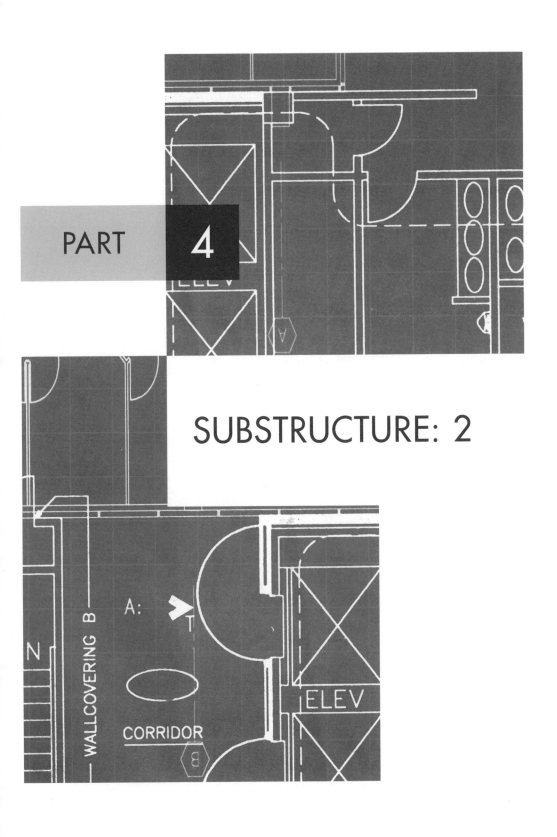

PART 4

SUBSTRUCTURE: 2

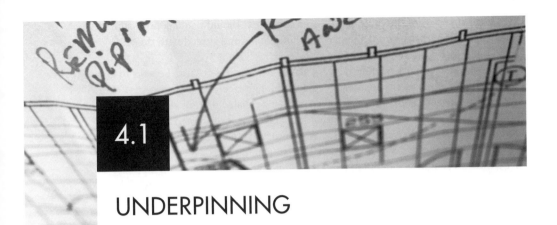

4.1

UNDERPINNING

The main objective of underpinning is to transfer the load carried by an existing foundation from its present bearing level to a new level at a lower depth. It can also be used to replace an existing weak foundation.

Underpinning may be necessary for one or more of the following reasons:

- As a remedy for:
  - uneven loading;
  - unequal resistance of the subsoil;
  - action of tree roots;
  - action of subsoil water;
  - cohesive soil settlement;
  - deterioration of foundation concrete.
- To increase the loadbearing capacity of a foundation, which may be required to enable an extra storey to be added to the existing structure or if a change of use would increase the imposed loadings.
- As a preliminary operation to lowering the adjacent ground level when constructing a basement at a lower level than the existing foundations of the adjoining structure or when laying deep services near to or below the existing foundations.

■■■■ SITE SURVEY AND PRELIMINARY WORKS

Before any underpinning is commenced the following surveying and preliminary work should be carried out:

1. Notice should be served to the adjoining owners, setting out in detail the intention to proceed with the proposed works and giving full details of any proposed temporary supports such as shoring.

2. A detailed survey of the building to be underpinned should be made, recording any defects or cracks, supplemented by photographs and agreed with the building owner where possible.
3. Glass slips or 'telltales' should be fixed across any vertical or lateral cracks to give a visual indication of any further movement taking place.
4. A series of check levels should be taken against a reliable datum; alternatively metal studs can be fixed to the external wall and their levels noted. These levels should be checked periodically as the work proceeds to enable any movements to be recorded and the necessary remedial action taken.
5. Permission should be obtained, from the adjoining owner, to stop up all flues and fireplaces to prevent the nuisance and damage that can be caused by falling soot.
6. If underpinning is required to counteract unacceptable settlement of the existing foundations an investigation of the subsoil should be carried out to determine the cause and to forecast any future movement so that the underpinning design will be adequate.
7. The loading on the structure should be reduced as much as possible by removing imposed floor loads and installing any shoring that may be necessary.

▨▨■ WALL UNDERPINNING

Traditional underpinning to walls is carried out by excavating in stages alongside and underneath the existing foundation, casting a new foundation, building up to the underside of the existing foundation in brickwork or concrete, and finally pinning between the old and new work with a rich dry mortar.

To prevent the dangers of fracture or settlement the underpinning stages or bays should be kept short and formed to a definite sequence pattern so that no two bays are worked consecutively. This will enable the existing foundation and wall to arch or span the void created underneath prior to underpinning. The number and length of the bays will depend upon the following factors:

■ total length of wall to be underpinned;
■ width of existing foundation;
■ general condition of existing substructure;
■ superimposed loading of existing foundation;
■ estimated spanning ability of existing foundation;
■ subsoil conditions encountered.

The generally specified maximum length for bays used in the underpinning of traditional wall construction is 1.500 m, with the proviso that at no time should the sum total of unsupported lengths exceed 25% of the total wall length.

Bays are excavated and timbered as necessary, after which the bottom of the excavation is prepared to receive the new foundation. To give the new foundation strip continuity, dowel bars are inserted at the end of each bay. Brick and concrete block underpinning is toothed at each end to enable the bonding to be continuous, whereas *in-situ* concrete underpinning usually has splice bars or dowels projecting to provide the continuity. Brickwork would normally be in a dense clay common brick bedded in 1:3 cement mortar laid in English bond for strength. Concrete blockwork should meet manufacturers' specifications for foundations and have a compressive strength of 7 N/mm^2, or preferably 10 N/mm^2. Concrete used in underpinning is usually specified as a 1:2:4/20 mm aggregate mix (C20) using rapid-hardening cement. The final pinning mix should consist of 1 part rapid-hardening cement to 3 parts of well graded fine aggregate from 10 mm down to fine sand with a water/cement ratio of 0.35. In both methods the projection of the existing foundation is cut back to the external wall line so that the loads are transmitted to the new foundation and not partially dissipated through the original foundation strip onto the backfill material (see Fig. 4.1.1).

▣■■ PRETEST METHOD OF UNDERPINNING

This method is designed to prevent further settlement of foundations after underpinning has been carried out by consolidating the soil under the new foundation before the load from the underpinning is applied. The perimeter of the wall to be underpinned is excavated in stages as described for wall underpinning, the new foundation strip is laid, and a hydraulic jack supporting a short beam is placed in the centre of the bay under the existing foundation. A dry mortar mix is laid between the top of the beam and the existing foundation, and before it has finally set the jack is extended to give a predetermined load on the new foundation, thus pretesting the soil beneath.

This process is repeated along the entire underpinning length until the whole wall is being supported by the hydraulic jacks. Underpinning is carried out using brickwork or concrete walling between the jacks, which are later removed and replaced with underpinning to complete the operation.

▣■■ JACK OF MIGA PILE UNDERPINNING

This is a method that can be used in the following circumstances:

■ Depth of suitable bearing capacity subsoil is too deep to make traditional wall underpinning practical or economic.
■ Where a system giving no vibration is required. It is worth noting that this method is also practically noiseless.
■ If a system of variable depth is required.
■ The existing foundation is structurally sound.

The system consists of short precast concrete pile lengths jacked into the ground until a suitable subsoil is reached (see Fig. 4.1.2). When the jack pile has reached the required depth the space between the top of the pile and underside of the existing foundation is filled with a pinned concrete cap. The existing foundation must be in a good condition, because in the final context it will act as a beam spanning over the piles. The condition and hence the spanning ability of the existing strip foundation will also determine the spacing of the piles.

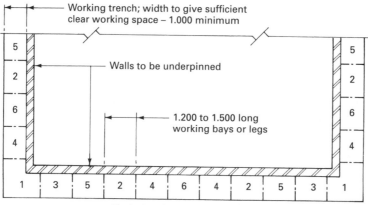

Typical underpinning schedule

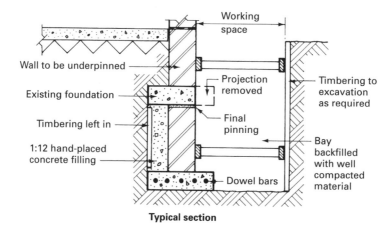

Typical section

Projection
cut back

Existing
foundation

New brickwork
toothed at ends

25 mm diameter
dowel bars

Timbering to
excavation
as required

Final pinning
carried out
after new wall
has settled

New foundation

Typical elevation

Figure 4.1.1   Traditional brick underpinning

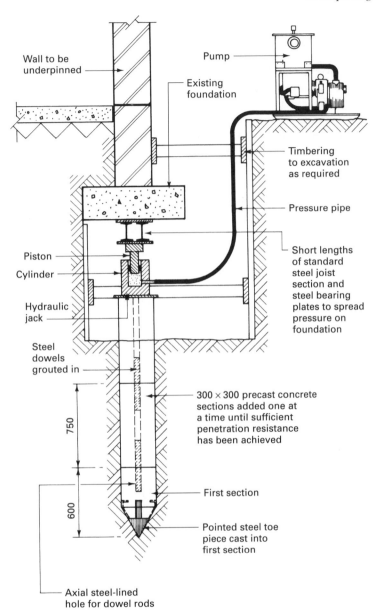

Wall to be underpinned

Pump

Existing foundation

Timbering to excavation as required

Pressure pipe

Piston

Cylinder

Hydraulic jack

Short lengths of standard steel joist section and steel bearing plates to spread pressure on foundation

Steel dowels grouted in

300 × 300 precast concrete sections added one at a time until sufficient penetration resistance has been achieved

750

600

First section

Pointed steel toe piece cast into first section

Axial steel-lined hole for dowel rods

Figure 4.1.2   Jack or miga pile underpinning

▨■■ NEEDLE AND PILES

If the wall to be underpinned has a weak foundation that is considered unsuitable for spanning over the heads of jack piles an alternative method giving the same degree of flexibility can be used. This method uses pairs of jacks or usually bored piles in conjunction with an *in-situ* reinforced concrete beam or needle placed above the existing foundation. The system works on the same principle as a dead shoring arrangement relying on the arching effect of bonded brickwork. If water is encountered when using bored piles a pressure pile can be used as an alternative. The formation of both types of pile is described in Chapter 4.2 on piling. Typical arrangements to enable the work to be carried out from both sides of the wall or from the external face only are shown in Fig. 4.1.3.

▨■■ PYNFORD STOOLING METHOD

The Pynford method of underpinning enables walls to be underpinned in continuous runs without the use of needles or raking shoring. The procedure is to cut away portions of brickwork, above the existing foundation, to enable precast concrete or steel stools to be inserted and pinned. The intervening brickwork can be removed, leaving the structure supported entirely on the stools. Reinforcing bars are threaded between and around the stools and caged to form the ring beam reinforcement. After the formwork has been placed and the beam cast, final pinning can be carried out using a well-rammed dry mortar mix (see Fig. 4.1.4). This method replaces the existing foundation strip with a reinforced concrete ring beam, from which other forms of deep underpinning can be carried out if necessary.

▨■■ CERFAX HOOPSAFE METHOD

This is most appropriate as remedial treatment where differential settlement can be identified. A limited amount of external excavation is needed to expose the substructural wall to a depth just above foundation level. Here, an *in-situ* concrete beam with purpose-made longitudinal voids created with plastic conduits is cast around the building's periphery. The small-diameter voids accommodate steel stressing tendons for post-tensioning to bind the walls into a solid unit.

Continuity provided by the post-tensioned beams integrates the substructural wall and compensates for weaker areas of subsoil. However, where areas or pockets of weakness in the ground can be identified, these should be stabilised by granular consolidation to complement the remedial treatment. This method is relatively fast when applied to regular plan shapes such as a square or rectangle. Offsets and extensions to buildings are less easy to peripherate. Figure 4.1.5 shows the principles of application.

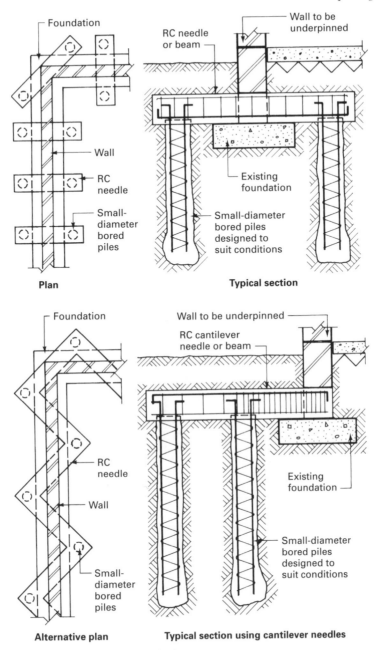

Figure 4.1.3 Needle and pile underpinning

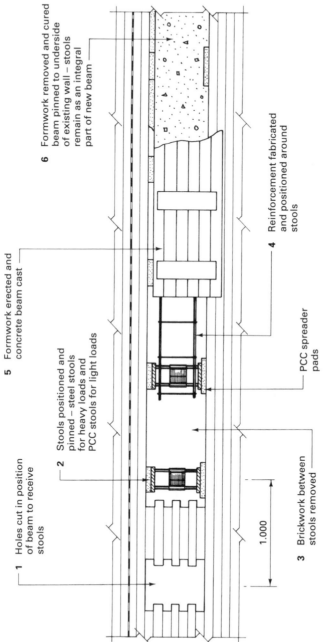

1   Holes cut in position
    of beam to receive
    stools

2   Stools positioned and
    pinned – steel stools
    for heavy loads and
    PCC stools for light loads

3   Brickwork between
    stools removed

4   Reinforcement fabricated
    and positioned around
    stools

5   Formwork erected and
    concrete beam cast

6   Formwork removed and cured
    beam pinned to underside
    of existing wall – stools
    remain as an integral
    part of new beam

    PCC spreader
    pads

    1.000

Figure 4.1.4 'Pynford' stooling method of underpinning

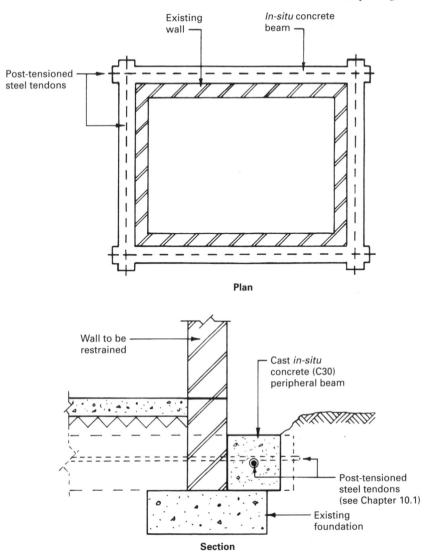

Plan

Section

Figure 4.1.5   Hoopsafe foundation support

▮▮▮ DOUBLE-ANGLE ROOT PILING

This is another low-cost modern alternative to traditional underpinning for stabilising existing substructural walls and foundations. Figure 4.1.6 shows the installation of reinforced concrete angle piles by air-flushed rotary percussion drill from inside and outside the building. The voids are lined with steel casings cut off at the surface prior to lowering of reinforcement and placing concrete. Spacings and depth of borings will depend on site conditions such as the occurrence of solid bearing strata and extent of structural damage. Although disruption to the

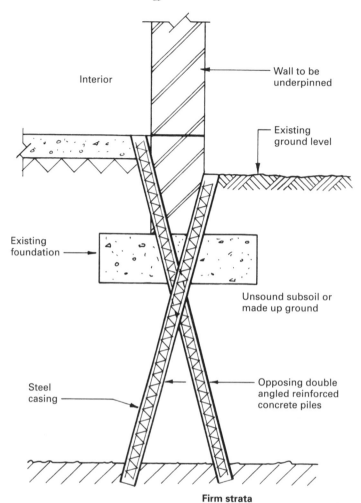

Figure 4.1.6   Angle piled foundation stabilisation

building's interior is most likely (unless external application only is considered adequate), the relatively short timescale and minimal excavation make for an economic process.

■ ■ ■ OTHER FORMS OF UNDERPINNING USING BEAMS

Stressed steel

This consists of standard universal beam sections in short lengths of 600 to 1,500 mm with a steel diaphragm plate welded to each end that is drilled to take high tensile torque bolts. Short lengths of wall are removed and the steel beam inserted. The joints between adjacent diaphragm plates are formed so that a small space occurs on the lower or tension side to give a predetermined camber when the bolts are tightened. Final pinning between the top of the stressed steel beam and the wall completes the operation.

Prestressed concrete

Short precast concrete blocks are inserted over the existing foundations as the brickwork is removed. The blocks are formed to allow for post-tensioning stressing tendons to be inserted, stressed and anchored to form a continuous beam. Final pinning completes the underpinning.

Note that all forms of beam underpinning can also be used to form a lintel or beam, in any part of a wall, prior to the formation of a large opening.

▨■■ FINAL PINNING

Although final pinning is usually carried out by ramming a stiff dry cement mortar mix into the space between the new underpinning work and the existing structure, alternative methods are available such as:

- **Flat jacks** These are circular or rectangular hollow plates of various sizes made from thin sheet metal, which can be inflated with high-pressure water for temporary pinning or, if work is to be permanent, with a strong cement grout. The increase in thickness of the flat jacks is approximately 25 mm.
- **Wedge bricks** These are special bricks of engineering quality of standard face length but with a one brick width and a depth equal to two courses. The brick is made in two parts, the lower section having a wide sloping channel in its top bed surface to receive the wedge-shaped and narrower top section. Both parts have a vertical slot through which the bedding grout passes to key the two sections together.

▨■■ UNDERPINNING COLUMNS

This is a more difficult task than underpinning a wall. It can be carried out on brick or stone columns by inserting a series of stools, casting a reinforced concrete base, and then underpinning by the methods previously described.

Structural steel or reinforced concrete columns must be relieved of their loading before any underpinning can take place. This can be achieved by variations of one basic method. A collar of steel or precast concrete members is fixed around the perimeter of the column. Steel collars are usually welded to the structural member, whereas concrete columns are usually chased to a depth of 25 to 50 mm to receive the support collar. The column loading is transferred from the collar to cross-beams or needles, which in turn transmit the loads to the ground at a safe distance from the proposed underpinning excavations. Cantilever techniques that transfer the loadings to one side of the structural member are possible provided sufficient kentledge and anchorage can be obtained (see Fig. 4.1.7). The underpinning of the column foundation can now be carried out by the means previously described.

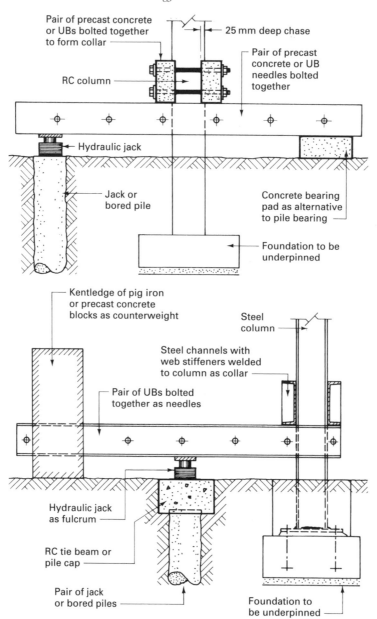

Figure 4.1.7   Typical column underpinning arrangements

PILED FOUNDATIONS

A pile can be loosely defined as a column inserted in the ground to transmit the structural loads to a lower level of subsoil. Piled foundations have been used in this context for hundreds of years and until the twentieth century were invariably of driven timber. Today a wide variety of materials and methods are available to solve most of the problems encountered when confronted with the need for deep foundations. It must be remembered that piled foundations are not necessarily the answer to all awkward foundation problems, but should be considered as an alternative to other techniques when suitable bearing-capacity soil is not found near the lowest floor of the structure.

The unsuitability of the upper regions of a subsoil may be caused by:

- low bearing capacity of the subsoil;
- heavy point loads of the structure exceeding the soil bearing capacity;
- presence of highly compressible soils near the surface such as filled ground and underlying peat strata;
- subsoils such as clay, which may be capable of moisture movement or plastic failure;
- high water table.

■■■ CLASSIFICATION

Piled foundations may be classified by the way in which they transmit their loads to the subsoils or by the way they are formed. Transmittance of loading to a lower level may be by:

- **End bearing**   The shafts of the piles act as columns, carrying the loads through the overlying weak subsoils to firm strata into which the pile toe has penetrated. This can be a rock stratum or a layer of firm sand or gravel that has been compacted by the displacement and vibration encountered during the driving.

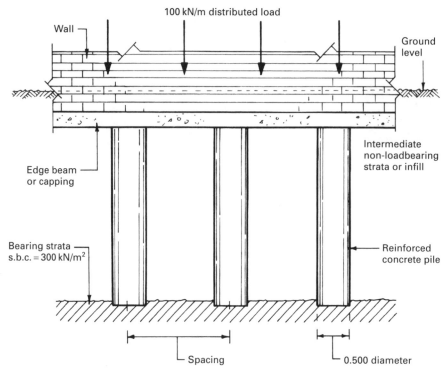

Figure 4.2.1 Pile spacing

The design of end bearing piled foundations can be determined by establishing the safe bearing capacity of the subsoil by sampling and labatory analysis, and relating this to the load distribution on each pile. For example, the wall shown in Fig. 4.2.1 carries a load of 100 kN/m onto piles bearing on subgrade of safe bearing capacity of 300 kN/m^2. Using piles of say 0.500 m diameter their spacing will be:

Pile base area = $3.142 \times (0.25)^2 = 0.20$ m^2
Each pile carries: $300 \times 0.2 = 60$ kN
The extent of wall on one pile = $100/60 = 1.670$ m
Therefore pile spacing is not more than 1.670 m.
Minimum permissible spacing is 3 × pile diameter (1.500 m) or 1.000 m (take greater value). Therefore calculated spacing is satisfactory.

■ **Friction** Any foundation imposes on the ground a pressure that spreads out to form a bulb of pressure as shown in Fig. 4.2.2. If a suitable loadbearing stratum cannot be found at an acceptable level, particularly in stiff clay soils, it is possible to use a pile to carry this bulb of pressure to a lower level where a higher bearing capacity is found. The friction or floating pile is supported mainly by the adhesion or friction action of the soil around the perimeter of the pile shaft. Vertical stress at various points in a pile can be calculated by multiplying the coefficients shown in Fig. 4.2.2 by the result of the pressure divided by the pile length squared, i.e. $K(P/L^2)$.

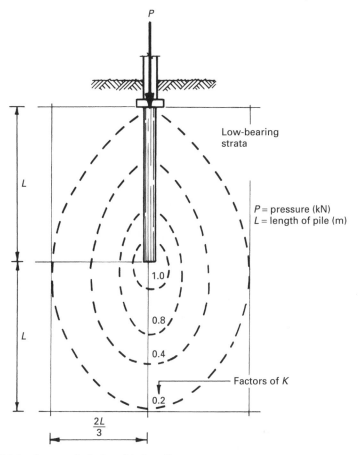

Figure 4.2.2   Pressure bulb for a friction pile

In most situations piles work on a combination of the two principles outlined above, but the major design consideration identifies the pile class.

Piles may be preformed and driven, thus displacing the soil through which they pass, and are therefore classified as **displacement** piles. Alternatively the soil can be bored out and subsequently replaced by a pile shaft, and such piles are classified as **replacement** piles.

■■■ DOWNDRAG

When piles are driven through weak soils such as alluvial or silty clay, peat layers or reclaimed land, the long-term settlement and consolidation of the ground can cause downdrag of the piles. This is in simple terms a transference of the downdrag load of consolidating soil onto a pile shaft. It is called negative skin friction because the resistance to this transfer of load is the positive skin friction of the pile shaft. Downdrag loads have been known to equal the design bearing capacity of the pile.

Three methods to counteract the effects of downdrag are possible:

- Increase the number of piles being used to carry the structural loads to accommodate the anticipated downdrag loads. This method may necessitate the use of twice the number of piles required to carry only the structural loadings.
- Increase the size of the piles designed for structural loads only: this can be expensive, because the cost of forming piles does not increase pro rata with its size.
- Design for the structural loads only and prevent the transference of downdrag loads by coating the external face of the shaft with a slip layer, thus reducing the negative skin friction. A special bituminous compound has been developed to fulfil this task and is applied in a 10 mm thick coat to that part of the shaft where negative skin friction can be expected. The toe of the pile is never coated, so that the end bearing capacity is not reduced. Care must be taken to ensure that the coating is not damaged during transportation, storage and driving, particularly in hot weather, when a reflective coating of lime wash may be necessary.

■ ■ ■ DISPLACEMENT PILES

This is a general term applied to piles that are driven, thus displacing the soil, and includes those piles that are preformed, partially preformed or are driven *in-situ* piles.

TIMBER PILES

These are usually of square sawn hardwood or softwood in lengths up to 12.000 m with section sizes ranging from 225 mm × 225 mm to 600 mm × 600 mm. They are easy to handle, and can be driven by percussion with the minimum of experience. Most timber piles are fitted with an iron or steel driving shoe, and have an iron ring around the head to prevent splitting due to impact. Although not particularly common, they are used in sea defences such as groynes and sometimes as guide piles for large trestles in conjunction with steel sheet piling. Loadbearing capacities can be up to 350 kN per pile, depending upon section size and/or species.

PRECAST CONCRETE PILES

These are used on medium to large contracts where soft soils overlying a firm stratum are encountered and at least 100 piles will be required. Lengths up to 18.000 m with section sizes ranging from 250 mm × 250 mm to 450 mm × 450 mm carrying loadings of up to 1,000 kN are generally economical for the conditions mentioned above. The precast concrete driven pile has little frictional bearing strength because the driving operation moulds the cohesive soils around the shaft, which reduces the positive frictional resistance.

Problems can be encountered when using this form of pile in urban areas for the following reasons:

- Transportation of the complete length of pile through narrow and/or congested streets.
- The driving process, which is generally percussion, and can set up unacceptable noise and/or vibrations.
- The fact that many urban sites are in themselves restricted or congested, thus making it difficult to manoeuvre the long pile lengths around the site.

PREFORMED CONCRETE PILES

These are available as reinforced precast concrete or prestressed concrete piles, but because of the problems listed above and the site difficulties that can be experienced in splicing or lengthening preformed piles their use has diminished considerably in recent years in favour of precast piles formed in segments or the partially preformed types of pile. Typical examples of the segmental type are West's 'Hardrive' and 'Segmental' piles. The **Hardrive** pile is composed of standard interchangeable precast concrete units of 10.000, 5.000 and 2.500 m lengths designed to carry working loads up to 800 kN. The pile lengths are locked together with four H-section pins located at the corners of an aligning sleeve (see Fig. 4.2.3). The **Segmental** pile is designed for lighter loading conditions of up to 300 kN, and is formed by joining the 1.000 m standard lengths together with spigot and socket joints up to a maximum length of 15.000 m. Special half-length head units are available to reduce wastage to a minimum (see Fig. 4.2.4).

STEEL PREFORMED PILES

These are used mainly in conjunction with marine structures and where overlying soils are very weak. Two forms are encountered: the hollow box pile, which is made from welded rolled steel sections to BS EN 10210-2: *Hot finished structural hollow sections of non-alloy and fine grain structural steels. Tolerances, dimensions and sectional properties*, and the universal bearing pile of conventional 'H' section to BS 4-1: *Structural steel sections. Specification for hot-rolled sections*. These piles are relatively light and therefore are easy to handle and drive. Splicing can be carried out by site welding to form piles up to 15.000 m long with loadbearing capacities up to 600 kN. Consideration must always be given to the need to apply a protective coating to the pile to guard against corrosion.

COMPOSITE PILES

These are sometimes referred to as partially preformed piles, and are formed by a method that combines the use of precast and *in-situ* concrete or steel and *in-situ* concrete. They are used mainly on medium to large contracts where the presence of running water or very loose soils would render the use of bored or preformed piles unsuitable. Composite piles provide a pile of readily variable length made from easily transported short lengths. Typical examples are 'Prestcore', West's 'Shell' and cased piles.

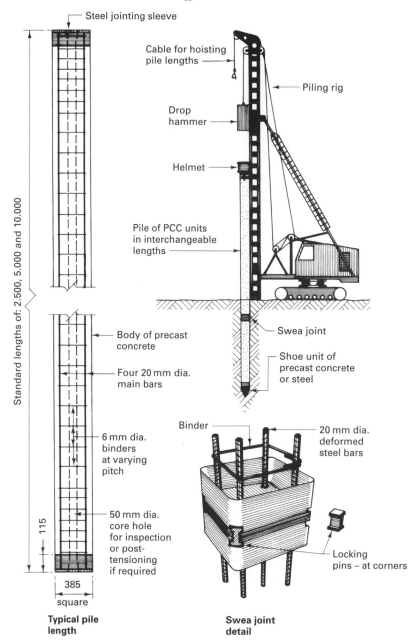

Steel jointing sleeve

Cable for hoisting
pile lengths

Piling rig

Drop
hammer

Helmet

Pile of PCC units
in interchangeable
lengths

Swea joint

Standard lengths of: 2.500, 5.000 and 10.000

Body of precast
concrete

Shoe unit of
precast concrete
or steel

Four 20 mm dia.
main bars

6 mm dia.
binders
at varying
pitch

50 mm dia.
core hole
for inspection
or post-
tensioning
if required

Binder

20 mm dia.
deformed
steel bars

Locking
pins – at corners

115

385
square

**Typical pile
length**

**Swea joint
detail**

Figure 4.2.3   West's Hardrive precast modular piles

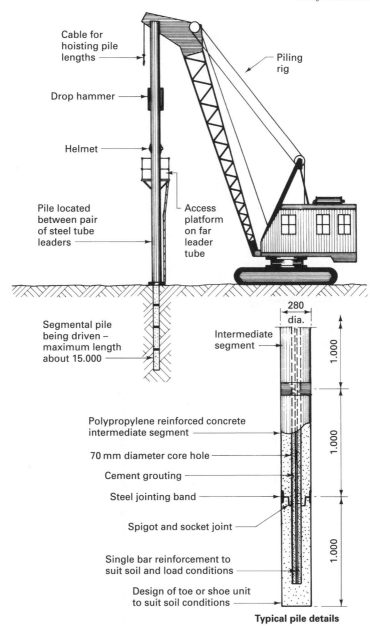

Cable for
hoisting pile
lengths

Piling
rig

Drop hammer

Helmet

Pile located
between pair
of steel tube
leaders

Access
platform
on far
leader
tube

Segmental pile
being driven –
maximum length
about 15.000

280
dia.

Intermediate
segment

1.000

Polypropylene reinforced concrete
intermediate segment

70 mm diameter core hole

Cement grouting

Steel jointing band

Spigot and socket joint

Single bar reinforcement to
suit soil and load conditions

Design of toe or shoe unit
to suit soil conditions

1.000

1.000

Typical pile details

Figure 4.2.4   West's Segmental piles

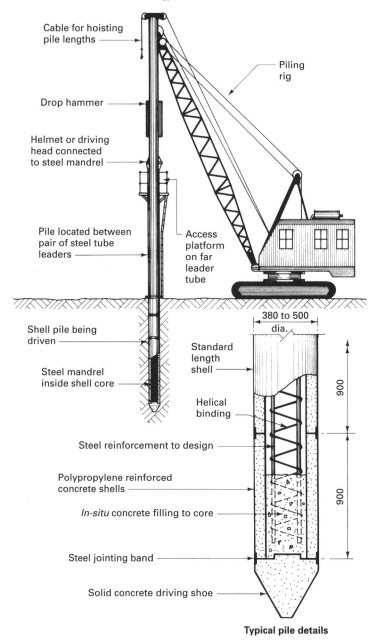

Cable for hoisting
pile lengths

Piling
rig

Drop hammer

Helmet or driving
head connected
to steel mandrel

Pile located between
pair of steel tube
leaders

Access
platform
on far
leader
tube

380 to 500
dia.

Shell pile being
driven

Standard
length
shell

Steel mandrel
inside shell core

Helical
binding

900

Steel reinforcement to design

Polypropylene reinforced
concrete shells

In-situ concrete filling to core

900

Steel jointing band

Solid concrete driving shoe

Typical pile details

Figure 4.2.5   West's Shell piles

The **Prestcore** pile is a composite pile formed inside a lined, bored hole and is, strictly speaking, a replacement pile and is therefore described in more detail in the following section on replacement piles.

The **Shell** pile is, however, a driven or displacement pile consisting of a series of precast shells threaded onto a mandrel and top-driven to the required set. After driving and removing the mandrel the hollow core can be inspected, a cage of reinforcement inserted, and the void filled with *in-situ* concrete. Lengths up to 60.000 m with bearing capacities within the range 500 to 1,200 kN are possible with this method. The precast concrete shells are reinforced by a patent system using fibrillated polypropylene film as a substitute for the traditional welded steel fabric (see Fig. 4.2.5).

Piles formed in this manner solve many of the problems encountered with waterlogged and soft substrata by being readily adaptable in length, the shaft can be inspected internally before the *in-situ* concrete is introduced, the flow of water or soil into the pile is eliminated, and the presence of corrosive conditions in the soil can be overcome by using special cements in the shell construction.

BSP CASED PILES

These are typical composite piles using steel and *in-situ* concrete. Cased piles are bearing piles consisting of a driven tube that is filled with *in-situ* concrete. The casing is manufactured from steel strip or plate, which is formed into a continuous helix with the adjoining edges butt-welded. These piles are usually driven into position by using an internal drop hammer operating within the casing. Usually a flat circular plate is welded to the base of the casing and a cushion plug of concrete with a very low water content is placed to a depth equal to $2\frac{1}{2}$ times the pile diameter directly on top of the plate shoe. Pile lengths are available up to 24.000 m as a single tube, but should extra length be required extension casings can be butt-welded on after the first length has been driven to a suitable depth.

Apart from situations such as jetties, where a fair amount of the pile is left projecting, cased piles do not require reinforcement except for splice bars at the top to bond the pile to a pile cap. The *in-situ* concrete is a standard 1:2:4 mix with a water/cement ratio of 0.4 to 0.5 (C20). A wide range of diameters from 250 to 600 mm are available with varying casing thicknesses to give working loads per pile ranging from 150 to 1,500 kN according to the type of subsoil. Typical cased pile details are shown in Fig. 4.2.6.

DRIVEN *IN-SITU* OR CAST-IN-PLACE PILES

These are an alternative to preformed displacement piles and are usually applied on medium to large contracts where subsoil conditions or loadings are likely to create variations in the lengths of pile required. They can be formed economically in diameters of 300 to 600 mm with lengths up to 18.000 m designed to carry loads of up to 1,300 kN. They generally require heavy piling rigs, and an open, level site where noise is unrestricted.

In some systems the tube that is used to form the pile shaft is top driven, and in others such as the Franki system driving is carried out by means of an internal

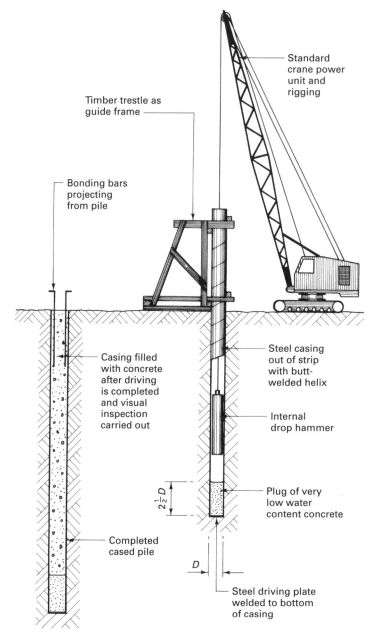

Standard
crane power
unit and
rigging

Timber trestle as
guide frame

Bonding bars
projecting
from pile

Casing filled
with concrete
after driving
is completed
and visual
inspection
carried out

Steel casing
out of strip
with butt-
welded helix

Internal
drop hammer

Completed
cased pile

$2\frac{1}{2}D$

Plug of very
low water
content concrete

D

Steel driving plate
welded to bottom
of casing

Figure 4.2.6   BSP cased piles

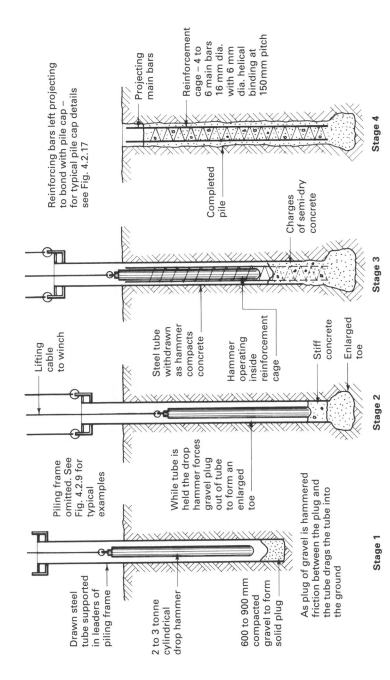

Figure 4.2.7 Franki driven *in-situ* piles

Reinforcing bars left projecting to bond with pile cap – for typical pile cap details see Fig. 4.2.17

Projecting main bars

Reinforcement cage – 4 to 6 main bars 16 mm dia. with 6 mm dia. helical binding at 150 mm pitch

Completed pile

Charges of semi-dry concrete

Stage 4

Lifting cable to winch

Steel tube withdrawn as hammer compacts concrete

Hammer operating inside reinforcement cage

Stiff concrete

Enlarged toe

Stage 2

Stage 3

Piling frame omitted. See Fig. 4.2.9 for typical examples

While tube is held the drop hammer forces gravel plug out of tube to form an enlarged toe

600 to 900 mm compacted gravel to form solid plug

As plug of gravel is hammered between the plug and the tube drags the tube into the ground

Drawn steel tube supported in leaders of piling frame

2 to 3 tonne cylindrical drop hammer

Stage 1

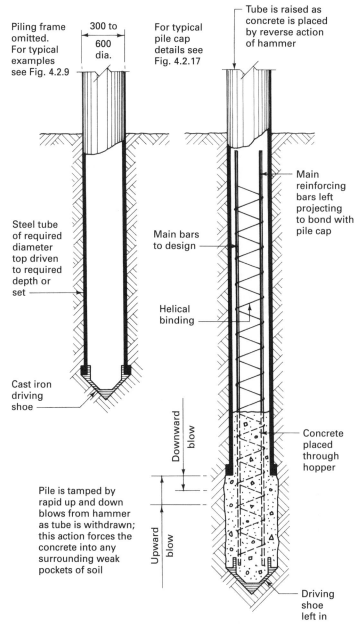

Piling frame omitted. For typical examples see Fig. 4.2.9

300 to 600 dia.

For typical pile cap details see Fig. 4.2.17

Tube is raised as concrete is placed by reverse action of hammer

Steel tube of required diameter top driven to required depth or set

Cast iron driving shoe

Main bars to design

Helical binding

Downward blow

Main reinforcing bars left projecting to bond with pile cap

Concrete placed through hopper

Pile is tamped by rapid up and down blows from hammer as tube is withdrawn; this action forces the concrete into any surrounding weak pockets of soil

Upward blow

Driving shoe left in

Figure 4.2.8   Vibro cast *in-situ* pile

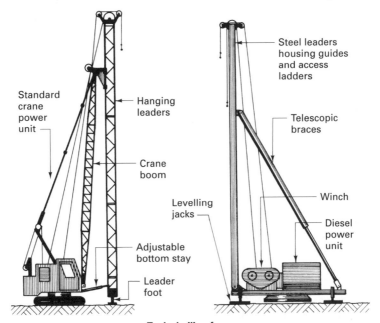

Typical piling frames

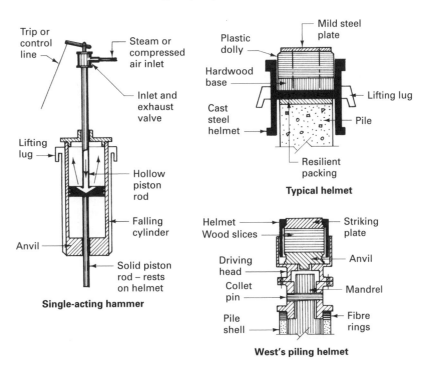

Single-acting hammer

Typical helmet

West's piling helmet

Figure 4.2.9 Piling frames and equipment

drop hammer working on a plug of dry concrete. *In-situ* concrete for the core is introduced into the lined shaft through a hopper or skip, and the concrete can be consolidated by impact of the internal drop hammer or by vibration of the tube as it is withdrawn (see Figs 4.2.7 and 4.2.8).

A problem that can be encountered with this form of pile is necking due to groundwater movement washing away some of the concrete, thus reducing the effective diameter of the pile shaft and consequently the cover of concrete over the reinforcement. If groundwater is present the other forms of displacement pile previously described should be considered.

PILEDRIVING

Displacement piles are generally driven into the ground by holding them in the correct position against the piling frame and applying hammer blows to the head of the pile. Exceptions are encountered, such as the cased pile shown in Fig. 4.2.6. The piling frame can be purpose made or an adaptation of a standard crane power unit. The basic components of any piling frame are the vertical member that houses the leaders or guides, which in turn support the pile and guide the hammer onto the head of the pile (see Fig. 4.2.9). Pile hammers come in a variety of types and sizes powered by gravity, steam, compressed air or diesel.

Drop hammers

These are blocks of cast iron or steel with a mass range of 1,500 to 8,000 kg and are raised by a cable attached to a winch. The hammer, which is sometimes called a monkey or ram, is allowed to fall freely by gravity onto the pile head. The free-fall distance is controllable, but generally a distance of about 1.200 m is employed. Drop hammers are slower than the following power hammers and may inflict more damage to the pile caps.

Single-acting hammers

Activated by steam or compressed air, these have much the same effect as drop hammers in that the hammer falls freely by gravity through a distance of about 1.500 m. Two types are available: in one case the hammer is lifted by a piston rod; in the other the piston is static and the cylinder is raised and allowed to fall freely (see Fig. 4.2.9). Both forms of hammer deliver a very powerful blow.

Double-acting hammers

These are activated by steam or compressed air, and consist of a heavy fixed cylinder in which there is a light piston or ram that delivers a large number of rapid light blows (90 to 225 blows per minute) in a short space of time, as opposed to the heavier blows over a longer period of the drop and single–acting hammers. The object is to try to keep the pile constantly on the move rather than being driven in a series of jerks. This type of hammer has been largely replaced by the diesel hammer and by vibration techniques.

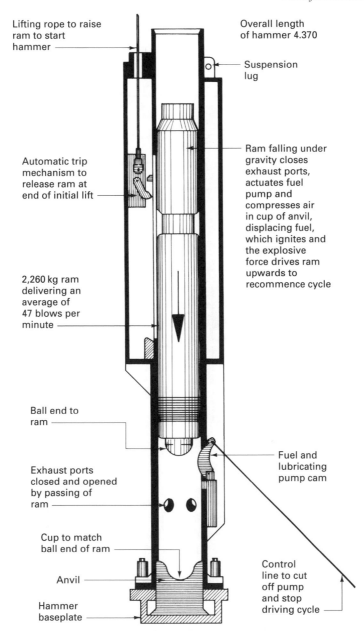

Lifting rope to raise ram to start hammer

Overall length of hammer 4.370

Suspension lug

Automatic trip mechanism to release ram at end of initial lift

Ram falling under gravity closes exhaust ports, actuates fuel pump and compresses air in cup of anvil, displacing fuel, which ignites and the explosive force drives ram upwards to recommence cycle

2,260 kg ram delivering an average of 47 blows per minute

Ball end to ram

Fuel and lubricating pump cam

Exhaust ports closed and opened by passing of ram

Cup to match ball end of ram

Anvil

Control line to cut off pump and stop driving cycle

Hammer baseplate

Figure 4.2.10   Typical diesel hammer details

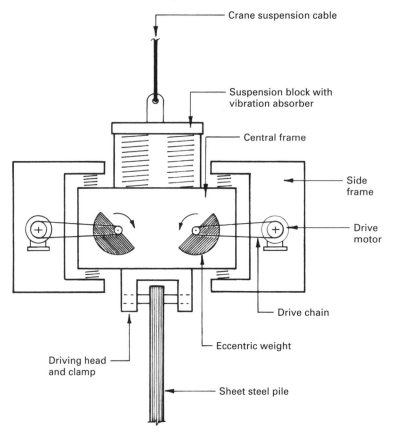

Figure 4.2.11   Operating principles of a vibrating hammer

Diesel hammers

These have been designed to give a reliable and economic method of piledriving. Various sizes giving different energy outputs per blow are available, but most deliver between 46 and 52 blows per minute. The hammer can be suspended from a crane or mounted in the leaders of a piling frame. A measured amount of liquid fuel is fed into a cup formed in the base of the cylinder. The air being compressed by the falling ram is trapped between the ram and the anvil, which applies a preloading force to the pile. The displaced fuel, at the precise moment of impact, results in an explosion that applies a downward force on the pile and an upward force on the ram, which returns to its starting position to recommence the complete cycle. The movement of the ram within the cylinder activates the fuel supply, and opens and closes the exhaust ports (see Fig. 4.2.10). Water- or air-cooled variations exist for popular application to vertical and inclined driving.

Vibration techniques

These can be used in driving displacement piles where soft clays, sands and gravels are encountered. They are notably efficient, with preformed steel pipe, 'H' and sheet pile sections. The equipment consists of a vibrating unit mounted on the pile head transmitting vibrations of the required frequency and amplitude down the length of the pile shaft. It achieves this with two eccentric rotors propelled in opposite directions to generate vertical vibrations in the pile. These are, in turn, transmitted to the surrounding soil, reducing its shear strength and enabling the pile to sink into the subsoil under its own weight and also that of the vibrator unit. Figure 4.2.11 shows the operating principles of the unit attached to the head of a sheet steel pile. To aid the driving process and to reduce the risk of damage to the pile during driving, water-jetting techniques can be used. Water is directed at the soil around the toe of the pile to loosen it and ease the driving process. The water pipes are usually attached to the perimeter of the pile shaft, and are therefore taken down with the pile as it is being driven. The water-jetting operation is stopped before the pile reaches its final depth so that the toe of the pile is finally embedded in undisturbed soil. The main advantage is the relatively silent operation. Additionally there is no damage to the pile cap, and high-voltage electricity may be used instead of combustible fuels.

To protect the heads of preformed piles from damage due to impact from hammers, various types of protective cushioned helmet are used. These can be either of a general nature, as shown in Fig. 4.2.9, or special purpose for a particular system, such as those used for West's piling.

■■■ REPLACEMENT PILES

Sometimes referred to as **bored piles**, these are formed by removing a column of soil and replacing it with *in-situ* concrete or, as in the case of composite piles, with precast and *in-situ* concrete. Replacement or bored piles are considered for sites where piling is being carried out in close proximity to existing buildings or where vibration and/or noise is restricted. The formation of this type of pile can be considered under three general classifications:

- percussion bored piles;
- rotary bored piles;
- prestcore piles.

PERCUSSION BORED PILES

These are suitable for small and medium-sized contracts of up to 300 piles in clay or gravel subsoils. Pile diameters are usually from 300 to 950 mm, designed to carry loads up to 1,500 kN. Apart from the common factor with all replacement piles that the strata penetrated can be fully explored, these piles can be formed by using a shear leg or tripod rig requiring as little as 1.800 m headroom.

A steel tube made up from lengths (1.000 to 1.400 m) screwed together is sunk by extracting the soil from within the tube liner using percussion cutters or balers according to the nature of the subsoil to be penetrated. The steel lining tube will usually sink under its own weight, but it can be driven in with slight pressure, normally applied by means of hydraulic jacks. When the correct depth has been reached, a cage of reinforcement is placed within the liner and the concrete is introduced. Tamping is carried out as the liner is extracted by using a winch or

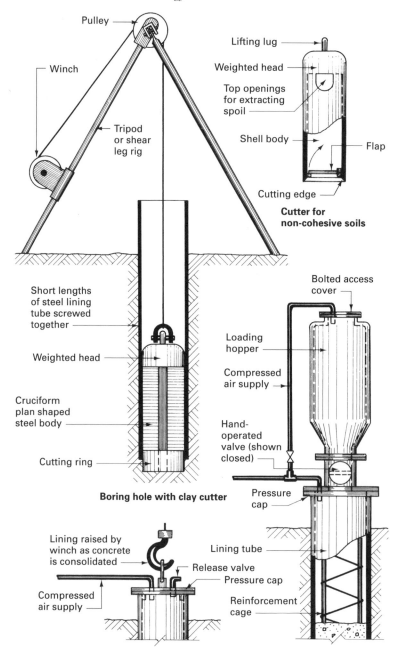

Pulley

Winch

Tripod
or shear
leg rig

Lifting lug

Weighted head

Top openings
for extracting
spoil

Shell body

Flap

Cutting edge

**Cutter for
non-cohesive soils**

Short lengths
of steel lining
tube screwed
together

Weighted head

Cruciform
plan shaped
steel body

Cutting ring

Bolted access
cover

Loading
hopper

Compressed
air supply

Hand-
operated
valve (shown
closed)

Pressure
cap

Boring hole with clay cutter

Lining raised by
winch as concrete
is consolidated

Compressed
air supply

Release valve
Pressure cap

Lining tube

Reinforcement
cage

Alternative methods of forming pressure piles

Figure 4.2.12   Typical percussion bored pile details

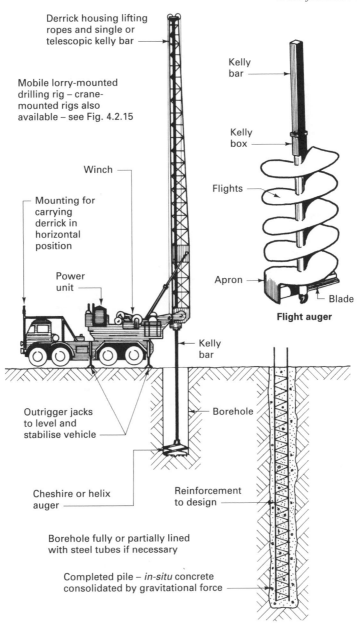

Derrick housing lifting ropes and single or telescopic kelly bar

Mobile lorry-mounted drilling rig – crane-mounted rigs also available – see Fig. 4.2.15

Winch

Mounting for carrying derrick in horizontal position

Power unit

Outrigger jacks to level and stabilise vehicle

Cheshire or helix auger

Borehole fully or partially lined with steel tubes if necessary

Completed pile – *in-situ* concrete consolidated by gravitational force

Kelly bar

Kelly box

Flights

Apron

Blade

Flight auger

Kelly bar

Borehole

Reinforcement to design

Figure 4.2.13 Typical rotary bored pile details

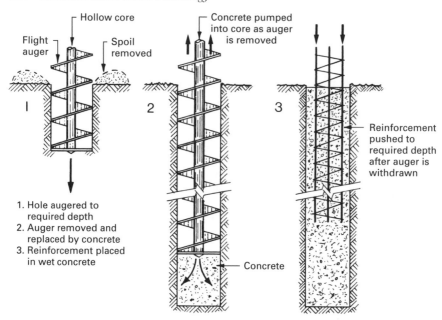

Figure 4.2.14   Grout injection piling

hydraulic jack operating against a clamping collar fixed to the top of the steel tube lining. An internal drop hammer can be used to tamp and consolidate the concrete, but usually compressed air is the method employed.

If waterlogged soil is encountered, a pressure pile is usually formed by fixing to the head of the steel liner an airlock hopper through which the concrete can be introduced and consolidated while the borehole remains under pressure in excess of the hydrostatic pressure of the groundwater (see Fig. 4.2.12).

ROTARY BORED PILES

These can range from the short bored pile used in domestic dwellings (see *Construction Technology*, Chapter 2.2) to the very-large-diameter piles used for concentrated loads in multi-storey buildings and bridge construction. The rotary bored pile is suitable for most cohesive soils, such as clay, and is formed using an auger, which may be operated in conjunction with the steel tube liner according to the subsoil conditions encountered.

Two common augers are in use. One is the Cheshire auger, which has $1\frac{1}{2}$ to 2 helix turns at the cutting end and is usually mounted on a lorry or tractor. The turning shaft or kelly bar is generally up to 7.500 m long and is either telescopic or extendable. The soil is cut by the auger, raised to the surface, and spun off the helix to the side of the borehole, from where it is removed from site (see Fig. 4.2.13). Alternatively, a continuous or flight auger can be used, where the spiral motion brings the spoil to the surface for removal from site. Flight augers are usually mounted on an adapted excavator or crane power unit.

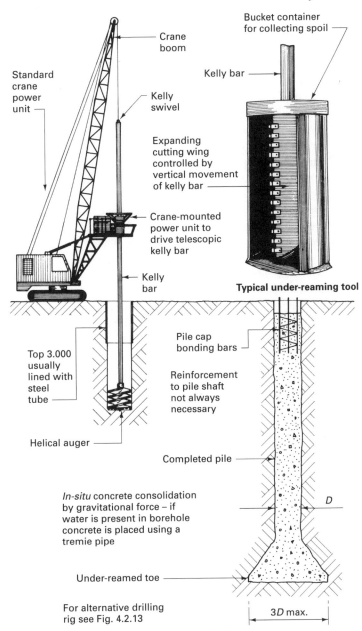

Bucket container
for collecting spoil

Crane
boom

Kelly bar

Standard
crane
power
unit

Kelly
swivel

Expanding
cutting wing
controlled by
vertical movement
of kelly bar

Crane-mounted
power unit to
drive telescopic
kelly bar

Typical under-reaming tool

Kelly
bar

Top 3.000
usually
lined with
steel
tube

Pile cap
bonding bars

Reinforcement
to pile shaft
not always
necessary

Helical auger

Completed pile

In-situ concrete consolidation
by gravitational force – if
water is present in borehole
concrete is placed using a
tremie pipe

Under-reamed toe

D

For alternative drilling
rig see Fig. 4.2.13

3*D* max.

Figure 4.2.15   Typical large-diameter bored pile details

A development of continuous or flight auger replacement piling uses a modified auger with an open-ended hollow central core. The borehole is excavated as described for a conventional auger but, on reaching the required depth, concrete is pumped into the hollow core as the auger is withdrawn. This practice is time efficient, as there is no need to case the pile or temporarily support the excavation. There is also little disturbance to the surrounding subsoil. When the auger is removed, a reinforcement cage is vibrated into the concrete to the required depth. To ease pumping and reinforcement placement, the concrete design may specify rounded aggregates and a higher than normal water/cement ratio. Therefore pile diameter and quantity of piles may be greater than with more conventional methods. This technique is also known as **grout injection piling** (see Fig. 4.2.14).

Large-diameter bored piles are usually considered to be those over 750 mm, and can be formed with diameters up to 2.600 m with lengths ranging from 24.000 to 60.000 m to carry loadings from 2,500 to 8,000 kN. They are suitable for use in stiff clays for structures having highly concentrated loadings, and can be lined or partially lined with steel tubes as required. The base or toe of the pile can be enlarged or under-reamed up to three times the shaft diameter to increase the bearing capacity of the pile. Reinforcement is not always required, and the need for specialists' knowledge at the design stage cannot be overemphasised. Compaction of the concrete, which is usually placed by a tremie pipe, is generally by gravitational force (see Fig. 4.2.15). Test loading of large–diameter bored piles can be very expensive, and if the local authority insists on test loading it can render this method uneconomic.

PRESTCORE PILES

These are a form of composite pile consisting of precast and *in-situ* concrete. The formation of the borehole is as described previously for the percussion bored pile, using light, easy to handle equipment requiring a low headroom or working height. The main advantage of this form of pile lies in the fact that the problem of necking is eliminated, which makes the system suitable for piling in waterlogged soils. The formation of a prestcore pile can be divided into four distinct stages:

1. **Boring**   A lined borehole formed by percussion methods using a tripod rig.
2. **Assembly**   Precast units that form the core of the pile are assembled on a special mandrel, and reinforcement is inserted before the core unit is lowered into position.
3. **Pressing the core**   The pile core is raised and lowered by means of a pneumatic winch attached to the head of the lining tube to consolidate the bearing stratum.
4. **Grouting**   The lining tube is withdrawn and the pile is grouted with the aid of compressed air to expel any groundwater.

The assembly arrangement and the unit details are shown in Fig. 4.2.16.

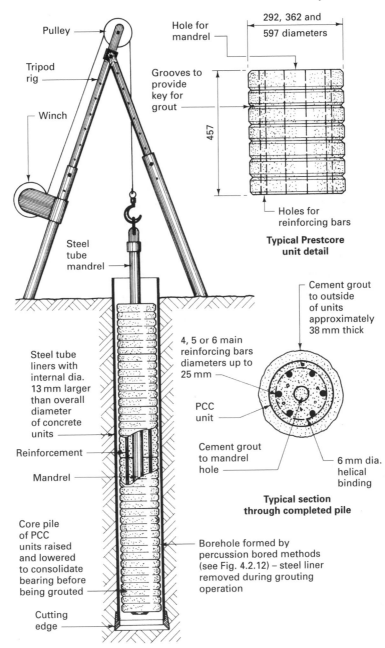

Pulley

Tripod rig

Winch

Hole for mandrel

Grooves to provide key for grout

292, 362 and 597 diameters

457

Holes for reinforcing bars

Typical Prestcore unit detail

Steel tube mandrel

Steel tube liners with internal dia. 13 mm larger than overall diameter of concrete units

Reinforcement

Mandrel

Core pile of PCC units raised and lowered to consolidate bearing before being grouted

Cutting edge

4, 5 or 6 main reinforcing bars diameters up to 25 mm

PCC unit

Cement grout to mandrel hole

Borehole formed by percussion bored methods (see Fig. 4.2.12) – steel liner removed during grouting operation

Cement grout to outside of units approximately 38 mm thick

6 mm dia. helical binding

Typical section through completed pile

Figure 4.2.16   BSP Prestcore bored pile

▨▪■ PILE TESTING

The main objective of forming a test pile is to confirm that the design and formation of the chosen pile type is adequate. It is always advisable to form at least one test pile on any piling contract and, indeed, many local authorities will insist upon this being carried out. The test pile must not be used as part of the finished foundations, and should be formed and tested in such a position that it will not interfere with the actual contract but is nevertheless truly representative of site conditions.

Test piles are usually overloaded by at least 50% of the design working load to near failure or to actual failure depending upon the test data required. Any loading less than total failure loading should remain in place for at least 24 hours. The test pile is bored to the required depth or driven to the required set (which is a predetermined penetration per blow or series of blows), after which it can be tested by one of the following methods:

- A grillage or platform of steel or timber is formed over the top of the test pile and loaded with a quantity of kentledge composed of pig iron or precast concrete blocks. A hydraulic jack is placed between the kentledge and the head of the pile and the test load gradually applied.
- Three piles are formed, and the outer two piles are tied across their heads with a steel or concrete beam. The object is to jack down the centre or test pile against the uplift of the outer piles. Unless two outer piles are available for the test this method can be uneconomic if special outer piles have to be bored or driven.
- Anchor-stressing wires are secured into rock or by some other means to provide the anchorage and uplift the resistance to a cross-beam of steel or concrete by passing the wires over the ends of the beam. The test load is applied by a hydraulic jack placed between the cross-member and the pile head, as described for the previous method.
- The constant rate of penetration test is a method whereby the pile under test is made to penetrate the ground at a constant rate, generally in the order of about 0.8 mm per minute. The load needed to produce this rate of penetration is plotted against a deflection or timebase. When no further load is required to maintain this constant rate of penetration, the ultimate bearing capacity has been reached.

Note that the safe load for a single pile cannot necessarily be multiplied by the number of piles in a cluster to obtain the safe load of a group of piles.

▨▪■ PILE CAPS

Apart from the simple situations such as domestic dwellings, or where large-diameter piles are employed, piles are not usually used singly but are formed into a group or cluster. The load is distributed over the heads of the piles in the group by means of a reinforced cast *in-situ* concrete pile cap. To provide structural continuity the reinforcement in the piles is bonded into the pile cap; this may necessitate the breaking out of the concrete from the heads of the piles to expose the reinforcement. The heads of piles also penetrate the base of the pile cap some 100 to 150 mm to ensure continuity of the members.

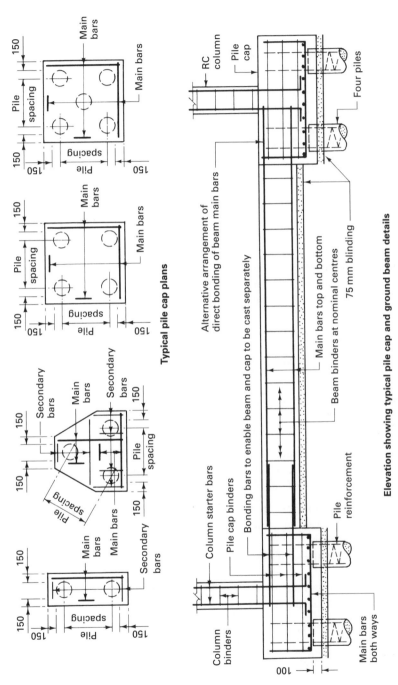

Typical pile cap plans

Elevation showing typical pile cap and ground beam details

Figure 4.2.17 Typical pile cap and ground beam details

Piles should be spaced at such a distance that the group is economically formed and at the same time any interaction between adjacent piles is prevented. Actual spacings must be selected upon subsoil conditions, but the usual minimum spacings are:

- **Friction piles**: 3 pile diameters or 1.000 m, whichever is greater.
- **End bearing piles**: 2 pile diameters or 750 mm, whichever is greater.

The plan shape of the pile cap should be as conservative as possible, and this is usually achieved by having an overhang of 150 mm. The Federation of Piling Specialists have issued the following guide table as to suitable pile cap depths, having regard to both design and cost requirements:

Pile diameter (mm)	300	350	400	450	500	550	600	750
Depth of cap (mm)	700	800	900	1,000	1,100	1,200	1,400	1,800

The main reinforcement is two-directional, formed in bands over the pile heads to spread the loads, and usually takes the form of a U-shaped bar suitably bound to give a degree of resistance to surface cracking of the faces of the pile cap (see typical details shown in Fig. 4.2.17).

In many piling schemes, especially where capped single piles are used, the pile caps are tied together with reinforced concrete tie beams. The beams can be used to carry loadings such as walls to the pile foundations (see Fig. 4.2.17).

▩▩■■ PILING CONTRACTS

The formation of piled foundations is a specialist's task, and so a piling contractor may be engaged to carry out the work by one of three methods:

- Nominated subcontractor.
- Direct subcontractor.
- Main contractor under a separate contract.

Piling companies normally supply the complete service of advising, designing and carrying out the complete site works, including setting out if required. When seeking a piling tender the builder should have and supply the following information:

- **site investigation report**, giving full details of subsoil investigations, adjacent structures, topography of site and any restrictions regarding headroom, noise or vibration limitations;
- **site layout drawings**, indicating levels, proposed pile positions and structural loadings;
- **contract data** regarding tender dates, contract period, completion dates, a detailed specification and details of any special or unusual contract clauses.

The appraisal of piling tenders is not an easy task, and the builder should take into account not only costs but also any special conditions attached to the submitted tender and the acceptance by the local authority of the proposed scheme and system.

Reference:
BS 8004: *Code of practice for foundations.*

4.3

SUBSOIL ANALYSIS AND FOUNDATIONS

Loadings in buildings consist of the combined dead weight of the structure plus the imposed loading from people, furnishings and the effect of snow and wind. These exert a downward pressure upon the soil on which the structure is founded, and this in turn promotes a reactive force in the form of an upward pressure from the soil. The structure is in effect sandwiched between these opposite pressures, and the design of the building must be able to resist the resultant stresses set up within the structural members and the general building fabric. The supporting subsoil must be able to develop sufficient reactive force to give stability to the structure to prevent failure due to unequal settlement and to prevent failure of the subsoil due to shear. To enable designers to select, design and detail a suitable foundation they must have adequate data regarding the nature of the soil on which the structure will be founded, and this is normally obtained from a planned soil investigation programme.

▨▨■■ SOIL INVESTIGATION

Soil investigation is specific in its requirements whereas site investigation is all-embracing, taking into account such factors as topography, location of existing services, means of access and any local restrictions. Soil investigation is a means of obtaining data regarding the properties and characteristics of subsoils by providing samples for testing or providing a means of access for visual inspection. The actual data required and the amount of capital that can be reasonably expended on any soil investigation programme will depend upon the type of structure proposed and on how much previous knowledge the designer has of a particular region or site.

The main methods of soil investigation can be enumerated as follows:

- Trial pits: small contracts where foundation depths are not likely to exceed 3.000 m.
- Boreholes: medium to large contracts with foundations up to 30.000 m deep.

TRIAL PITS

These present a relatively cheap method of obtaining soil data by enabling easy visual inspection of the soil stratum in its natural condition. The pits can be hand- or machine-excavated to a plan size of 1.200 m × 1.200 m and spaced at centres to suit the scope of the investigation. A series of pits set out on a 20.000 m grid would give a reasonable coverage of most sites. The pits need to be positioned so that the data obtained are truly representative of the actual conditions, but not in such a position where their presence could have a detrimental effect on the proposed foundations. In very loose soils, or soils having a high water table, trial pits can prove to be uneconomical because of the need for pumps and/or timbering to keep the pits dry and accessible. The spoil removed will provide disturbed samples for testing purposes, whereas undisturbed samples can be cut and extracted from the walls of the pit.

BOREHOLES

These enable disturbed or undisturbed samples to be removed for analysis and testing, but undisturbed samples are sometimes difficult to obtain from soils other than rock or cohesive soils. The core diameter of the samples obtained varies from 100 to 200 mm according to the method employed in extracting the sample. Disturbed samples can be obtained by using a rotary flight auger or by percussion boring in a manner similar to the formation of small-diameter bored piles using a tripod or shear leg rig. Undisturbed samples can be obtained from cohesive soils using 450 mm long × 100 mm diameter sampling tubes, which are driven into the soil to collect the sample; upon removal the tube is capped, labelled and sent off to a laboratory for testing. Undisturbed rock samples can be obtained by core drilling with diamond-tipped drills where necessary.

■ ■ ■ CLASSIFICATION OF SOILS

Soils may be classified by any of the following methods:

- physical properties;
- geological origin;
- chemical composition;
- particle size.

It has been established that the physical properties of soils can be closely associated with their particle size, both of which are of importance to the foundation engineer, architect or designer. All soils can be defined as being coarse-grained or fine-grained, each resulting in different properties.

COARSE-GRAINED SOILS

These would include sands and gravels having a low proportion of voids, negligible cohesion when dry, high permeability and slight compressibility, which takes place almost immediately upon the application of load.

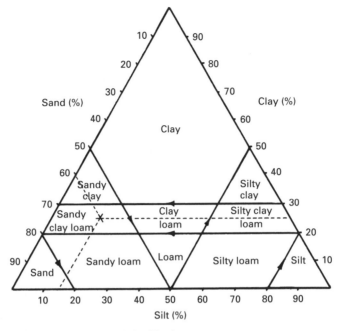

Figure 4.3.1 Triangular chart, soil classification

FINE-GRAINED SOILS

These include the cohesive silts and clays having a high proportion of voids, high cohesion, very low permeability and high compressibility, which takes place slowly over a long period of time.

There are of course soils that can be classified in between the two extremes described above. BS 1377: *Methods of test for soils for civil engineering purposes* divides particle sizes as follows:

- Clay particles: less than 0.002 mm.
- Silt particles: between 0.002 and 0.06 mm.
- Sand particles: between 0.06 and 2 mm.
- Gravel particles: between 2 and 60 mm.
- Cobbles: between 60 and 200 mm.

Soils that are composed mainly of clay, silt and sand can be classified on the triangular chart shown in Fig. 4.3.1. Given a soil sample analysis of 60% sand, 15% silt and 25% clay, it can be identifed as sandy clay loam.

Note: Silt is fine particles of sand, easily suspended in water. Loam is very fine particles of clay, easily dissolved in water.

The silt, sand and gravel particles are also further subdivided into fine, medium and coarse with particle sizes lying between the extremes quoted above.

Fine-grained soils such as clays are difficult to classify positively by their particle size distribution alone, and therefore use is made of the reaction of these soils to a change in their moisture content. If the moisture content is high, the volume is consequently large and the soil is basically a suspension of clay particles in water. As the moisture content decreases, the soil passes the liquid limit and becomes plastic. The liquid limit of a soil is defined as 'the moisture content at which a soil passes from the plastic state to the liquid state', and this limit can be determined by the test set out in BS 1377.

Further lowering of the moisture content will enable the soil to pass the plastic limit and begin to become a solid. The plastic limit of a soil is reached when a 20 g sample just fails to roll into a 3 mm diameter thread when rolled between the palm of the hand and a glass plate. When soils of this nature reach the solid state the volume tends to remain constant, and any further decrease in moisture content will only alter the appearance and colour of the sample. It must be clearly understood that the change from one definable state to another is a gradual process and not a sudden change.

■ ■ ■ SHEAR STRENGTH OF SOILS

The resistance that can be offered by a soil to the sliding of one portion over another or its shear strength is of importance to the designer because it can be used to calculate the bearing capacity of a soil and the pressure it can exert on such members as timbering in excavations. Resistance to shear in a soil under load depends mainly upon its particle composition. If a soil is granular in form, the frictional resistance between the particles increases with the load applied, and consequently its shear strength also increases with the magnitude of the applied load. Conversely, clay particles, being small, develop no frictional resistance, and therefore clay's shear strength will remain constant whatever the magnitude of the applied load. Intermediate soils such as sandy clays normally give only a slight increase in shear strength as the load is applied.

To ascertain the shear strength of a particular soil sample the triaxial compression test as described in BS 1377 is usually employed for cohesive soils. Non-cohesive soils can be tested in a shear box, which consists of a split box into which the sample is placed and subjected to a standard vertical load while a horizontal load is applied to the lower half of the box until the sample shears.

COMPRESSIBILITY

Another important property of soils that must be ascertained before a final choice of foundation type and design can be made is compressibility, and two factors must be taken into account:

- the rate at which compression takes place;
- the total amount of compression when full load is applied.

When dealing with non-cohesive soils such as sands and gravels the rate of compression will keep pace with the construction of the building, and therefore when the structure is complete there should be no further settlement if the soil remains in the same state. A soil is compressed when loaded by the expulsion of air and/or water from the voids, and by the natural rearrangement of the particles. In cohesive soils the voids are very often completely saturated with water, which in itself is nearly incompressible, and therefore compression of the soil can take place only by the water moving out of the voids, thus allowing settlement of the particles. Expulsion of water from the voids within cohesive soils *can* occur, but only at a very slow rate, mainly because of the resistance offered by the plate-like particles of the soil through which it must flow. This gradual compressive movement of a soil is called **consolidation**. Uniform settlement will not normally cause undue damage to a structure, but uneven settlement can cause progressive structural damage.

STRESSES AND PRESSURES

The above comments on shear strength and compressibility clearly indicate that cohesive soils present the most serious problems when giving consideration to foundation choice and design. The two major conditions to be considered are:

- shearing stresses;
- vertical pressures.

Shearing stress

The maximum stress under a typical foundation carrying a uniformly distributed load will occur on a semicircle whose radius is equal to half the width of the foundation, and the isoshear line value will be equal to about one-third the applied pressure (see Fig. 4.3.2). The magnitude of this maximum pressure should not exceed the shearing resistance value of the soil.

Vertical pressure

This acts within the mass of the soil upon which the structure is founded, and should not be of such a magnitude as to cause unacceptable settlement of the structure. Vertical pressures can be represented on a drawing by connecting together points that have the same value, forming what are termed **pressure bulbs**. Most pressure bulbs are plotted up to a value of 0.2 of the pressure per unit area, which is considered to be the limit of pressure that could influence settlement of the structure. Typical pressure bulbs are shown in Figs 4.3.2 and 4.3.3. A comparison of these typical pressure bulbs will show that generally vertical pressure decreases with depth; the 0.2 value will occur at a lower level under strip foundations than under rafts, isolated bases and bases in close proximity to one another, which form combined pressure bulbs. The pressure bulbs illustrated in Figs 4.3.2 and 4.3.3 are based on the soil being homogeneous throughout the depth under consideration. As, in reality, this is not always the case, it is important that soil investigation is carried out at least to the depth of the theoretical pressure bulb. Great care must be taken where underlying strata of highly compressible soil are encountered to ensure that these are not overstressed if cut by the anticipated pressure bulb.

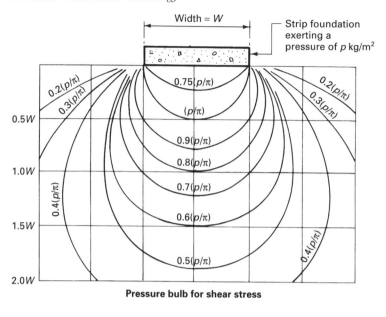

Pressure bulb for shear stress

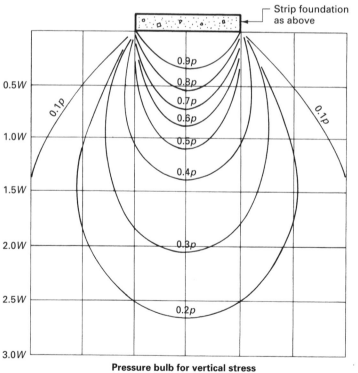

Pressure bulb for vertical stress

Figure 4.3.2   Strip foundations: typical pressure bulbs

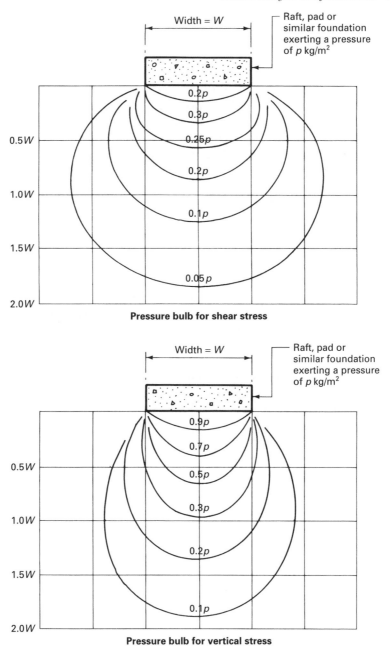

Figure 4.3.3   Raft or similar foundations: typical pressure bulbs

Contact pressure

It is very often incorrectly assumed that a foundation that is uniformly loaded will result in a uniform contact pressure under the foundation. This would only be true if the foundation was completely flexible, such as the bases to a pin-jointed frame. The actual contact pressure under a foundation will be governed by the nature of the soil and the rigidity of the foundation, and because in practice most large structures have a rigid foundation the contact pressure distribution is not uniform. In cohesive soils there is a tendency for high stresses to occur at the edges: this is usually reduced slightly by the yielding of the clay soil. Non-cohesive soils give rise to a parabolic contact pressure distribution with increasing edge pressures as the depth below ground level of the foundation increases. When selecting the basic foundation format consideration must be given to the concentration of the major loads over the position where the theoretical contact pressures are at a minimum to obtain a balanced distribution of contact pressure (see Fig. 4.3.4).

Plastic failure

This is a form of failure that can occur in cohesive soils if the ultimate bearing capacity of the soil is reached or exceeded. As the load on a foundation is increased, the stresses within the soil also increase until all resistance to settlement has been overcome. Plastic failure, which can be related to the shear strength of the soil, occurs when the lateral pressure being exerted by the wedge of relatively undisturbed soil immediately below the foundation causes a plastic shear failure to develop, resulting in a heaving of the soil at the sides of the foundation moving along a slip circle or plane. In practice this movement tends to occur on one side of the building, causing it to tilt and settle (see Fig. 4.3.5). Plastic failure is likely to happen when the pressure applied by the foundation is approximately six times the shear strength of the soil.

APPROXIMATE BEARING CAPACITIES FOR DIFFERENT SUBSOILS

Clay

Very stiff boulder clays and hard clays	$420–650 \ \text{kN/m}^2$
Stiff and sandy clays	$220–420 \ \text{kN/m}^2$
Firm and sandy clays	$110–220 \ \text{kN/m}^2$
Soft clays	$55–110 \ \text{kN/m}^2$
Very soft clays	$<55 \ \text{kN/m}^2$

Sand

Compact graded sands and gravels	$430–650 \ \text{kN/m}^2$
Loose graded sands and gravel	$220–430 \ \text{kN/m}^2$
Compact sands of consistent grade	$220–430 \ \text{kN/m}^2$
Loose sands of consistent grade	$110–220 \ \text{kN/m}^2$
Silts	$55–110 \ \text{kN/m}^2$

Note: Figures given are for dry sands. If saturated, the bearing capacity is about one half of that given.

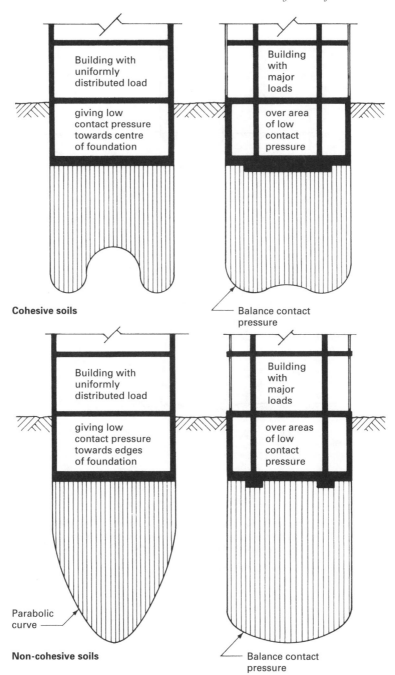

Figure 4.3.4 Typical contact pressures

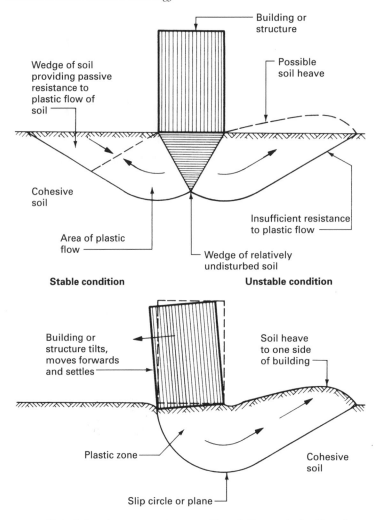

Figure 4.3.5   Plastic failure of foundations

Rocks

Sound igneous and gneissic rocks	$10{,}700 \text{ kN/m}^2$
Dense lime and sandstones	$4{,}300 \text{ kN/m}^2$
Slate and schists	$3{,}200 \text{ kN/m}^2$
Dense shale, mudstones and soft sandstone	$2{,}200 \text{ kN/m}^2$
Clay shales	$1{,}100 \text{ kN/m}^2$
Dense chalk	650 kN/m^2
Thin bedded limestone and sandstone	*
Broken or shattered rocks and soft chalk	*

* To be assessed *in situ*, as may be unstable.

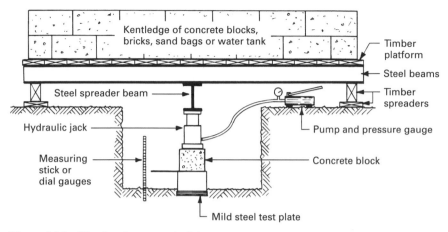

Figure 4.3.6 Plate bearing test: principles

PLATE BEARING TEST

In addition to subjecting subsoil to the previously mentioned site and laboratory tests, trial pit investigation provides an opportunity to load the soil physically in positions where it is anticipated that the foundations will be. The equipment shown in Fig. 4.3.6 replicates the design situation at a proportional reduction, i.e. a proportionally smaller foundation bearing area is loaded proportionally less. Data are recorded as measured in settlement (mm) and soil bearing capacity (kN/m^2).

The procedure is to jack between the steel test plate and an immovable load or kentledge hydraulically until the required pressure or resistance is achieved. Thereafter, pressure increase and corresponding settlement are measured, usually at 12 and 24 hour intervals. Loading is applied in increments of about one-fifth of the anticipated ultimate load per 24 hour interval. The results are plotted graphically, as shown in Fig. 4.3.7. The safe design load should be the ultimate load divided by a factor of safety of at least 3.

Example
If the foundation design is to be a pad of base dimensions 1,400 mm × 1,400 mm loaded to 40 tonnes, a proportionally reduced test plate of 700 mm × 700 mm (1/4 area) may be used to carry 10 tonnes (40 × 1/4). Therefore 10 tonnes × safety factor 3 = 30 tonnes (approx. 300 kN) ultimate load to be applied.

■■■ FOUNDATION TYPES

There are many ways in which foundations can be classified, but one of the most common methods is by form, resulting in four basic types:

- **Strip foundations**   Light loadings, particularly in domestic buildings (see Chapter 2.2 of *Construction Technology*). Heavier loadings can sometimes be founded on a reinforced concrete strip foundation, as described and illustrated in Chapter 2.3 of *Construction Technology*.

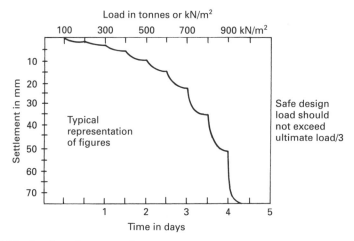

Figure 4.3.7   Load–settlement graph

- **Raft foundations**   Light loadings, average loadings on soils with low bearing capacities and structures having a basement storey (see Chapter 2.3 of *Construction Technology*).
- **Pad or isolated foundations**   A common method of providing the foundation for columns of framed structures and for the supporting members of portal frames (see Chapter 2.3 of *Construction Technology*, Chapter 6.2 of this volume and the summary below).
- **Piled foundations**   A method for structures where the loads have to be transmitted to a point at some distance below the general ground level (see Chapter 4.2 of this volume).

PAD FOUNDATIONS

Pad foundations are usually square on plan, unless obstructions or limitations such as the site boundary impede the spread in one direction. A rectangular base is generally acceptable in these situations, although the foundation area and amount of reinforcement may need to increase to accommodate the irregular transfer of loads. Pad foundations can be of mass concrete, but are usually reinforced to limit the amount of excavation and concrete fill, and to increase the load-carrying potential. Pads are used mainly to support columns and piers or as a base for heavy manufacturing machinery. Reinforcement in the pad is tied to projecting vertical starter bars for column reinforcement (see Chapter 2.3 of *Construction Technology*).

   The area of pad required is obtained by dividing the load on the subsoil (column imposed and dead load + the foundation weight) by the safe bearing capacity of the subsoil.

Example

A 40 tonne column and foundation load on a subsoil of safe bearing capacity of 200 kN/m².

40 tonne is approximately 400 kN ($40 \times 9.81 = 392.4$ kN).

Therefore $400 \div 200 = 2$ m² base area.

If square, the foundation base will be $\sqrt{2} = 1.400$ m $\times$ 1.400 m.

On sites where space is restricted and column loadings are high, pad foundations can be sufficiently large to be economically constructed in combination with an adjacent pad. In some situations it may be more practical and economic to combine several pads into a continuous foundation, as described in the next subsection.

As a guide to the thickness of an individual pad foundation, the following formula for punching shear* may be used as an estimate:

$$\text{Thickness of pad base} = \frac{W}{\text{Perimeter of column} \times 2.5 \times \text{Max. shear stress}}$$

where W = total dead and imposed load (column + foundation).

Max. shear stress of concrete varies, depending on specification: typically 690 kN/m².

Perimeter of a square column of say, 0.300 m sides = 1.200 m.

$$\text{Thickness of pad base} = \frac{400}{1.2 \times 2.5 \times 690} = 0.193 \text{ m, i.e. } 200 \text{ mm.}$$

* Punching shear is the term used to describe the situation where the side of a column meets its support pad, i.e. immediately underneath the column as shown in Fig. 4.3.8.

COMBINED FOUNDATIONS

When designing a foundation the area of the base must be large enough to ensure that the load per unit area does not exceed the bearing capacity of the soil upon which it rests. The ideal situation when considering column foundations is to have a square base of the appropriate area, with the column positioned centrally. Unfortunately this ideal condition is not always possible. When developing sites in urban areas the proposed structure is very often required to abut an existing perimeter wall, with the proposed perimeter columns in close proximity to the existing wall. Such a situation would result in an eccentric column loading if conventional isolated bases were employed. One method of overcoming this problem is to place the perimeter columns on a reinforced concrete continuous column foundation in the form of a **strip**. The strip is designed as a beam with the columns acting as point loads, which will result in a negative bending moment occurring between the columns, requiring top tensile reinforcement in the strip.

If sufficient plan area cannot be achieved by using a continuous column foundation a **combined column foundation** could be considered. This form

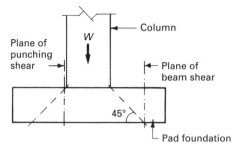

Figure 4.3.8   Proximity of shear in foundation pads

of foundation consists of a reinforced concrete slab placed at right angles to the line of columns, linking together an outer or perimeter column to an inner column. The base slab must have sufficient area to ensure that the load per unit area will not cause overstressing of the soil, leading to unacceptable settlement. To alleviate the possibility of eccentric loading the centres of gravity of the columns and slab should be designed to coincide (see Fig. 4.3.9). Combined foundations of this type are suitable for a pair of equally loaded columns or where the outer column carries the lightest load. The effect of the column loadings on the slab foundation is similar to that described above for continuous column foundations where a negative bending moment will occur between the columns.

In situations where the length of the slab foundation is restricted or where the outer column carries the heavier load the slab plan shape can be in the form of a **trapezium** with the widest end located nearest to the more heavily loaded column (see Fig. 4.3.9). As with the rectangular base, the trapezoidal base will have negative bending moments between the columns, and to eliminate eccentric loading the centres of gravity of columns and slab should coincide.

Alternative column foundations to the combined and trapezoidal forms described above are the **balanced base foundations**, which can be designed in one of two basic forms:

■ Cantilever foundations.
■ Balanced base foundations.

These foundations are usually specified where it is necessary to ensure that no pressure is imposed on an existing substructure or adjacent service such as a drain or sewer. A **cantilever foundation** can take two forms, both of which use a cantilever beam, which in one case acts about a short fulcrum column positioned near the existing structure, whereas in the alternative version the beam cantilevers beyond a slab base (see Fig. 4.3.10).

A **balanced base foundation** can be considered where it is possible to place the inner columns directly onto a base foundation. These columns are usually eccentric on the inner base, causing within the base a tendency to rotate. This movement must be resisted or balanced by the inner column base and the connecting beam (see Fig. 4.3.11).

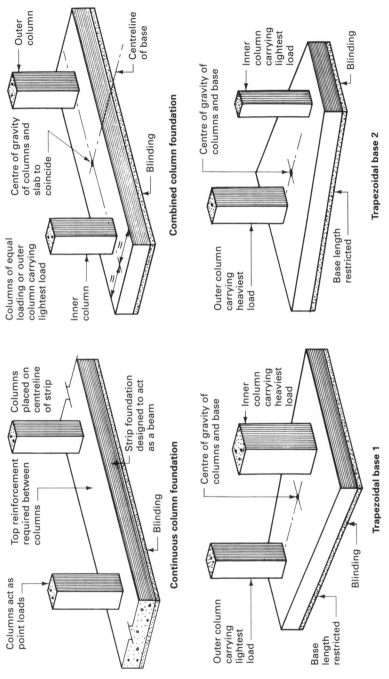

Figure 4.3.9 Typical combined foundations

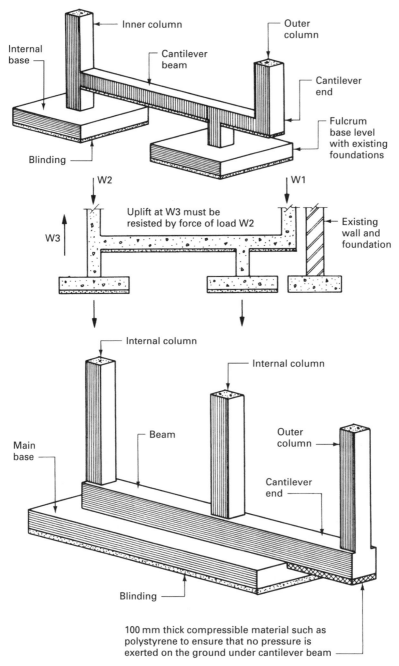

Figure 4.3.10　Typical cantilever bases

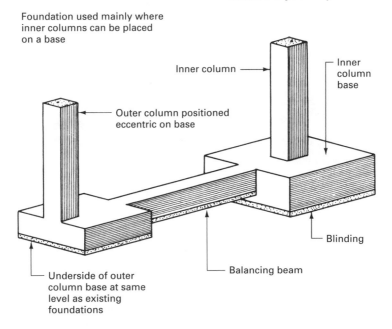

Foundation used mainly where inner columns can be placed on a base

Inner column

Inner column base

Outer column positioned eccentric on base

Blinding

Underside of outer column base at same level as existing foundations

Balancing beam

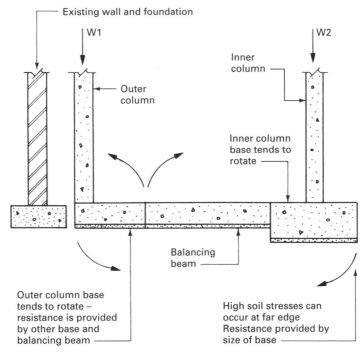

Existing wall and foundation

W1

W2

Inner column

Outer column

Inner column base tends to rotate

Balancing beam

Outer column base tends to rotate – resistance is provided by other base and balancing beam

High soil stresses can occur at far edge Resistance provided by size of base

Figure 4.3.11   Typical balanced base foundation

DEEP BASEMENTS

The construction of shallow single-storey basements and methods of waterproofing are considered in Chapter 2.6 of *Construction Technology* and Chapter 3.2 of this volume.

One of the main concerns of the contractor and designer when constructing deep and/or multi-storey basements is the elimination as far as practicable of the need for temporary works such as timbering, which in this context would be extensive, elaborate and costly. The solution is to be found in the use of diaphragm walling as the structural perimeter wall by constructing this ahead of the main excavation activity or alternatively by using a reinforced concrete land sunk open caisson if the subsoil conditions are favourable (see Fig. 3.3.10).

Diaphragm walls can be designed to act as pure cantilevers in the first instance, but this is an expensive and usually unnecessary method. Therefore the major decision is one of providing temporary support until the floors, which offer lateral restraint, can be constructed or, alternatively, providing permanent support if the deep basement is to be constructed free of intermediate floors. Diaphragm walls can take the form of *in-situ* reinforced concrete walling installed using the bentonite slurry trench system (see Fig. 3.2.7), contiguous piling techniques as described in Chapter 3.1, or precast concrete units as described below.

■ ■ ■ PRECAST CONCRETE DIAPHRAGM WALLS

The main concept of precast concrete diaphragm walls is based on the principles of *in-situ* reinforced concrete walling installed using a bentonite slurry-filled trench but with the advantages obtained by using factory-produced components. The wall is constructed by forming a bentonite slurry-filled trench of suitable width and depth, and inserting into this the precast concrete panel or posts and panels according to the system being employed. As a normal bentonite slurry mix would not effectively seal the joints between the precast concrete components, a special mixture of bentonite and cement with a special retarder additive to control setting time is used. The precast units are placed into position within this mixture, which will set sufficiently within a few days to enable excavation to take place within the basement area right up to the face of the diaphragm wall. To ensure that a clean wall face is exposed upon excavation it is usual practice to coat the proposed exposed wall faces with a special compound to reduce adhesion between the faces of the precast concrete units and the slurry mix.

The usual formats for precast concrete diaphragm walls are either simple tongue and groove jointed panels or a series of vertical posts with precast concrete infill panels, as shown in Fig. 4.4.1. If the subsoil is suitable it is possible to use a combination of precast concrete and *in-situ* reinforced concrete to form a diaphragm wall by installing precast concrete vertical posts or beams at suitable centres and linking these together with an *in-situ* reinforced concrete wall constructed in 2.000 m deep stages as shown in Fig. 4.4.2. The main advantage of this composite walling is the greater flexibility in design.

■■■ CONSTRUCTION METHODS

Four basic methods can be used in the construction of deep basements without the need for timbering or excessive subsoil support, and these can be briefly described as follows:

- A series of lattice beams are installed so that they span between the tops of opposite diaphragm walls, enabling the walls to act initially as propped cantilevers, receiving their final lateral restraint when the internal floors have been constructed (see Fig. 4.4.3).
- The diaphragm walls are exposed by carrying out the excavation in stages, and ground anchors are used to stabilise the walls as the work proceeds. This is a very popular method where no lateral restraint in the form of floors is to be provided (see Fig. 4.4.3). The technique and details of ground anchor installations are given in conjunction with prestressing systems in Chapter 10.1.
- After the perimeter diaphragm walls have been constructed, the ground floor slab and beams are cast, providing top edge lateral restraint to the walls. An opening is left in this initial floor slab through which operatives, materials and plant can pass to excavate the next stage and cast the floor slab and beams. This method can be repeated until the required depth has been reached (see Fig. 4.4.4).
- The centre area between the diaphragm walls can be excavated, leaving an earth mound or berm around the perimeter to support the walls while the lowest basement floor is constructed. Slots to accommodate raking struts acting between the wall face and the floor slab are cut into the berm; final excavation and construction of the remainder of the basement can take place working around the raking struts (see Fig. 4.4.4).

■■■ WATERPROOFING BASEMENTS

Standard methods for waterproofing basements such as the application of membranes, using dense monolithic concrete structural walls and tanking

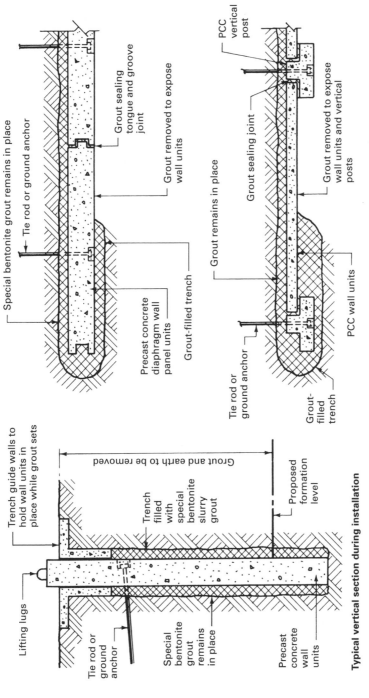

Special bentonite grout remains in place

Tie rod or ground anchor

Grout sealing tongue and groove joint

Grout removed to expose wall units

Precast concrete diaphragm wall panel units

Grout-filled trench

Grout remains in place

Grout sealing joint

Grout removed to expose wall units and vertical posts

PCC vertical post

Tie rod or ground anchor

Grout-filled trench

PCC wall units

Trench guide walls to hold wall units in place while grout sets

Grout and earth to be removed

Trench filled with special bentonite slurry grout

Proposed formation level

Lifting lugs

Tie rod or ground anchor

Special bentonite grout remains in place

Precast concrete wall units

Typical vertical section during installation

Figure 4.4.1   Typical precast concrete panel diaphragm wall details

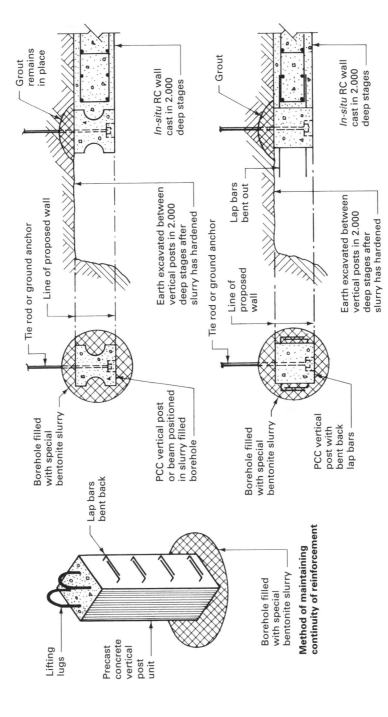

Figure 4.4.2   Typical precast and *in-situ* concrete diaphragm wall details

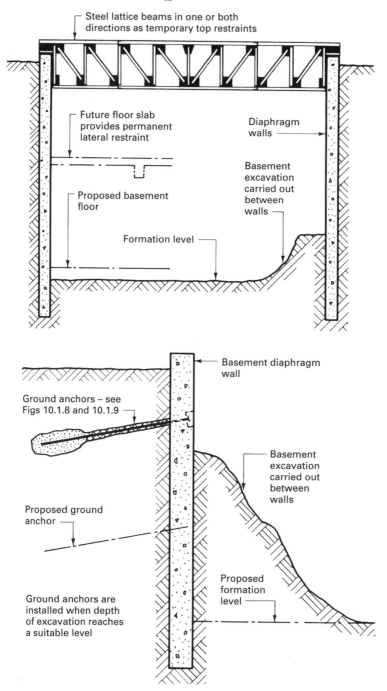

Figure 4.4.3   Deep basement construction methods: 1

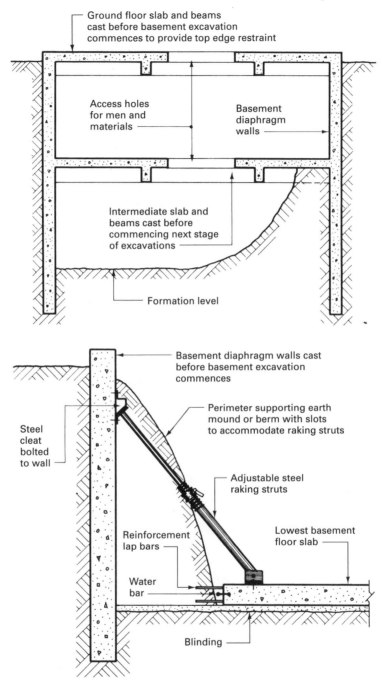

Figure 4.4.4   Deep basement construction methods: 2

techniques using mastic asphalt, are covered in the context of single-storey basements (see Chapter 2.6 of *Construction Technology*). Another method that can be used for waterproofing basements is the drained or cavity tanking system using special floor tiles produced by Atlas Stone Products Ltd.

DRAINED CAVITY SYSTEM

This system of waterproofing is suitable for basements of any depth where there are no intermediate floors, or where such floors are so constructed that they would not bridge the cavity. The basic concept of this system is very simple, in that it accepts that it is possible to have a small amount of water seepage or moisture penetration through a dense concrete structural perimeter wall, and therefore the best method of dealing with any such penetration is to allow it to be collected and drained away without it reaching the interior of the building. This is achieved by constructing an inner non-loadbearing wall to create a cavity and laying a floor consisting of precast concrete Dryangle tiles over the structural floor of the basement. This will allow any water trickling down the inside face of the outer wall to flow beneath the floor tiles, where it can be discharged into the surface water drains or, alternatively, pumped into the drains if these are at a higher level than the invert of the sump. Typical details are shown in Fig. 4.4.5.

The concrete used to form the structural basement wall should be designed as if it were specified for waterproof concrete, which, it must be remembered, is not necessarily vapourproof. All joints should be carefully designed, detailed and constructed, including the fixing of suitable water bars (see Fig. 2.6.1, *Construction Technology*). It must be emphasised that the drained cavity system is designed to deal only with seepage that may occur, and therefore the use of a dense concrete perimeter wall is an essential component in the system. The dense concrete used in the outer wall needs to be well vibrated to obtain maximum consolidation, and this can usually be achieved by using poker vibrators.

The *in-situ* concrete used in any form of basement construction can be placed by using chutes, pumps or tremie pipes. The placing of *in-situ* concrete using pumps is covered in Chapter 2.5, dealing with contractors' plant, but the placing of concrete by means of a tremie pipe is considered in the section on placing concrete below ground.

■■■ BUOYANCY OR TANKED BASEMENT

Where site space permits, the building dead and imposed loading may be distributed over sufficient basement raft foundation base area to prevent undue settlement. Multi-storey buildings on relatively small sites of restricted foundation bearing area can still be constructed, even where the subsoil composition is unsuited to piling techniques, or where for practical reasons piling is not viable. In these situations it may be possible to design the basement as a huge tank on which the building effectively 'floats'.

In principle, the concept is to counterbalance the mass of subsoil removed for the basement with the mass of the building in its place. Very careful design calculations

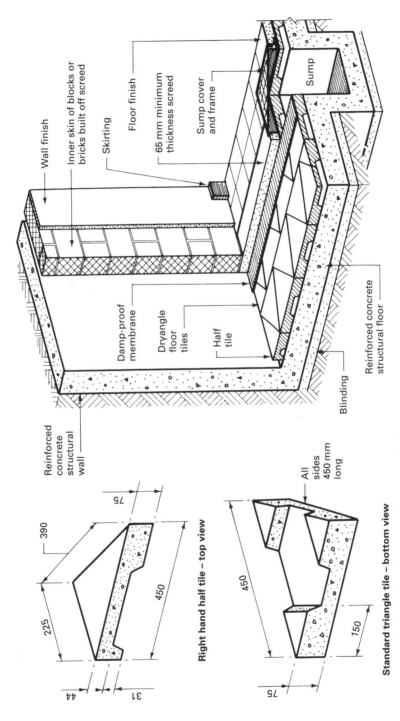

Wall finish

Inner skin of blocks or
bricks built off screed

Skirting

Floor finish

65 mm minimum
thickness screed

Sump cover
and frame

Sump

Reinforced
concrete
structural
wall

Damp-proof
membrane

Dryangle
floor
tiles

Half
tile

Reinforced concrete
structural floor

Blinding

390

225

75

450

44

31

Right hand half tile – top view

All
sides
450 mm
long

450

150

75

Standard triangle tile – bottom view

Figure 4.4.5   Drained cavity system of waterproofing basements

will be required to ensure the balance; to avoid excessive and eccentric loadings and to allow for extraneous imposed loading including wind and snow. Generally, the deeper the excavation the more compressed the subsoil will be and the greater its bearing capacity. The opportunity to excavate at depth will normally enhance the loading potential considerably, although when exposing clay subsoil there can be a problem of it swelling as the soil above is removed. A substantial loading of *in-situ* concrete will control this, the weight of which must be incorporated into the balancing calculations.

▨ ▨ ▨ PLACING CONCRETE BELOW GROUND

A **tremie pipe** can be used to place *in-situ* concrete below the ground or water level where segregation of the mix must be avoided, and can be used in the construction of piled foundations, basements and diaphragm walls. For work below ground level a rigid tube of plastic or metal can be used; alternatively a flexible PVC tube can be employed, this often being referred to as **elephant trunking**. In all cases the discharge end of the tremie pipe is kept below the upper level of the concrete being placed and the upper end of the pipe is fitted with a feed hopper attachment to receive the charges of concrete. As the level of the placed concrete rises, the tube is shortened by removing lengths of pipe as required. The tremie pipe and its attached hopper head must be supported as necessary throughout the entire concrete-placing operation.

Placing concrete below water level by means of a tremie pipe, very often suspended from a crane, requires more care and skill than placing concrete below ground level. The pipe should be of adequate diameter for the aggregates being used, common tube diameters being 150 and 200 mm. It should be watertight and have a smooth bore. The operational procedure is as follows:

1. The tremie pipe is positioned over the area to be concreted, and lowered until the discharge end rests on the formation level.
2. A rabbit or travelling plug of plastic or cement bags is inserted into the top of the pipe to act as a barrier between the water and concrete. The weight of the first charge concrete will force the plug out of the discharge end of the tube.
3. When filled with concrete the tremie pipe is raised so that the discharge end is just above the formation level to allow the plug to be completely displaced and thus enable the concrete to flow.
4. Further concrete charges are introduced into the pipe and allowed to discharge within the concrete mass already placed, the rate of flow being controlled by raising and lowering the tremie pipe.
5. As the depth of the placed concrete increases, the pipe is shortened by removing tube sections as necessary.

When placing concrete below the water level by the tremie pipe method, great care must be taken to ensure that the discharge end of the pipe is not withdrawn clear of the concrete being placed, as this could lead to a weak concrete mix by water mixing with the concrete. One tremie pipe will cover an area of approximately 30 m², and if more than one tremie pipe is being used simultaneous placing is usually recommended.

■■■ SULPHATE-BEARING SOILS

A problem that can occur when using concrete below ground level is the deterioration of the set cement due to sulphate attack by naturally occurring sulphates such as calcium, magnesium and sodium sulphates, which are found particularly in clay soils. The main factors that influence the degree of attack are:

- amount and nature of sulphate in the soil;
- level of water table and amount of groundwater movement;
- form of construction;
- type and quality of concrete.

The permanent removal of the groundwater in the vicinity of the concrete structure is one method of limiting sulphate attack, but this is very often impracticable and/or uneconomic. The only real alternative is to use a fully compacted concrete of low permeability. Once the chemical reaction between the sulphates and the set cement has taken place, further deterioration can come about only when fresh sulphates are brought into contact with the concrete by groundwater movement. Therefore concrete which is situated above the water table is not likely to suffer to a great extent.

Generally, precast concrete components suffer less than their *in-situ* counterparts because of the better control possible during casting under factory conditions. Basement walls and similar structural members usually have to resist lateral water pressure on one side only, which increases the risk of water penetration, and if evaporation can occur from the inner face, sulphate attack can take place throughout the wall thickness.

Ordinary Portland cement manufactured to the recommendations of BS EN 197-1: *Cement. Composition, specifications and conformity criteria for common cements*, usually contains a significant amount of tricalcium aluminate, which is reacted upon by sulphates in the soil, resulting in an expansion of the set cement, causing a breakdown of the concrete. Sulphate-resisting Portland cement manufactured to the recommendations of BS 4027: *Specification for sulphate-resisting Portland cement*, has the proportion of tricalcium aluminate present limited to a maximum of 3.5%, which will give the set cement a considerable resistance to sulphate attack. The minimum cement content and maximum free water/cement ratio of a concrete to be placed in a sulphate-bearing soil are related to the concentration of sulphate present in the soil, and readers seeking further information are advised to consult the following Building Research Establishment publications:

- BRE Report BR 164: *Sulphate resistance of buried concrete.*
- BRE Report BR 279: *Sulphate and acid attack on concrete in the ground: recommended procedures for soil analysis.*
- BRE Information Paper IP 15/92: *Assessing the risk of sulphate attack on concrete in the ground.*
- BRE Digest 276: *Hardcore.*
- BRE Special Digest 1: *Concrete in aggressive ground.*

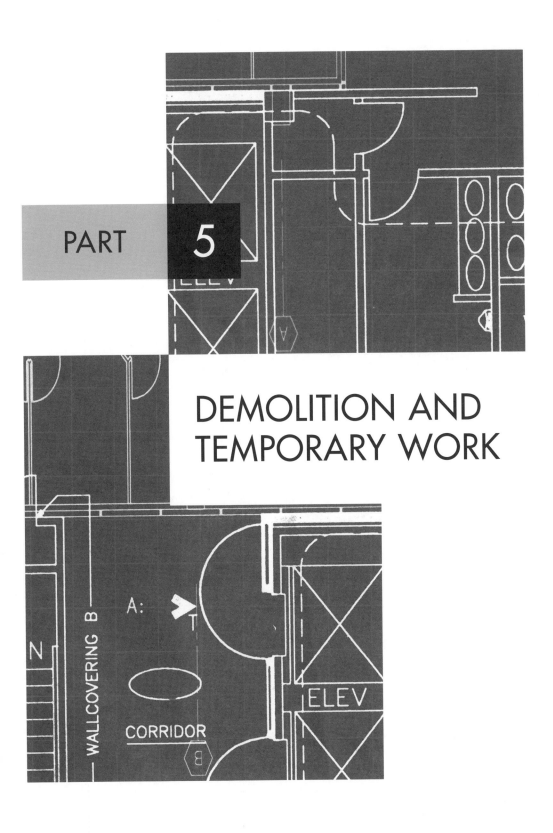

PART 5

DEMOLITION AND TEMPORARY WORK

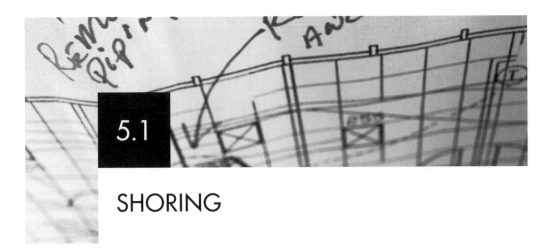

5.1

SHORING

All forms of shoring are temporary supports applied to a building or structure. The assembly and dismantling of a shoring system must comply with the requirements of the Construction (Health, Safety and Welfare) Regulations 1996, particularly Regulations 9 and 10. These relate to work on structures where there is a perceived risk of accidental collapse and danger to any person on or adjacent to the site. Relevant clauses relating to risk assessment also occur in the Work at Height Regulations 2005. Although shoring is an interim measure while a permanent support structure is built, it can remain in place for a considerable time. Therefore it is no less important than other aspects of construction. Regardless of the time it is to remain in place, it must be subject to planned safety procedures and competent supervision during assembly, dismantling and any associated demolition.

The following situations may justify application of shoring:

■ to give support to walls that are dangerous or are likely to become unstable because of subsidence, bulging or leaning;
■ to avoid failure of sound walls caused by the removal of an underlying support, such as where a basement is being constructed near a sound wall;
■ during demolition works to give support to an adjacent building or structure;
■ to support the upper part of a wall during formation of a large opening in the lower section of the wall;
■ to give support to a floor or roof to enable a support wall to be removed and replaced by a beam.

Structural softwood is the usual material used for shoring members; its strength to weight ratio compares favourably with that of structural steel, and its adaptability is superior to that of steel. Shoring arrangements can also be formed by coupling together groups of scaffold tubulars.

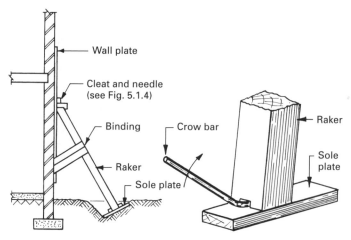

Figure 5.1.1   Single raking shore and base tightening process

▨ ▧ ■ SHORING SYSTEMS

There are three basic shoring systems:

- dead shoring;
- raking shoring;
- flying shoring.

Each shoring system has its own function to perform and is based upon the principles of a perfectly symmetrical situation. The simplest example is shown in Fig. 5.1.1. A single raker and short wall plank or binding combine with a sole plate and needle and cleat at the top. The sole plate is set in the ground and inclined at a slightly acute angle to the raker. This allows the raker to be levered and tightened in place with a crowbar before securing with cleats and dogs.

Where flying shores are required, the disposition of buildings does not always lend itself to symmetrical arrangements. Some asymmetrical examples are shown in Fig. 5.1.2. See Figs 5.1.6 and 5.1.7 for details of regularly disposed flying shore components.

DEAD SHORING

This type of shoring is used to support dead loads that act vertically downwards. In its simplest form it consists of a vertical prop or shore leg with a head plate, sole plate and some means of adjustment for tightening and easing the shore. The usual arrangement is to use two shore legs connected over their heads by a horizontal beam or needle. The loads are transferred by the needle to the shore legs and hence down to a solid bearing surface. It may be necessary to remove pavings and cut holes in suspended timber floors to reach a suitable bearing surface; if a basement is encountered, a third horizontal member called a transom will be necessary, because it is impracticable to manipulate a shore leg through two storeys. A typical example of this situation is shown in Fig. 5.1.3.

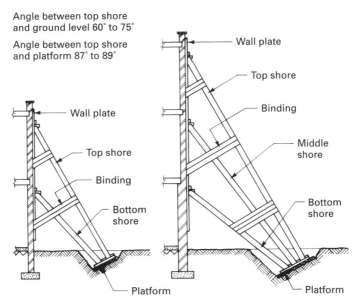

Angle between top shore
and ground level 60° to 75°

Angle between top shore
and platform 87° to 89°

Wall plate

Top shore

Binding

Middle
shore

Bottom
shore

Platform

Wall plate

Top shore

Binding

Bottom
shore

Platform

Typical raking shore arrangements (see also Fig. 5.1.5)

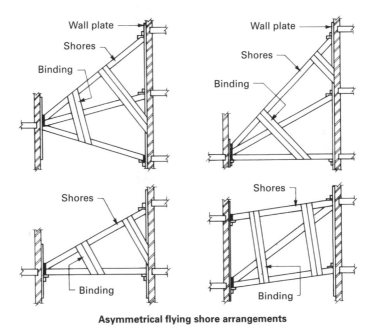

Wall plate

Shores

Binding

Wall plate

Shores

Binding

Shores

Binding

Shores

Binding

Asymmetrical flying shore arrangements

Figure 5.1.2 Shoring arrangements

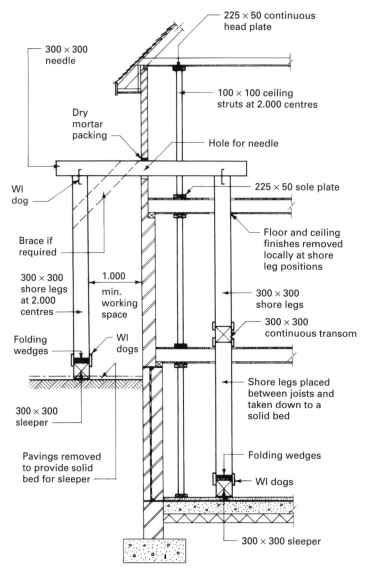

225 × 50 continuous head plate

300 × 300 needle

100 × 100 ceiling struts at 2.000 centres

Dry mortar packing

Hole for needle

WI dog

225 × 50 sole plate

Floor and ceiling finishes removed locally at shore leg positions

Brace if required

300 × 300 shore legs at 2.000 centres

1.000 min. working space

WI dogs

300 × 300 shore legs

300 × 300 continuous transom

Folding wedges

Shore legs placed between joists and taken down to a solid bed

300 × 300 sleeper

Pavings removed to provide solid bed for sleeper

Folding wedges

WI dogs

300 × 300 sleeper

Cross bracing, longitudinal bracing and hoardings to be fixed as necessary

Figure 5.1.3   Dead shoring

The sequence of operations necessary for a successful dead shoring arrangement is as follows:

1. Carry out a thorough site investigation to determine:
   - number of shores required, by ascertaining possible loadings and window positions;
   - bearing capacity of soil and floors;
   - location of underground services that may have to be avoided or bridged.
2. Fix ceiling struts between suitable head and sole plates to relieve the wall of floor and roof loads. The struts should be positioned as close to the wall as practicable.
3. Strut all window openings within the vicinity of the shores to prevent movement or distortion of the opening. The usual method is to place timber plates against the external reveals and strut between them; in some cases it may be necessary to remove the window frame to provide sufficient bearing surface for the plates.
4. Cut holes through the wall slightly larger in size than the needles.
5. Cut holes through ceilings and floors for the shore legs.
6. Position and level sleepers on a firm base, removing pavings if necessary.
7. Erect, wedge and secure shoring arrangements.

Upon completion of the builder's work it is advisable to leave the shoring in position for at least seven days before easing the supports, to ensure the new work has gained sufficient strength to be self-supporting.

RAKING SHORING

This shoring arrangement transfers the floor and wall loads to the ground by means of sloping struts or rakers. It is very important that the rakers are positioned correctly so that they are capable of receiving maximum wall and floor loads. The centreline of the raker should intersect with the centrelines of the wall or floor bearing; common situations are detailed in Fig. 5.1.4. One raker for each floor is required and ideally should be at an angle of between 40° and 70° with the horizontal; therefore the number of rakers that can be used is generally limited to three. A four-storey building can be shored by this method if an extra member, called a **rider**, is added (see Fig. 5.1.5).

The operational sequence for erecting raking shoring is as follows:

1. Carry out site investigation as described for dead shoring.
2. Mark out and cut mortises and housings in wall plate.
3. Set out and cut holes for needles in external wall.
4. Excavate to a firm bearing subsoil and lay grillage platform and sole plate.
5. Cut and erect rakers, commencing with the bottom shore. A notch is cut in the heel so that a crowbar can be used to lever the raker down the sole plate and thus tighten the shore (see Fig. 5.1.1). The angle between sole plate and shores should be at its maximum about 89° to ensure that the tangent point is never reached and not so acute that levering is impractical.
6. Fix cleats, distance blocks, binding and if necessary cross-bracing over the backs of the shores.

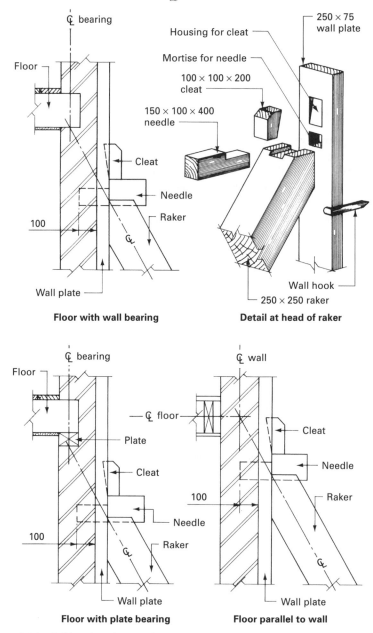

Figure 5.1.4 Raking shore intersections

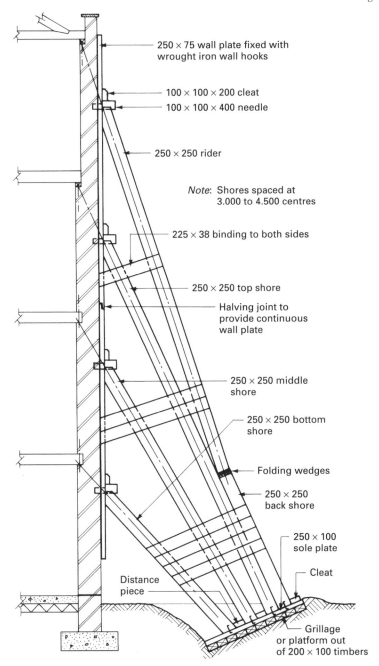

250 × 75 wall plate fixed with
wrought iron wall hooks

100 × 100 × 200 cleat
100 × 100 × 400 needle

250 × 250 rider

Note: Shores spaced at
3.000 to 4.500 centres

225 × 38 binding to both sides

250 × 250 top shore

Halving joint to
provide continuous
wall plate

250 × 250 middle
shore

250 × 250 bottom
shore

Folding wedges

250 × 250
back shore

250 × 100
sole plate

Cleat

Distance
piece

Grillage
or platform out
of 200 × 100 timbers

Figure 5.1.5 Typical multiple raking shore

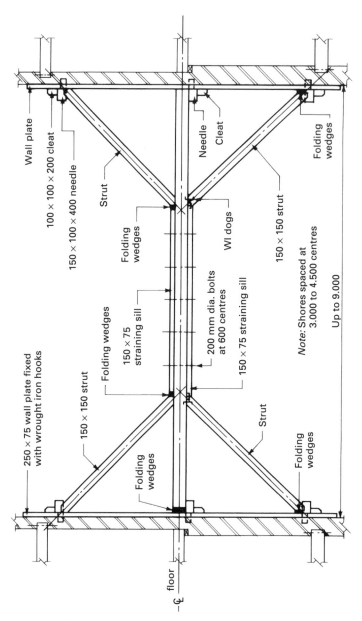

Wall plate

100 × 100 × 200 cleat

150 × 100 × 400 needle

Strut

Needle

Cleat

Folding wedges

Folding wedges

150 × 150 strut

Folding wedges

150 × 75 straining sill

200 mm dia. bolts at 600 centres

WI dogs

150 × 75 straining sill

Note: Shores spaced at 3.000 to 4.500 centres

Up to 9.000

250 × 75 wall plate fixed with wrought iron hooks

150 × 150 strut

Folding wedges

Strut

Folding wedges

℄ floor

Figure 5.1.6   Typical single flying shore

FLYING SHORING

These shores fulfil the same functions as a raking shore but have the advantage
of providing a clear working space under the shoring. They can be used between
any parallel wall surfaces provided the span is not in excess of 12.000 m, when the
arrangement would become uneconomic. Short spans up to 9.000 m usually have a
single horizontal member, whereas the larger spans require two horizontal shores to
keep the section sizes within the timber range commercially available (see Figs 5.1.6
and 5.1.7).

It is possible with all forms of shoring to build up the principal members from
smaller sections by using bolts and timber connectors, ensuring all butt joints are
well staggered to give adequate rigidity. This is in effect a crude form of laminated
timber construction.

The site operations for the setting out and erection of a flying shoring system are
similar to those listed for raking shoring.

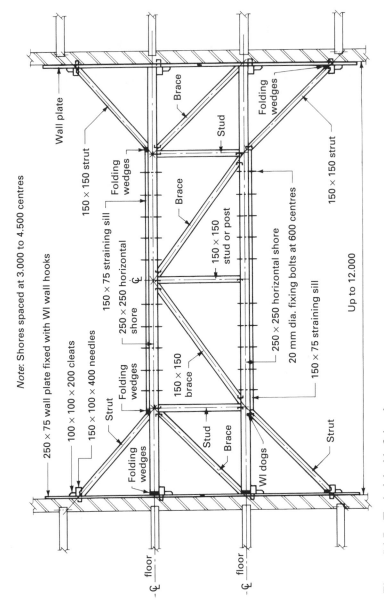

Note: Shores spaced at 3.000 to 4.500 centres

Wall plate

250 × 75 wall plate fixed with WI wall hooks

100 × 100 × 200 cleats

150 × 100 × 400 needles

Strut

150 × 150 strut

150 × 75 straining sill

Folding wedges

250 × 250 horizontal shore

Folding wedges

150 × 150 brace

Folding wedges

Brace

Stud

Folding wedges

Brace

150 × 150 stud or post

250 × 250 horizontal shore

20 mm dia. fixing bolts at 600 centres

150 × 75 straining sill

150 × 150 strut

Up to 12.000

Stud

Brace

WI dogs

Strut

℄ floor

℄ floor

Figure 5.1.7 Typical double flying shore

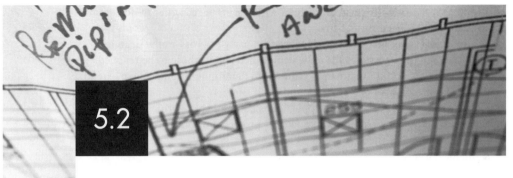

DEMOLITION

This is a skilled and sometimes dangerous operation and unless of a very small nature should be entrusted to a specialist contractor. Demolition of a building or structure can be considered under two headings:

- taking down or removals: partial demolition of a structure;
- demolition: complete removal of a structure.

Before any taking down or demolition is commenced it is usual to remove carefully all saleable items such as copper, lead, steel fittings, domestic fittings, windows, doors and frames (see subsection on salvaging).

Taking down requires a thorough knowledge of building construction and design so that loadbearing members and walls can be correctly identified and adequately supported by struts, props and suitable shoring. Most partial demolition works will need to be carried out manually using hand tools such as picks and hammers. These operations usually relate to the removal of small parts of a building, such as brickwork to form a new opening or roofwork alterations to create dormer windows.

■■■ PRELIMINARY CONSIDERATIONS

SURVEY

Before any works of demolition are started, a detailed survey and examination should be made of the building or structure and its curtilage. Photographs of any existing defects on adjacent properties should be taken, witnessed and stored in a safe place. The relationship and condition of adjoining properties that may be affected by the demolition should also be considered and noted, taking into account the existence of easements, wayleaves, party rights and boundary walls.

- **Roofs and framed structures**   Check whether proposed order of demolition will cause unbalanced thrusts to occur.
- **Walls**   Check whether these are loadbearing, party or cross-walls. Examine condition and thickness of walls to be demolished and of those to be retained.
- **Basements**   Careful examination required to determine whether these extend under public footpaths or beyond boundary of site.
- **Cantilevers**   Check nature of support to balconies, heavy cornices and stairs.
- **Services**   These may have to be sealed off, protected or removed, and could include any or all of the following:
    - drainage runs;
    - electricity cables;
    - gas mains and service pipes;
    - water mains and service pipes;
    - telephone cables above and below ground level;
    - radio and television relay cables;
    - district heating mains.

A careful survey of the whole site is advisable to ensure that any hazardous, flammable or explosive materials such as oil drums and gas cylinders are removed before the demolition work commences. If the method of construction of the existing structure is at all uncertain all available drawings should be carefully studied and analysed; alternatively a detailed survey of the building should be conducted under the guidance of an experienced surveyor.

INSURANCE

Insurance companies and their underwriters will regard demolition work as particularly hazardous. Any aspect of demolition is normally contracted out to specialists, as it may not be included under a general contractor's insurance policy. Even minor work such as removals may not be included. Therefore, where demolition or removals form part of a work contract, general builders should ensure that adequate insurance is in place to cover all risks, including claims from operatives and other parties to the work. Insurance will also be required for any other third party risk, including cover for any claim for loss or damage to property, business, public utilities and local authority maintained roads and paving.

SALVAGING

This is otherwise known as salving or recycling of materials. Old building materials can often be sold, as they are sometimes preferred to new materials. For example, when constructing an extension to an existing building, salvaged roof tiles if matched well will immediately mellow in and be less contrasting than new. They will also be less expensive than new, unless they are rare because of discontinued manufacture. Other examples of building items of value include fireplaces, roofing slates, stairways, London stock bricks and useful lengths of structural timber.

Architectural salvage is now a significant business, responsible for the preservation and reuse of much of the featurework from our old buildings.

When quoting for demolition work, contractors will consider the salvage potential. However, there will be a calculated balance between the time and cost factors for careful reclamation, transporting, cleaning, preparing and marketing old materials for resale, against simply demolishing without regard. The latter will still require transportation of surplus materials from site, as well as attracting tipping costs at licensed receivers.

HOARDINGS

A site survey will determine whether the location justifies special protection for adjoining properties, public highways or other places used by the general public. Consultation with the local authority will be required to determine their requirements for temporary works, not least the health and safety aspects. The local authority usually requires a formal application, licensing fees and possibly financial deposits against damages, particularly if the work abuts a public thoroughfare and highway. Some examples of standard boarded and fan hoardings are included under the section on security and protection in Chapter 1.2 of *Construction Technology*.

ASBESTOS SURVEY

In the UK there are many laws and regulations relating to the use and handling of asbestos. This is understandable, because inhaled asbestos fibres may be carcinogenic, in time manifesting as a form of cancer of the lungs. Much of our existing building stock contains large quantities of asbestos. It was originally used in ignorance of its latent effect on personal health, often sprayed onto steelwork as fire protection, used as a pipe insulation and in board form as a cladding or lining. Therefore, before altering or demolishing buildings, it is essential that an asbestos appraisal is undertaken to assess whether there is a risk to operatives. The Health and Safety at Work etc. Act is the principal statute, under which there are numerous supplementary regulations and advisory papers, including:

- The Asbestos (Licensing) Regulations 1983. These require a licence to be held by an employer carrying out work on sprayed and other asbestos-finished coatings, and where asbestos insulating board is used.
- The Control of Asbestos at Work Regulations 2002. These relate to all work with asbestos, and affect anyone liable to exposure. The regulations have a specific requirement for identification and management of asbestos products in non-domestic buildings. This provides for maintenance of records documenting up-to-date specialist surveys for location, condition and assessment of exposure risk.
- See also the Health and Safety Executive (HSE) publications *Methods for the Determination of Hazardous Substances* (MDHS 100); *Surveying, sampling and assessment of asbestos containing materials – 2001*. Contact www.hsebooks.co.uk.

Depending on the intended course of action, asbestos surveys can be categorised generally, as:

- **A presumptive survey**   This involves a visual inspection of the premises to locate and identify all materials that could contain asbestos materials. This survey is mainly for the benefit of building owners and facilities managers. They will have the responsibility for ensuring that asbestos products are maintained in a manner harmless to the building occupants.
- **A sampling survey**   This survey is usually provided as a supplement to a presumptive survey for positively identifying the nature of asbestos in suspected materials. Where possible asbestos-containing materials are known to be located in inaccessible areas, they are assumed to be a health risk. *Note*: Some established types of asbestos now banned under the Asbestos (Prohibition) Regulations 1999 are named as chrysotile, amphibole, amosite and crocidolite.
- **Sampling and identification survey**   This type of survey is required before demolition and refurbishment work commences. It can be used as a basis for specialist contractors to tender for asbestos removal. Accessibility to all parts of the building is essential to obtain material samples for laboratory analysis and for destruction testing on site.

Note: The Personal Protective Equipment Regulations 1992 and the Control of Substances Hazardous to Health Regulations 2002 require employers to ensure that their employees are appropriately dressed and protected when working with or in the presence of materials posing a potential health risk. For asbestos surveys and removals, dress requirements will necessitate total isolation by means of disposable outer protective clothing and respiratory equipment.

▧■■ STATUTORY NOTICES

Under the Public Health Act 1961, the local authority for the area in England or Wales must be notified before any demolition work can be started by the building owner or their agent. In the Inner London area there are by-laws relating to the demolition of buildings, and they require notification to be deposited with the district surveyor. In Scotland a warrant is required from the building authority of the burgh or county in which the proposed demolition work is to take place. It is also necessary in Great Britain to inform the Health and Safety Executive of the local authority or the appropriate area authority applicable under the Health and Safety at Work etc. Act 1974.

Prior to the commencement of any demolition work the building owner or their agent must notify the public utilities companies, i.e. the gas, electricity, water and drainage authorities, the telecommunications services, and the associated companies responsible for other installations such as radio and cable television relay lines. Note that it is the contractor's responsibility to ensure that all services and other installations have been rendered safe or removed by the authority or company concerned.

■■■ SUPERVISION AND SAFETY

Regulations 7 to 11 of the Construction (Health, Safety and Welfare) Regulations 1996 relate specifically to safety measures applied to buildings subject to dismantling and demolition. This high-risk activity must be undertaken in a safe manner, requiring thorough planning and organisation. It must be supervised by a competent person to ensure that no one is exposed to unnecessary danger. The supervisor must be experienced in the type of demolition concerned, and where more than one contractor is involved each must appoint a competent supervisor.

A systematic planned approach is endorsed by Regulation 6 of the Work at Height Regulations 2005. Specific reference is made to an employer's and self-employed person's responsibility to undertake a risk assessment under Regulation 3 of the Management of Health and Safety at Work Regulations 1999.

FURTHER PLANNING AND RISK CONSIDERATIONS

■ Method of construction of the building and the structural interdependence and integrity of elements of construction.

■ Type of materials used, their strengths and weaknesses, their potential as a health and safety hazard (see subsection on asbestos survey) and the possibility for reclamation and reuse (see subsection on salvaging).

■ Whether the demolition is full or partial. Occasionally, facades are retained as part of a preservation order: therefore temporary supports may be required (see Chapter 5.1 on shoring).

■ Need to isolate the effects of demolition. This may include measures to prevent dust and debris spilling onto a public thoroughfare and into the curtilage of adjacent buildings. Means for protection to be considered, e.g. dust sheeting and hoardings.

■ Exposure of contaminants. In addition to hazardous materials used in the construction process, there may be contamination of the land from previous occupancies. Consider extraction of soil samples for laboratory analysis.

■ Location and isolation of all existing services to the site and inform relevant supplier. See listing on page 271.

■ Local planning authority and Health and Safety Executive requirements and restrictions on hours and days of work. Also, determine acceptable noise and dust levels.

■ Handling and disposal methods for debris. See section on rubble chutes and skips.

■■■ METHODS OF DEMOLITION

There are several methods of demolition, and the choice is usually determined by:

■ **Type of structure** For example, two-storey framed structure, reinforced concrete chimney.

■ **Type of construction**   For example, masonry wall, prestressed concrete, structural steelwork.
■ **Location of site**   For example, a detached building on an isolated site, which is defined as a building on a site where the minimum distance to the boundary is greater than twice the height of the building to be demolished. A confined site is one where not all the boundaries are at a distance exceeding twice the height of the building to be demolished.

Every site is unique, and it must be assessed with particular regard to selection of demolition technique. Methods vary, and the procedures given below are included for general guidance.

HAND DEMOLITION

This involves the progressive demolition of a structure by operatives using hand-held tools; lifting appliances may be used to hoist and lower members or materials once they have been released. Buildings are usually demolished, by this method, in the reverse order to that of their construction storey by storey. Debris should be allowed to fall freely only where the horizontal distance from the point of fall to the public highway or an adjoining property is greater than 6.000 m, or half the height from which the debris is dropped, whichever is the greater. In all other cases a chute or skip should be used.

PUSHER ARM DEMOLITION

This is a method of progressive demolition using a machine fitted with a steel pusher arm exerting a horizontal thrust on to the building fabric. This method should be used only when the machine can be operated from a firm level base with a clear operating base of at least 6.000 m. The height of the building should be reduced by hand demolition if necessary to ensure that the height above the pusher arm does not exceed 600 mm. The pusher arm should not be overloaded, and generally should be operated from outside the building. An experienced operator is required, who should work from within a robust cab capable of withstanding the impact of flying debris and fitted with shatterproof glass cab windows. Where this method of demolition is adopted in connection with attached buildings, the structure to be demolished should first be detached from the adjoining structure by hand demolition techniques.

DELIBERATE COLLAPSE DEMOLITION

This method involves the removal of key structural members, causing complete collapse of the whole or part of the building. Expert engineering advice should be obtained before this method is used. It should be used only on detached isolated buildings on reasonably level sites so that the safety to personnel can be carefully controlled.

DEMOLITION BALL TECHNIQUES

Due to potential difficulties of control over the swinging and slewing of a heavy weight, this method is restricted under UK safety guidance. Where it is acceptable, the following techniques could be used:

- vertical drop;
- swinging in line with the jib;
- slewing jib.

Whichever method is used, a skilled operator is essential.

The use of a demolition ball from a normal-duty mobile crane should be confined to the vertical drop technique only. A heavy-duty machine such as a convertible dragline excavator should be used for the other techniques, but in all cases an anti-spin device should be attached to the hoist rope. It is advisable to reduce the length of the crane jib as the demolition work proceeds, but at no time should the jib head be less than 3.000 m above the part of the building being demolished.

Pitched roofs should be removed by hand demolition down to wall plate level and at least 50–70% of the internal flooring removed to allow for the free fall of debris within the building enclosure. Demolition should then proceed progressively storey by storey.

Demolition ball techniques should not be used on buildings over 30.000 m high, because the fall of debris is uncontrollable. Attached buildings should be separated from the adjoining structure by hand demolition to leave a space of at least 6.000 m, or half the height of the building, whichever is the greater; a similar clear space is required around the perimeter of the building to give the machine operating space.

WIRE ROPE PULLING DEMOLITION

Only steel wire ropes should be used, and the size should be adequate for the purpose, but in no case less than 38 mm circumference. The rope should be firmly attached at both ends and the pulling tension gradually applied. No person should be forward of the winch or on either side of the rope within a distance of three-quarters of the length between the winch and the building being demolished.

If, after several pulls, the method does not cause collapse of the structure it may have been weakened and should not therefore be approached, but demolished by an alternative means such as a demolition ball or a pusher arm. A well-anchored winch or heavy vehicle should be used to apply the pulling force, great care being taken to ensure that the winch or the vehicle does not lift off its mounting, wheels or tracks.

DEMOLITION BY EXPLOSIVES

This is a specialist method where charges of explosives are placed within the fabric of the structure and detonated to cause partial or complete collapse. It should never be attempted by a building contractor without the advice and supervision of an expert.

OTHER METHODS

Where site conditions are not suitable for the use of explosives, the following specialist methods can be considered:

■ **Gas expansion burster**   A steel cylinder containing a liquefied gas, which expands with great force when subjected to an electric charge, is inserted into a prepared cavity in the fabric to be demolished. On being fired, the expansion of the cylinder causes the fabric mass to be broken into fragments.

■ **Hydraulic burster**   This consists of a steel cylinder with a number of pistons, which are forced out radially under hydraulic pressure.

■ **Thermal reaction**   A structural steel member that is to be cut out and removed is surrounded by a mixture of a metal oxide and a reducing agent. This covering is ignited, usually by an electric current, which results in the liberation of a large quantity of heat, causing the steel to become plastic. A small force such as a wire rope attached to a winch will be sufficient to cause collapse of the member.

■ **Thermic lance**   This method involves a steel tube, sometimes packed with steel rods, through which oxygen is passed. The tip of the lance is preheated by conventional means to melting point (approximately 1,000 °C) when the supply of oxygen is introduced. This sets up a thermochemical reaction giving a temperature of around 3,500 °C at the reaction end, which will melt all the materials normally encountered, causing very little damage to surrounding materials.

The dangers and risks encountered with any demolition works cannot be overemphasised, and all builders should seek the advice of and employ specialist contractors to carry out all but the simple demolition tasks.

See also BS 6187: *Code of practice for demolition*.

▨▨■■ RUBBLE CHUTES AND SKIPS

RUBBLE CHUTES

In principle and original concept, these are reputed to derive from bottomless refuse containers strung together with a central core of rope. Purpose-made developments are produced from reinforced rubber, usually in effective lengths of 1.000 m (actual unit length is 1.100 m), to suspend collectively from a building or scaffold system and provide safe conveyance of waste materials to a suitable discharge point or skip receptacle.

Proprietary chute units typically taper from 510 mm diameter down to 380 mm diameter and have internal ribbing to resist wear and abrasion. Each unit is fitted with two uppermost brackets on opposing sides for chain link retention and fixing of adjacent sections. Figure 5.2.1 shows a typical standard unit with variations having a reinforcing ring (use every sixth section), side entry and hopper attachments. Figure 5.2.2 shows the application to a building, with material discharge to a skip placed at ground level.

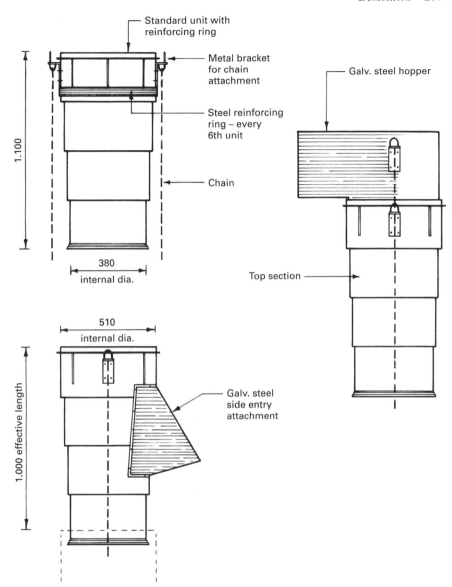

Figure 5.2.1   Typical rubble chute details

SKIPS OR DUMPSTERS

Skips are manufactured from mild steel sheet with welded box sections and angles to provide ribbed reinforcement and rigidity. They are produced in a wide range of sizes as rubbish/demolition material receptacles to suit all applications from minor domestic work to major demolition contracts. Skip capacities can be categorised:

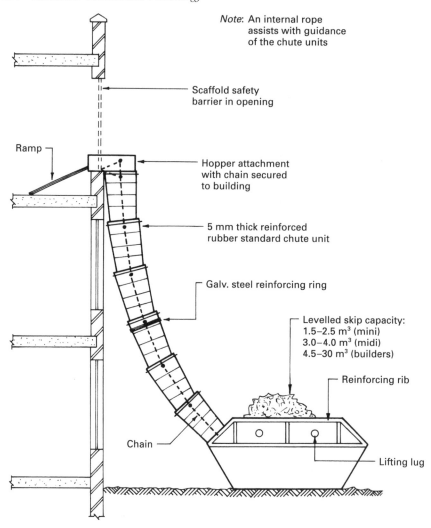

Note: An internal rope assists with guidance of the chute units

Scaffold safety barrier in opening

Ramp

Hopper attachment with chain secured to building

5 mm thick reinforced rubber standard chute unit

Galv. steel reinforcing ring

Levelled skip capacity:
1.5–2.5 m³ (mini)
3.0–4.0 m³ (midi)
4.5–30 m³ (builders)

Reinforcing rib

Chain

Lifting lug

Figure 5.2.2   Rubble chute and skip

- mini: 1.5 to 2.5 m³
- midi: 3.0 to 4.0 m³
- builders: 4.5 to 30 m³

Skips are hired out on the basis of a delivery and removal charge, plus a nominal rent for each day the skip remains on site. The removal charge will reflect the fee levied to the skip contractor for tipping at licensed fill sites. The local authority highways department and the local police must be consulted if the skip is to occupy all or part of the road and/or pavement. Permission to encroach onto the highway may be granted subject to suitable lighting being provided, an alternative temporary walkway, if required, and payment of a fee and deposit against damage to the highway. Provision of temporary traffic control lights may also be necessary.

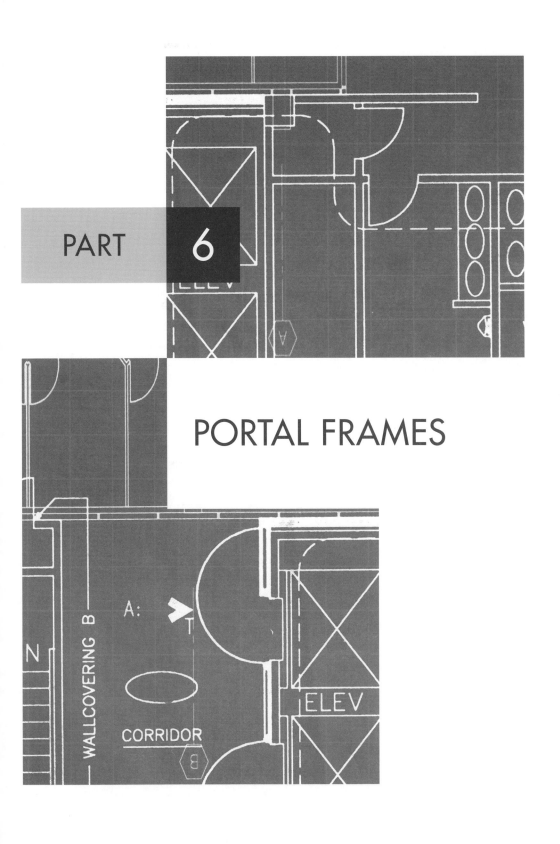

PART 6

PORTAL FRAMES

6.1

PORTAL FRAME THEORY

A portal frame may be defined as a continuous or rigid frame that has the basic characteristic of a rigid or restrained joint between the supporting member or column and the spanning member or beam. The object of this continuity of the portal frame is to reduce the bending moment in the spanning member by allowing the frame to act as one structural entity, thus distributing the stresses throughout the frame.

If a conventional simply supported beam was used (over a large span) an excessive bending moment would occur at mid-span, which would necessitate a deep heavy beam or a beam shaped to give a large cross-section at mid-span. Alternatively a deep cross-member of lattice struts and ties could be used. The main advantage of the simply supported frame lies in the fact that the column loading is for all practicable purposes axial, and therefore no bending is induced into the supporting members, which may well ease design problems because it would be statically determinate, but does not necessarily produce an economic structure. Furthermore, the use of a portal frame eliminates the need for a lattice of struts and ties within the roof space, giving a greater usable volume to the structure and generally a more pleasing internal appearance.

The transfer of stresses from the beam to the column in rigid frames will require special care in the design of the joint between the members; similarly the horizontal thrust and/or rotational movement at the foundation connection needs careful consideration. Methods used to overcome excessive forces at the foundation are:

- reliance on the passive pressure of the soil surrounding the foundation;
- inclined foundations so that the curve of pressure is normal to the upper surface, thus tending to induce only compressive forces;
- a tie bar or beam between opposite foundations;
- introducing a hinge or pin joint where the column connects to the foundation.

▨▩■ HINGES

Portal frames of moderate height and span are usually connected directly to their foundation bases, forming rigid or unrestrained joints. The rotational movement caused by wind pressures tending to move the frames and horizontal thrusts of the frame loadings is generally resisted by the size of the base and the passive earth pressures. When the frames start to exceed 4.000 m in height and 15.000 m in span the introduction of a hinged or pin joint at the base connection should be considered.

A hinge is a device that will allow free rotation to take place at the point of fixity but at the same time will transmit both load and shear from one member to another. They are sometimes called pin joints, unrestrained joints or non-rigid joints. Because no bending moment is transmitted through a hinged joint the design is simplified by the structural connection becoming statically determinate. In practice it is not always necessary to provide a true 'pivot' where a hinge is included, but to provide just enough movement to ensure that the rigidity at the connection is low enough to overcome the tendency of rotational movement.

Hinges can be introduced into a portal frame design at the base connections and at the centre or apex of the spanning member, giving three basic forms of portal frame:

- **Fixed or rigid portal frame**   All connections between frame members are rigid. This will give bending moments of lower magnitude more evenly distributed than other forms. Used for small to medium-size frames where the moments transferred to the foundations will not be excessive.
- **Two-pin portal frame**   In this form of frame, hinges are used at the base connections to eliminate the tendency of the base to rotate. The bending moments resisted by the supporting members will be greater than those encountered in the rigid portal frame. Main use is where high base moments and weak ground conditions are encountered.
- **Three-pin portal frame**   This form of frame has hinged joints at the base connections and at the centre of the spanning member. The effect of the third hinge is to reduce the bending moments in the spanning member but to increase deflection. To overcome this latter disadvantage a deeper beam must be used or, alternatively, the spanning member must be given a moderate pitch to raise the apex well above the eaves level. Two other advantages of the three-pin portal frame are that the design is simplified because the frame is statically determinate and, on site, they are easier to erect, particularly when preformed in sections.

A comparison of the bending moment diagrams for roof loads of the three forms of portal frame with a simply supported beam are shown in Fig. 6.1.1.

Another form of rigid frame is the **arch rib frame**, which is not strictly a portal frame as it has no supporting members. The main design objective is to design the arch to follow the curve of pressure, thus creating a state of no bending when subjected to a uniformly distributed load. Any moments encountered with this form of frame are generally those induced by wind pressures. Hinges may be used in the same positions for the same reasons as described above for the

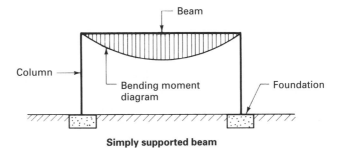

Simply supported beam

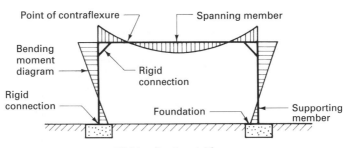

Rigid or fixed portal frame

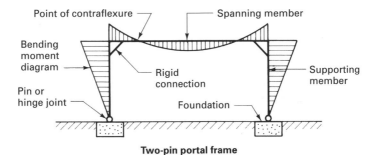

Two-pin portal frame

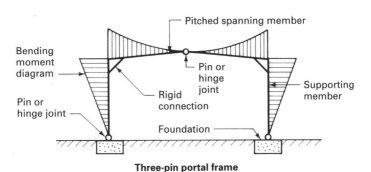

Three-pin portal frame

Figure 6.1.1 Portal frames: comparison of bending moments

conventional portal frames. The arch rib rigid frame is very often used where laminated timber is the structural material.

Most portal frames are made under factory-controlled conditions off site, which gives good dimensional and quality control but can create transportation problems. To lessen this problem and that of site erection, splices may be used. These can be positioned at the points of contraflexure (see Fig. 6.1.1), at the junction between spanning and supporting members and at the crown or apex of the beam. Most hinges or pin joints provide a point at which the continuity of fabrication is broken.

Portal frames constructed of steel, concrete or timber can take the form of the usual roof profiles used for single- or multi-span buildings such as flat, pitched, northlight, monitor and arch. Examples of northlight and monitor roofs are shown in Chapter 11.1. The frames are generally connected over the spanning members with purlins designed to carry and accept the fixing of lightweight insulated roof coverings or deckings. The walls can be of similar material fixed to sheeting rails attached to the supporting members or alternatively clad with brick or infill panels.

CONCRETE PORTAL FRAMES

Concrete portal frames are invariably manufactured from high-quality precast concrete suitably reinforced. In common with all precast concrete components for buildings, rapid advances in design and use were made after the Second World War, mainly because of the shortage of steel and timber that prevailed at that time. Now that the use of steel and timber is less restricted, reinforced concrete portal frames for new work is less common. In the main the use of precast concrete portal frames is confined to low-pitch (4° to 22$\frac{1}{2}$°) single-span frames, but two-storey and multi-span frames are available, giving a wide range of designs from only a few standard components.

The frames are generally designed to carry a lightweight (35 kg/m² maximum) roof sheeting or decking fixed to precast concrete purlins. Most designs have an allowance for snow loading of up to 150 kg/m² in addition to that allowed for the dead load of the roof covering. Wall finishes can be varied and intermixed because they are non-loadbearing and therefore have to provide only the degree of resistance required for fire, thermal and sound insulation, act as a barrier to the elements, and resist positive and negative wind pressures. Sheet claddings are fixed in the traditional manner, using hook bolts and purlins; sheet wall claddings are fixed in a similar manner to sheeting rails of precast concrete or steel spanning between or over the supporting members. Brick or block wall panels either of solid or cavity construction can be built off a ground beam constructed between the foundation bases; alternatively they can be built off the reinforced concrete ground floor slab. Remember that all such claddings must comply with any relevant building regulations, which will vary depending on whether the end use is for industrial, commercial, storage or agricultural purposes.

▉▉▉ FOUNDATIONS AND FIXINGS

The foundations for a precast concrete portal frame usually consist of a reinforced concrete isolated base or pad designed to suit loading and ground-bearing conditions. The frame can be connected to the foundations by a variety of methods:

- **Pocket connection**   The foot of the supporting member is located and housed in a void or pocket formed in the base so that there is an all-round clearance of 25 mm to allow for plumbing and final adjustment before the column is grouted into the foundation base.
- **Baseplate connection**   A steel baseplate is welded to the main reinforcement of the supporting member, or alternatively it could be cast into the column using fixing lugs welded to the back of the baseplate. Holding-down bolts are cast into the foundation base; the erection and fixing procedure follows that described for structural steelwork (see Chapter 10.5 of *Construction Technology*).
- **Pin joint or hinge connection**   A special base or bearing plate is bolted to the foundation, and the mechanical connection is made when the frames are erected (see Fig. 6.2.4).

The choice of connection method depends largely upon the degree of fixity required and the method adopted by the manufacturer of the particular system.

▨▨■■ ADVANTAGES

The main advantages to using precast concrete portal frames are as follows:

- Factory production will result in accurate and predictable components because the criteria for design, quality and workmanship recommended in BS 8110: *Structural use of concrete* can be more accurately controlled under factory conditions than casting components *in situ*.
- Most manufacturers produce a standard range of interchangeable components that, within the limitations of their systems, gives a well-balanced and flexible design range covering most roof profiles, single-span frames, multi-span frames and lean-to roof attachments. By adopting this limited range of members the producers of precast portal frames can offer their products at competitive rates coupled with reasonable delivery periods.
- Maintenance of precast concrete frames is not usually required unless the building owner chooses to paint or clad the frames.
- Precast dense concrete products have an inherent resistance to fire, and therefore it is not usually necessary to provide further fire-resistant treatment. However, the amount of reinforcement concrete cover will vary depending on the fire resistance required – see Approved Document B, Building Regulations to ascertain purpose grouping. For industrial premises the fire resistance is unlikely to be less than 2 hours, which will justify a minimum of 35 and 50 mm concrete cover for column and beam components respectively.
- The wind resistance of precast concrete portal frames to both positive and negative pressures is such that wind bracing is not usually required.
- Where members of the frame are joined or spliced together the connections are generally mechanical (nut and bolt), and therefore the erection and jointing can be carried out by semi-skilled operatives.

■ In most cases the foundation design, setting out and construction can be carried out by the portal frame supplier or their nominated subcontractor.

Typical details of single-span frames, multi-span frames, cladding supports, splicing and hinges are shown in Figs 6.2.1 to 6.2.4.

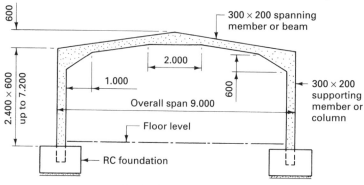

600

2.400 × 600
up to 7.200

300 × 200 spanning
member or beam

2.000

1.000

600

300 × 200
supporting
member or
column

Overall span 9.000

Floor level

RC foundation

Typical frame outline

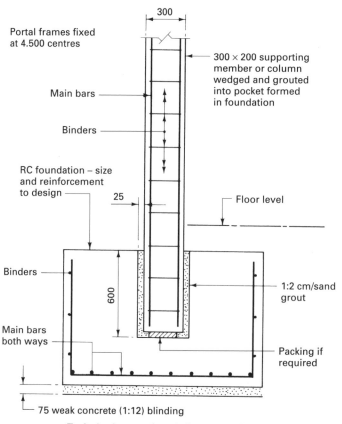

300

Portal frames fixed
at 4.500 centres

300 × 200 supporting
member or column
wedged and grouted
into pocket formed
in foundation

Main bars

Binders

RC foundation – size
and reinforcement
to design

25

Floor level

Binders

600

1:2 cm/sand
grout

Main bars
both ways

Packing if
required

75 weak concrete (1:12) blinding

Typical column to foundation connection

Figure 6.2.1   Typical single-span PCC frame

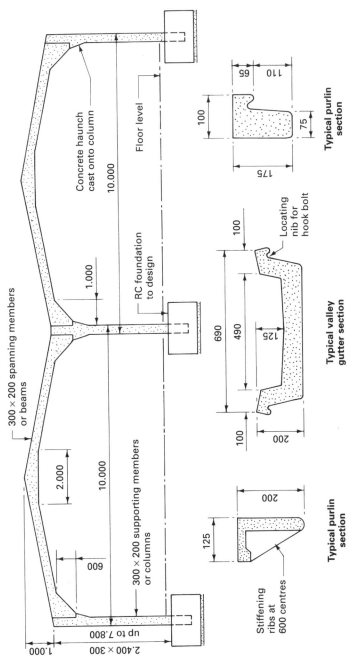

Figure 6.2.2 Typical multi-span precast concrete portal frame

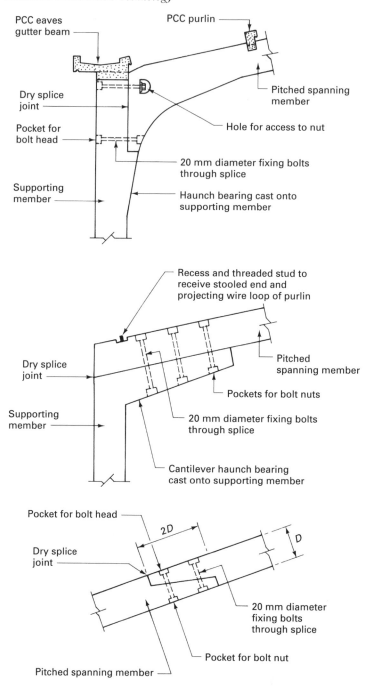

Figure 6.2.3   Typical splice details for PCC portal frames

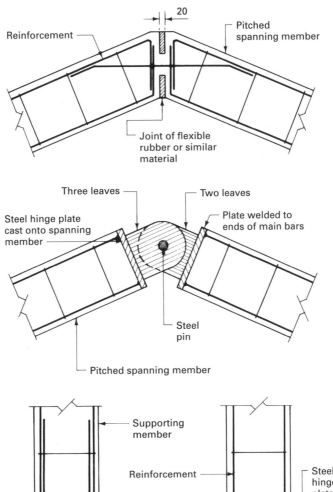

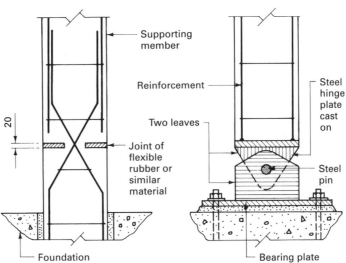

Figure 6.2.4 Typical hinge details for PCC portal frames

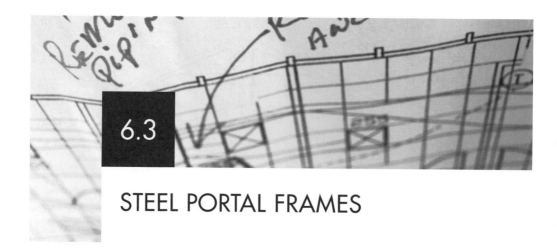

6.3

STEEL PORTAL FRAMES

Steel portal frames can be fabricated from standard universal beam, column and box sections. Alternatively a lattice construction of flats, angles or tubulars can be used. Most forms of roof profiles can be designed and constructed, giving a competitive range when compared with other materials used in portal frame construction. The majority of systems employ welding techniques for the fabrication of components, which are joined together on site using bolts or welding. An alternative system uses special knee joint, apex joint and base joint components, which are joined on site to square-cut standard beam or column sections supplied by the main contractor or by the manufacturer producing the jointing pieces.

The frames are designed to carry lightweight roof coverings of the same loading conditions as those given previously for precast concrete portal frames. Similarly, wall claddings can be of the same specification as for precast concrete portal frames and fixed in the same manner. Part L2 of the Building Regulations must be observed if thermal insulation requirements apply (may not be applicable for unheated storage-only facilities). If the building's use, irrespective of the framing material, is for an industrial process, the structure will have to comply with Part B to the Building Regulations: *Fire Safety*. The roof may also require purpose-made vents to relieve smoke logging and to reduce temperatures to ease firefighting access (see Chapter 11.1).

■ ■ ■ FOUNDATIONS AND FIXINGS

The foundation is usually a reinforced concrete isolated base or pad foundation designed to suit loading and ground-bearing conditions. The connection of the frame to the foundation can be by one of three basic methods:

■ **Pocket connection** The foot of the supporting member is inserted and grouted into a pocket formed in the concrete foundation, as described for precast concrete portal frames. To facilitate levelling some designs have gussets welded to the flanges of the columns, as shown in Fig. 6.3.1.

- **Baseplate connection**   Traditional structural steelwork column to foundation connection using a slab or a gusset base fixed to a reinforced concrete foundation with cast-in holding-down bolts (see Chapter 10.5 of *Construction Technology*).
- **Pin or hinge connection**   Special bearing plates designed to accommodate true pin or rocker devices are fixed by holding-down bolts to the concrete foundation to give the required low degree of rigidity at the connection.

■ ■ ■ ADVANTAGES

The main advantages of factory-controlled production are a standard range of manufacturer's systems, a frame of good wind resistance, and the fact that the ease of site assembly using semi-skilled operatives attributed to precast concrete portal frames can be equally applied to steel portal frames. A further advantage of steel is that, generally, the overall dead load of a steel portal frame is less than that of a comparable precast concrete portal frame. However, steel has the disadvantage of being a corrosive material, which will require a long-life protection of a patent coating or regular protective maintenance, generally by the application of coats of paint. Steel has a lower fire resistance than precast concrete, but if the frame is for a single-storey building structural fire protection may not be required under the Building Regulations (see Approved Document B). It depends very much on the purpose group or function of the building and the amount of space it occupies. However, the building's insurers usually require more effective fire protection than that determined by current building legislation, and may apply the Loss Prevention Certification Board's *Loss Prevention Standards (Fire and Security)*. Typical details of steel portal frames, cladding fixings, splicing and hinges are shown in Figs 6.3.1 to 6.3.3.

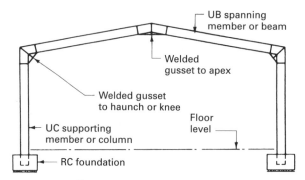

Typical steel portal frame profile

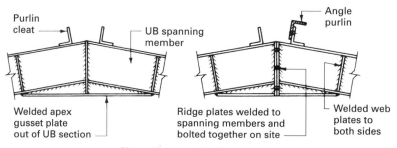

Alternative apex details

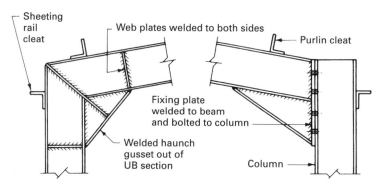

Alternative knee joint details

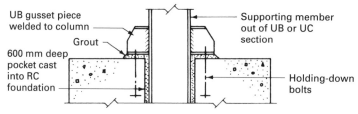

Pocket foundation connection

Figure 6.3.1   Typical steel portal frame details

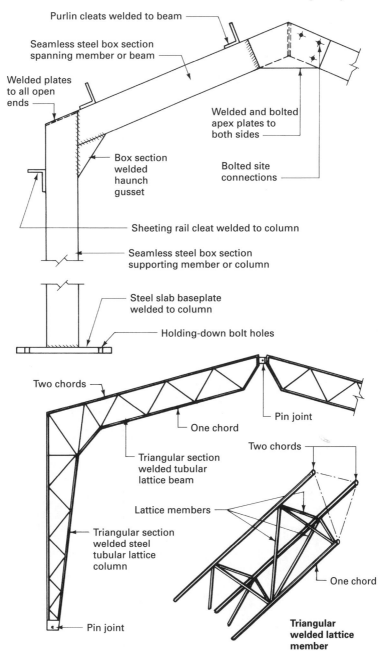

Purlin cleats welded to beam

Seamless steel box section spanning member or beam

Welded plates to all open ends

Welded and bolted apex plates to both sides

Box section welded haunch gusset

Bolted site connections

Sheeting rail cleat welded to column

Seamless steel box section supporting member or column

Steel slab baseplate welded to column

Holding-down bolt holes

Two chords

One chord

Pin joint

Triangular section welded tubular lattice beam

Two chords

Lattice members

Triangular section welded steel tubular lattice column

One chord

Pin joint

Triangular welded lattice member

Figure 6.3.2   Typical steel portal frames

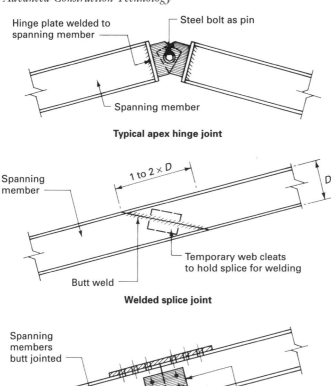

Hinge plate welded to spanning member

Steel bolt as pin

Spanning member

Typical apex hinge joint

Spanning member

1 to 2 × D

D

Temporary web cleats to hold splice for welding

Butt weld

Welded splice joint

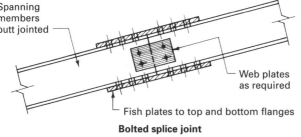

Spanning members butt jointed

Web plates as required

Fish plates to top and bottom flanges

Bolted splice joint

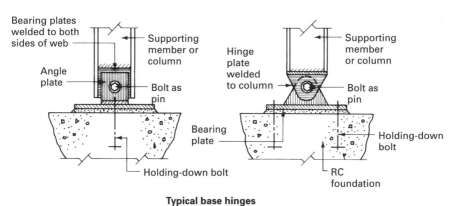

Bearing plates welded to both sides of web

Supporting member or column

Angle plate

Bolt as pin

Holding-down bolt

Hinge plate welded to column

Supporting member or column

Bolt as pin

Bearing plate

Holding-down bolt

RC foundation

Typical base hinges

Figure 6.3.3   Steel portal frames: splices and hinges

6.4

TIMBER PORTAL FRAMES

Timber portal frames can be manufactured by several methods, which produce a light, strong frame of pleasing appearance that renders them suitable for buildings such as churches, halls and gymnasiums where clear space and appearance are important. The common methods used are glued laminated portal frames, plywood-faced portal frames, and timber portal frames using solid members connected together with plywood gussets.

■ ■ ■ GLUED LAMINATED PORTAL FRAMES

The main objective of forming a laminated member consisting of glued layers of thin-section timber members is to obtain an overall increase in strength of the complete component compared with that which could be expected from a similar-sized solid section of a particular species of timber. This type of portal frame is usually manufactured by a specialist firm, because the jigs required would be too costly for small outputs. The selection of suitable-quality softwoods of the right moisture content is also important for a successful design. In common with other timber portal frames, these can be fully rigid, two- or three-pin structures.

Site work is simple, consisting of connecting the foot of the supporting member to the metal shoe fixing or to a pivot housing bolted to the concrete foundation and connecting the joint at the apex or crown with a bolt fixing or a hinge device. Most glued laminated timber portal frames are fabricated in two halves, which eases transportation problems and gives maximum usage of the assembly jigs. The frames can be linked together at roof level with timber purlins and clad with a lightweight sheeting or decking; alternatively, they may be finished with traditional roof coverings. Any form of walling can be used in conjunction with these frames provided such walling forms comply with any of the applicable Building Regulations. Typical details are shown in Fig. 6.4.1.

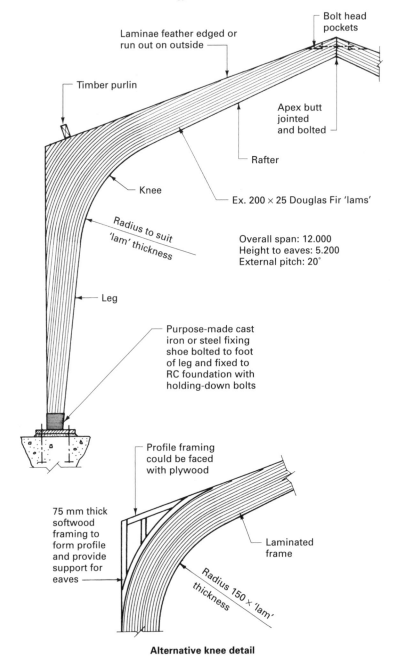

Laminae feather edged or run out on outside

Bolt head pockets

Timber purlin

Apex butt jointed and bolted

Rafter

Knee

Ex. 200 × 25 Douglas Fir 'lams'

Radius to suit 'lam' thickness

Overall span: 12.000
Height to eaves: 5.200
External pitch: 20°

Leg

Purpose-made cast iron or steel fixing shoe bolted to foot of leg and fixed to RC foundation with holding-down bolts

Profile framing could be faced with plywood

75 mm thick softwood framing to form profile and provide support for eaves

Laminated frame

Radius 150 × 'lam' thickness

Alternative knee detail

Figure 6.4.1   Typical glued laminated portal frame

▨▪▪ PLYWOOD-FACED PORTAL FRAMES

These frames are suitable for small halls, churches and schools with spans in the region of 9.000 m. The portal frames are in essence boxed beams, consisting of a skeleton core of softwood members faced on both sides with plywood, which takes the bending stresses. The hollow form of the construction enables electrical and other small services to be accommodated within the frame members. Design concepts, fixing and finishes are as given above for glued laminated portal frames. Typical details are shown in Fig. 6.4.2.

▨▪▪ SOLID TIMBER AND PLYWOOD GUSSETS

These frames were developed to provide a simple and economic timber portal frame for clear-span buildings using ordinary tools and basic skills. The general concept of this form of frame varies from the two types of timber portal frame previously described in that no gluing is used, the frames are spaced close together (600, 900 and 1200 mm centres), and they are clad with a plywood sheath so that the finished structure acts as a shell, giving a lightweight building that is very rigid and strong. The frames can be supplied in two halves and assembled by fixing the plywood apex gussets on site before erection, or they can be supplied as a complete frame ready for site erection.

The foundation for this form of timber portal frame consists of a ground beam or, alternatively, the frames can be fixed to the edge of a raft slab. A timber spreader or sole plate is used along the entire length of the building to receive and distribute the thrust loads of the frames. Connection to this spreader plate is made by using standard galvanised steel joist hangers or by using galvanised steel angle cleats. Standard timber windows and doors can be inserted into the side walls by trimming in the conventional way and infilling where necessary with studs, noggins and rafters. Typical details are shown in Fig. 6.4.3.

The advantages of all timber portal frame types are as follows:

- constructed from readily available materials at an economic cost;
- light in weight;
- easy to transport and erect;
- can be trimmed and easily adjusted on site;
- protection against fungal and/or insect attack can be by impregnation, or surface application;
- pleasing appearance either as a natural timber finish or painted.

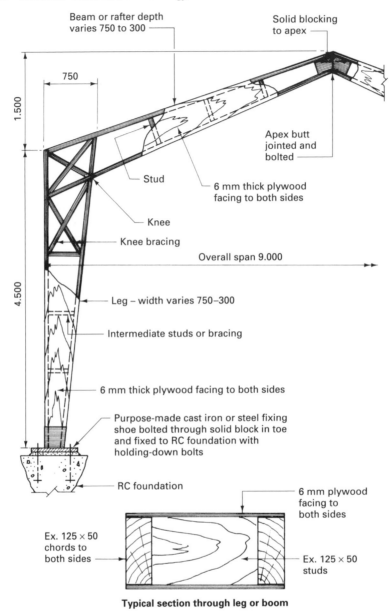

Beam or rafter depth
varies 750 to 300

Solid blocking
to apex

750

1.500

Apex butt
jointed and
bolted

Stud

6 mm thick plywood
facing to both sides

Knee

Knee bracing

Overall span 9.000

4.500

Leg – width varies 750–300

Intermediate studs or bracing

6 mm thick plywood facing to both sides

Purpose-made cast iron or steel fixing
shoe bolted through solid block in toe
and fixed to RC foundation with
holding-down bolts

RC foundation

6 mm plywood
facing to
both sides

Ex. 125 × 50
chords to
both sides

Ex. 125 × 50
studs

Typical section through leg or boom

Figure 6.4.2   Typical plywood-faced portal frame

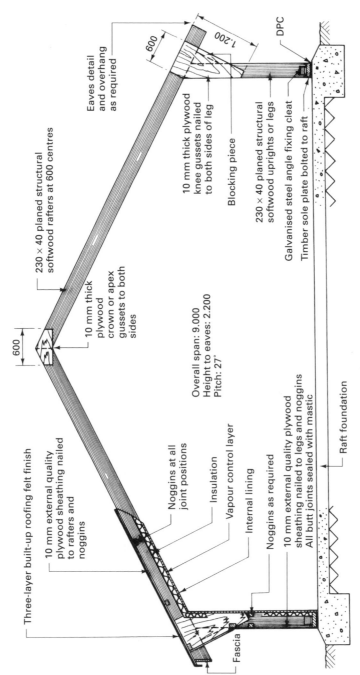

Three-layer built-up roofing felt finish

10 mm external quality plywood sheathing nailed to rafters and noggins

Noggins at all joint positions

Insulation

Vapour control layer

Internal lining

Noggins as required

10 mm external quality plywood sheathing nailed to legs and noggins
All butt joints sealed with mastic

Fascia

Eaves detail and overhang as required

230 × 40 planed structural softwood rafters at 600 centres

10 mm thick plywood crown or apex gussets to both sides

Overall span: 9.000
Height to eaves: 2.200
Pitch: 27°

600

600

1.200

600

DPC

10 mm thick plywood knee gussets nailed to both sides of leg

Blocking piece

230 × 40 planed structural softwood uprights or legs

Galvanised steel angle fixing cleat

Timber sole plate bolted to raft

Raft foundation

Figure 6.4.3   Typical solid timber and plywood gusset portal frame

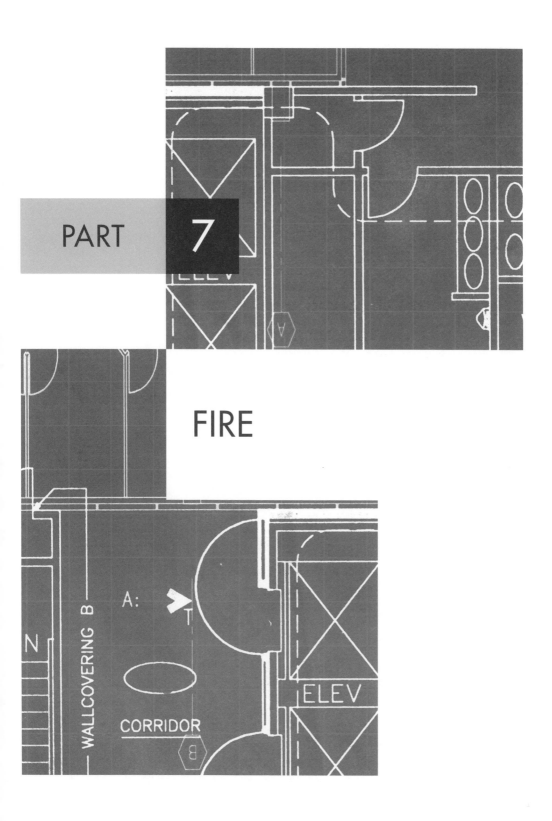

PART 7

FIRE

THE PROBLEM OF FIRE

Fire has always been an essential element of our technological advancement, providing heat, energy and light. Today, fire continues to be of great benefit to our well-being if it is controlled. If allowed to start and spread without strict control, it can be one of the greatest hazards with potential for destruction on a wide scale. Early civilisations considered fire to be a natural element like air and water. Later experimenters found that the residue of a burnt fuel (ash) weighed less than the fuel before it was burnt, and concluded that some substance was removed during the combustion period: this they called 'phlogiston' after the Greek word *phlogistos*, meaning 'inflammable'. The doctrine of phlogistics was overthrown by a French chemist Antoine Lavoisier (1743–94), who became known as the father of modern chemistry.

Lavoisier discovered by his researches and experiments that air consists of one-fifth oxygen and that the other main gas, nitrogen, accounted for the bulk of the remaining four-fifths. He showed that oxygen played an important part in the process of combustion, and that nitrogen does not support combustion. This discovery of the true nature of fire led to the conclusion that fire is a chemical reaction whereby atoms of oxygen combine with other atoms such as carbon and hydrogen, releasing water, carbon dioxide and energy in the form of heat. The chemical reaction will only start at a suitable temperature, which varies according to the substance or fuel involved. During combustion, gases will be given off, some of which are more inflammable than the fuel itself and therefore ignite and appear as flames, giving light, which is due to tiny particles being heated to a point at which they glow. Smoke is an indication of incomplete combustion and can give rise to deposits of solid carbon commonly known as soot.

From the discovery of the true nature of fire and processes of combustion it can be concluded that there are three essentials to all fires:

- **Fuel**   Generally any organic material is suitable.
- **Heat**   Correct temperature to promote combustion of a particular fuel. Heat can be generated deliberately, which is termed **ignition**, or it can be spontaneous when the fuel itself ignites.
- **Oxygen**   Air is necessary to sustain and support the combustion process.

The above is often referred to as the **triangle of fire**: remove any one of the three essentials and combustion cannot take place. This fact provides the whole basis for fire prevention, fire protection and firefighting. If non-combustible materials were used in the construction and furnishing of buildings, fires would not develop. This method is far too restrictive on the designer and builder: therefore combustible materials are used and protected with layers or coverings of non-combustible materials, e.g. plasterboard linings to wooden frames and combustible insulants. For guidance on minimum periods of fire resistance to structural elements, see Tables A1 and A2 in Approved Document B to the Building Regulations. There is also provision for materials of limited combustibility, e.g. applications to stair construction and roof decking, provided the extent of exposure is restricted. See Table A7 in Approved Document B to the Building Regulations and associated references to BS 476-11: *Method for assessing the heat emission from building products.* Firefighters try to remove one side of the fire triangle; to remove the fuel is not generally practicable, but by using a cooling agent such as water the heat can be reduced to a safe level, or alternatively by using a blanketing agent the supply of oxygen can be cut off and the fire extinguished.

Annual statistics for occurrence of fires and types of building affected are produced by the Office for National Statistics (ONS) and the Office of the Deputy Prime Minister (ODPM). Further data are also produced by the fire service and insurance companies. The presentation of data varies between the different sources, some providing figures for the UK as a whole and others for the constituent countries. The overall number of annual fatalities from fire (not just in buildings) in the UK averages in excess of 500 persons in recent years. Fatalities from fires in dwellings exceed 300 on average. The proportion of deaths attributed mainly to dwellings exceeds 90% of the total for all buildings.

Residential buildings are the most vulnerable, accounting for over 60,000 incidents annually or about 65% of all fires. In excess of 3,000 fires in dwellings are attributed to children playing with fire, i.e. matches and other easily ignitable home gadgets. Another high-risk area is commercial distribution, where large quantities of inflammable goods are held in store. One of the main causes of fires in non-domestic buildings is faulty electrical equipment and wiring. This is not an indication that the plant or installation is poor, but that regular maintenance, renewal and routine checks must be carried out and be documented by qualified and experienced personnel. In residential buildings the same provision should be made for electrical and gas services.

Given the impressive fire suppression record for sprinklers in commercial and industrial premises, building insurers and the fire prevention industry and professions have been considering and promoting the installation of sprinkler systems in residential accommodation. The government has responded with inclusion of these in Part B (Fire safety) of the Building Regulations. Initially, this guidance is limited to dwelling houses, residential care homes and in flats over 30m in height.

The annual cost of fires in buildings is difficult to assess. It probably runs into figures expressed in billions of pounds. The personal loss is impossible to value. The cost to manufacturers and employers in terms of loss of goodwill, loss of production, effect of fatalities and injuries to employees and the delay in returning to full production or working are almost incalculable. Therefore it can be seen that the seriousness of fires within buildings cannot be overstated.

Obviously designers and builders alike cannot be held responsible for the actions or non-actions of the occupants of the buildings they create, but they can ensure that these structures are designed and constructed in such a manner that they give the best possible resistance to the action of fire should it occur.

The precautions that can be taken within buildings to prevent a fire occurring – or, if it should occur, of containing it within the region of the outbreak, providing a means of escape for people in the immediate vicinity, and fighting the fire – can be studied under three headings:

- structural fire protection;
- means of escape in case of fire;
- firefighting.

Structural fire protection is generally known as a **passive** measure of fire protection, and is incorporated within the design specification of each element of construction. Firefighting and fire control mechanisms such as sprinklers, fire alarms, fusible link-operated shutters and doors are very much **active**. These are mostly integrated with the services of a building, and usually considered in the context of that subject. Therefore, apart from superficial reference, no depth of study of this specialism is included in this text. Readers are recommended to refer to Chapter 9 of *Building Services Technology, and Design*, published by Pearson Education/Chartered Institute of Building, for further information on firefighting and fire control services.

STRUCTURAL FIRE PROTECTION

The purpose of structural fire protection is to ensure that during a fire the temperature of structural members or elements does not increase to a figure at which their strength would be adversely affected. Additionally, when considering the features in layout and/or construction that are intended to reduce the effects of a fire, it can be established that containment of fire spread within buildings by such measures as compartmentation, as well as providing physical means for preventing it spreading to other buildings e.g. cavity barriers, is also of paramount importance. It is not practicable or possible to give an element complete protection in terms of time: therefore elements are given a fire resistance for a certain period of time that it is anticipated will give sufficient delay to the spread of fire, ultimate collapse of the structure, time for persons in danger to escape and to enable firefighting to be commenced.

These periods of time vary depending on the function or purpose group and occupancy of the building, and the size/height of the top floor above ground and the depth of basement. See tables in Appendix A to Approved Document B of the Building Regulations for specific requirements.

Before a fire-resistance period can be determined it is necessary to consider certain factors:

- fire load intensity of the building (amount of combustible material per m² of floor area);
- behaviour of materials under fire conditions;
- behaviour of combinations of materials under fire conditions;
- Building Regulation requirements as laid down in Part B.

■■■■ FIRE LOAD

Buildings can be graded as to the amount of overall fire resistance required by taking into account the following:

- size of building (floor area and height from ground to highest floor);
- use of building (purpose group);
- fire load.

The fire load is an assessment of potential fire severity based on the combustibility of materials within a building. Precise calculation of fire severity is impractical. It is therefore a broad estimate derived from building usage (purpose grouping) and its contents. This load is expressed as the amount of heat that would be generated per unit area by the complete combustion of its contents and combustible members, and is given a calorific value of joules per square metre. Note that the numerical grade is equivalent to the minimum number of hours' fire resistance that should be given to the elements of the structure.

- **Grade 1**   Low fire load, not more than 1,150 MJ/m^2. Typical buildings within this grade are flats, offices, restaurants, hotels, hospitals, schools, museums and public libraries.
- **Grade 2**   Moderate fire load, 1,150 to 2,300 MJ/m^2. Typical examples are retail shops, factories and workshops.
- **Grade 4**   High fire load, 2,300 to 4,600 MJ/m^2. Typical examples are certain types of workshop and warehouses.

When deciding the grade, no account is taken of the effects of any permanent fire protection installations such as sprinkler systems. The above principles are incorporated in the Building Regulations and, in particular, in Part B.

▨ ▨ ▨ ■ FIRE RESISTANCE

In addition to the previous assessment, fire resistance is the ability of a component or element of construction in a building to satisfy specific criteria from the relevant parts of BS 476: *Fire tests on building materials and structures*. These are usually measurements of the effects of temperature, either radiant or naked flame, over an exposure time, and may incorporate values for loadbearing capacity, integrity and insulation. Components can also be studied as separate entities with regard to their behaviour when subjected to the intense heat encountered during a fire and their ability to support fire spread over their exposed surfaces.

Structural steel is not considered to behave well under fire conditions, although its surface fire spread is negligible. As the fire progresses and the temperature of steel increases there is an actual gain in the ultimate strength of mild steel. This gain in strength decreases back to normal over the temperature range of 250 to 400 °C. The decrease in strength continues, and by the time the steel temperature has reached 550 °C it will have lost most of its useful strength. As the rise in temperature during the initial stages of a fire is rapid, this figure of 550 °C can be reached very quickly. If the decrease in strength results in the collapse of a member, the stresses it was designed to resist will be redistributed: this could cause other members to be overstressed, and progressive collapse could occur. Also, the high degree of thermal movement in steel may cause disturbance to, or loss of, bearing support, and this, too, will contribute to redistributed loadings and possibly progressive structural collapse.

Reinforced concrete structural members have good fire-resistance properties, and being non-combustible do not contribute to the spread of flame over their surfaces. It is possible, however, under the intense and prolonged heat of a fire, that the bond between the steel reinforcement and the concrete will be broken. This generally results in spalling of the concrete, which decreases both the protective cover of the concrete over the steel and the cross-sectional area. As for structural steel members, this can result in a redistribution of stresses leading to overloading of certain members, culminating in progressive collapse.

Timber, strange as it may seem, behaves very well structurally under the action of fire. This is due to its slow combustion rate, the strength of its core failure remaining fairly constant. The ignition temperature of timber is low (250–300 °C), but during combustion the timber chars at about 0.5–1.0 mm per minute, depending on the species and extent of heat and flame. The layer of charcoal so formed slows down the combustion rate of the core. Although its structural properties during a fire are good, timber, being an organic material and therefore combustible, will spread fire over its surface, which makes it unsuitable in most structural situations. Intumescent paints will provide a limited resistance to fire, but more successful protection is achieved by nailing and wire-binding plasterboard to the surface and finishing with a board finish plaster.

From the above brief considerations it is obvious that designers and builders need to have data on the performance, under the conditions of fire, of materials and especially combinations of materials forming elements. Such information is available in BS 476: *Fire tests on building materials and structures*. The BS is divided into parts that relate to the various fire tests applied to building materials and structures.

■■■ FIRE TESTS ON BUILDING MATERIALS AND STRUCTURES – BS 476

BS 476 consists of 17 parts numbered intermittently between 3 and 33. Omissions account for withdrawals such as Part 1, which was replaced by Part 8, which, in turn, was replaced by Parts 20 to 23. The following is a summary:

Part 3   *Classification and method of test for external fire exposure to roofs*

A series of tests for grading roof structures in terms of time for:

- resistance to external penetration by fire;
- distance of spread of flame over the external surface under certain conditions.

The tests are applied to a specimen of roof structure not less than 1.500 m × 1.200 m, which represents the actual roof construction including at least one specimen of any joints used and complete with any lining that is an integral part of the construction.

Three tests are applied:

- Preliminary ignition test.
- Fire penetration test.
- Roof surface spread of flame test.

After testing, the specimen or form of roof construction can be graded by a double letter designation in the range A–D. An AA designation is the most acceptable and DD the least. The initial letter relates to the time of fire penetration; the second letter is a measure of the surface spread of flame:

First letter (penetration)
A. No penetration within 1 hour.
B. Specimen penetrated in not less than $^1/_2$ hour.
C. Specimen penetrated in less than $^1/_2$ hour.
D. Specimen penetrated in the preliminary flame test.

Second letter (spread of flame)
A. No spread of flame.
B. Not more than 533 mm spread.
C. More than 533 mm spread.
D. Specimens that continue to burn for 5 minutes after withdrawal of the test flame or spread more than 381 mm across the region of burning in the preliminary flame test.

Specimens can be tested from a 45° inclination down to 1° (effectively flat), and may be prefixed EXT.S or EXT.F accordingly. If during the test any dripping from the underside of the specimen, any mechanical failure and/or development of holes is observed, a suffix 'X' is added to the designation thus:

EXT.S.AB–EXT.S.ABX

The Building Regulations, Approved Document B *Fire safety*, Sections 10 (Volume 1) and 14 (Volume 2): Roof coverings, has adopted the double letter classification. It is shown below compared with European Standard BS EN 13501: *Fire classification of construction products and building elements.*

Building Regulations classification	*BS EN 13501-5 classification*
AA, AB, AC	B_{ROOF}(t4)
BA, BB, BC	C_{ROOF}(t4)
CA, CB, CC	D_{ROOF}(t4)
AD, BD, CD	E_{ROOF}(t4)
DA, DB, DC, DD	F_{ROOF}(t4)

Notes:
The suffix (t4) indicates test number 4 in the Standard.
Classes A1 and A2 also occur in BS EN 13501. Products tested to these classifications satisfy the same criteria as Class B, and have no significant contribution to the fire load or fire growth.

Part 4   Non-combustibility test for materials

Generally, organic materials are combustible, whereas inorganic materials are non-combustible, which can be defined as a material not capable of undergoing combustion. For the purposes of BS 476, materials used in the construction and finishing of buildings or structures are classified as 'non-combustible' or 'combustible' according to their behaviour in the non-combustibility test.

A material is classified by this test as non-combustible if none of the three specimens tested:

■ causes the temperature reading from either of the two thermocouples used to rise by 50 °C or more above the initial furnace temperature;
■ is observed to flame continuously for 10 seconds or more inside the furnace.

Otherwise the material shall be deemed combustible.

Note that mixtures of organic and inorganic materials have a different behaviour pattern from that of the individual materials, and therefore such combinations must be tested and classified accordingly.

Part 6   Method of test for fire propagation for products

The scope of this test is to provide a means of comparing the contribution of combustible materials to the growth of fire. It is intended mainly for assessing the fire performance of linings to internal walls and ceilings. Specimens are given a fire propagation index (I) ranging from 0 (non-combustible) to 100 in a descending order according to a curve of rise in temperature plotted against time from ignition.

Part 7   Method of test to determine the classification of the surface spread of flame of products

This test is used to measure the lateral spread of flame for essentially flat materials, including composites that may be applied as exposed wall or ceiling finishes.

The sample for testing must have any surfacings or coatings applied in the usual manner. The 270 mm × 885 mm sample is fixed in the holder of the apparatus and subjected to radiant heat from the test furnace. During the first minute of the test a luminous gas flame is applied to the furnace end of the specimen. After 1 minute it is extinguished and at intervals of 1.5 minutes and 10 minutes the extent of flaming along the material specimen is recorded. The sample is placed into one of four classes, set out in Table 7.2.1.

Approved Document B2: Sections 3 (Volume 1) and 6 (Volume 2) specifies a higher class than class 1, called class 0, for wall and ceiling linings. This is defined as a non-combustible material throughout or, if the surface material is tested in

Table 7.2.1   BS 476-7, Table 1

Classification	Flame spread at $1^1/_2$ minutes		Final flame spread	
	Limit (mm)	Limit for one specimen in sample (mm)	Limit (mm)	Limit for one specimen in sample (mm)
Class 1	165	+25	165	+25
Class 2	215	+25	455	+45
Class 3	265	+25	710	+75
Class 4		Exceeding class 3 limits		

accordance with BS 476-6, it shall have an index (I) not exceeding 12 and a sub-index (i_1) not exceeding 6.

Part 10   *Guide to the principles and application of fire testing*

This relates to the general principles and methods of applying fire tests to building products, components and elements of construction for other parts of BS 476. It incorporates a general philosophy from BS 6336: *Guide to the development of fire tests, the presentation of test data and the role of tests in hazard assessment*, and terminology/definitions from BS 4422: *Fire. Vocabulary.*

Part 11   *Fire tests on building materials and structures*

This part of the British Standard is used to assess the heat emission from building materials when subjected to insertion in a furnace at 750 °C. The test is appropriate for reasonably homogeneous materials: therefore veneers and superficially treated products are not normally suited to this test unless the components are tested individually. Cylindrical specimens 45 mm in diameter × 50 mm high are prepared with a 2 mm diameter hole for location of a thermocouple. The test provides comparable results for duration of flaming and determination of whether a material is 'non-combustible' or of 'limited combustibility' (see Building Regulations, Approved Document B, Appendix A (A6 to A8)).

Part 12   *Method of test for ignitability of products by direct flame impingement*

This is used to establish the response of materials when subjected to impingement of flames having varying size and heat intensity. Several specimen sizes are used, ranging from 100 mm × 150 mm up to 500 mm × 750 mm, with a range of flame applications to the sample surfaces and edges. The materal reactions are measured over exposures of 1 second to 180 seconds. Results are letter designated:

I = Ignition occurred
T = Transient ignition
N = No ignition
W = Flaming or glowing to material edge or within 10 seconds of ignition source removal

Part 13   *Method of measuring the ignitability of products subjected to thermal irradiance*

This test is particularly useful for measuring the ignition characteristics of sheet materials, composites or assemblies less than 70 mm thick when used horizontally and subject to adjacent levels of thermal irradiance. Specimens 165 mm square are subjected to irradiance at five levels from 10 to 50 kW/m^2 for 15 minutes or until the specimen ignites. Test results and observations record time of ignition, glowing decomposition, melting, surface spalling, foaming, cracking, expansion and/or contraction. A gas pilot light can also be used to test ignitability of any gases leaving the sample.

Part 15   *Method for measuring the rate of heat release of products*

Specimens of sheet materials of up to 50 mm thickness are prepared in squares of 100 mm and subjected to radiant heat from a 5 kW cone-shaped electric heater (cone calorimeter). Data are recorded up to 2 minutes after flaming commences or until 60 minutes elapses. Test data in terms of duration and heat energy release rate provide for material comparisons.

PARTS 20 TO 24

These include provision for determining the fire resistance of various parts/elements/components of construction. The term 'fire resistance' relates to a complete element of building construction and not to the individual materials of which that element is composed. The tests enable elements of construction to be assessed according to their ability to retain their composition, to resist the passage of flame and hot gases, and to provide the necessary resistance to heat transmission in a defined pressure environment. This is measured against retention of fire separating and/or loadbearing facilities over a specified time.

Part 20   *Method for determination of the fire resistance of elements of construction (general principles)*

The general principles outlined in this part of BS 476 include details for preparing specimens, i.e. the number and size for specific tests and descriptions of the apparatus for specimen support and exposure. Procedures for effecting the various tests are also provided. Details of criteria for performance are shown measured in terms of **integrity** (loadbearing capacity and fire containment) and **insulation** (thermal transmittance) with reference to test report format and guidance on application of method. Test results are given in minutes from the start of the test until failure occurs. Likewise, the fire resistance is given in minutes for each element of construction described in Parts 21 to 24.

Part 21   *Method for determination of the fire resistance of loadbearing elements of construction*

This document defines the application to beams, columns, floors, flat roofs and walls.

Beams

The specimen is to be full size or have a minimum span of 4.000 m and be located to simulate actual site conditions. If the beam is exposed to fire on three faces, associated construction as in practice shall be included in the specimen.

Fire resistance

The test specimen is deemed to have failed when it can no longer support its design loading. This occurs if:

1. The deflection exceeds $L/20$, where L is the clear span of the specimen in mm.
2. The rate of deflection (mm/min) calculated at 1 minute intervals commencing 1 minute from the heat application exceeds the limit set by $L^2/9{,}000d$, where d is the distance from the top of the structural section to the bottom of the design tension zone in mm.

Take (1) or (2), whichever is exceeded first.

Columns

The specimen is to be full size or to have a minimum length of 3.000 m and be loaded to simulate actual site conditions. The specimen is heated on all exposed faces.

Fire resistance

This is the time taken for the specimen to no longer support the test load, i.e. to show a noticeable change in the rate of deformation.

Floors and flat roofs

The specimen is to be full size or a minimum of 3.000 m wide × 4.000 m span and be loaded to simulate actual site conditions. If a ceiling is intended to add to the fire resistance, it must be included with the test specimen. This also applies to any construction or expansion joints, light diffusers or any other components integral with the actual installation. The specimen is heated from one side by a furnace to produce a positive pressure at standard heating conditions (see Part 20) until failure occurs or the test is terminated.

Fire resistance

Loading as for columns, plus criteria to assess integrity and insulation.

- **Integrity**   A 100 mm × 100 mm × 20 mm thick cotton wool pad of mass 3 to 4 g is held over the centre of any crack through which flames and gases can pass. The pad is held 25 mm from and parallel to the crack for a period of 10 seconds to determine whether hot gases can cause ignition. The observation is repeated at frequent intervals. Alternatively, gap gauges can be used where a crack occurs, i.e. a 6 mm gauge penetrating into the furnace for a distance exceeding 150 mm, or a 25 mm gauge penetrating through the specimen and into the furnace.
- **Insulation**   The unexposed face of elements having a separating function is observed at intervals of not more than 5 minutes. Failure is deemed to occur if:
  - Mean temperature rises more than 140 °C above initial temperature.
  - Point temperature rises more than 180 °C above initial temperature.

See also BS 476-20, clauses 10.3 and 10.4.

Walls

The specimens must include provision for any mechanical joints and be tested from both sides. They must be full size or a minimum of 3.000 m × 3.000 m and loaded to simulated actual site conditions.

Fire resistance

Time taken to failure by any one of three observations:

- noticeable change in the rate of deformation;
- loss of integrity (as for floors and flat roofs);
- loss of insulation (as for floors and flat roofs).

Part 22   Methods for determination of fire resistance of non-loadbearing elements of construction

This is applicable to partitions, door sets and shutters (fully insulated, partially insulated and uninsulated), ceiling membranes and glazing.

Partitions

Specimens to be at least 3.000 m × 3.000 m and tested from one or either side. They must include any provision for mechanical joints such as built-in facilities for expansion and contraction.

Fire resistance

Failure time for integrity and insulation (see Part 21).

Door sets and shutters

Specimens are the same size as for partitions and are tested from both sides.

Fire resistance (fully insulated)

Integrity and insulation as previously described and/or sustained flaming, i.e. a visible flame occurring for at least 10 seconds.

Fire resistance (part insulated)

As for fully insulated except the cotton wool pad requirement for integrity only applies to areas other than non-insulating materials such as glazing.

Fire resistance (uninsulated)

Test for integrity only applies, but not the cotton wool pad requirement.

Ceilings

The specimen to be at least 4.000 m × 3.000 m and exposed to test from the underside only.

Fire resistance

For integrity and insulation as previously described.

Glazed elements
The specimen to be at least 3.000 m × 3.000 m and exposed to test from both sides.

Fire resistance
For integrity and insulation as previously described.

Part 23   Methods for determination of the contribution of components to the fire resistance of a structure

This part is in two sections:

1. The contribution made by components such as suspended ceilings when protecting steel beams.
2. The use of intumescent materials/seals with single-acting latched timber fire-resisting door assemblies.

Suspended ceilings protecting steel beams
The specimen to be full size or a minimum of 4.000 m × 3.000 m, with the underside only exposed to the furnace. The steel beams used to support the specimen have the top flange covered with a lightweight concrete floor at least 130 mm thick. Provision for any light fittings or similar outlets must be made within the specimen.

Fire resistance
The limit of effective protection is deemed to have been reached when any of the following are exceeded:

1. With an unloaded specimen, the beam temperature at any point attains 400 °C.
2. For a loaded specimen, see reference to Part 21: Beams, take (1) or (2).

Intumescent seals
This test is suitable for evaluating the effect of intumescent materials/seals between door to frame assemblies for doors with fire resistance specified at up to 1 hour (see BS 476: Part 22 and Chapter 7.1 of *Construction Technology*). Full-size specimens are exposed to the furnace on one side only, as described in Part 22.

Fire resistance
As previously described for integrity, or failure of the seal due to the presence of continuous flaming on the unexposed door face.

Part 24   Method for determination of the fire resistance of ventilation ducts

This provides information relating to the ability of vertical and horizontal ventilation or air-conditioning ducting to resist fire spread from one fire compartment to another, in the absence of fire dampers. Test specimens of ducting should be full size with perimeter fully exposed in the furnace. Thermocouples are strategically located throughout the duct length. Junctions with compartment walls and floors are suitably fire stopped as in standard installation practice (see Fig. 7.2.3).

Fire resistance

Failure of the specimen by one or more of the following:

- **Stability**   Duct collapses as retaining devices can no longer hold it.
- **Insulation**   When the exposed surface of the duct outside the furnace exceeds an average of 140 °C, or has a point value of 180 °C.
- **Integrity**   When cracks or holes allowing flames or hot gases to pass appear at the duct surface outside the furnace, i.e. sustained flaming occurs for at least 10 seconds. Measurements can also be taken using the cotton wool pad.

PARTS 31 TO 33

Part 31   *Methods for measuring smoke penetration through door sets and shutter assemblies*

Section 31.1: *Methods of measurement under ambient temperature conditions* provides guidance on smoke penetration potential through door sets by applying pressurised air tests and taking measurements of the air leakage rate. Its purpose is for smoke control only, and it has no reference to fire resistance (see Part 22). Full-size specimens are subjected to fan-delivered variable air pressures between the two door faces, in increments up to 100 Pa. Air penetration and leakage is measured in m^3/h. Doors fitted with patent peripheral brush seals can be expected to provide greater resistance to the simulated air flow. These will have both improved draught and smoke penetration resistance by comparison with unprotected doors. See also Building Regulations, Approved Document B: Appendix B, Table B1, where fire doors for certain situations (suffixed S) are required to have an air leakage rate not exceeding 3 m^3/h at 25 Pa.

Part 32   *Guide to full-scale fire tests within buildings*

Part 33   *Full-scale room test for surface products*

As indicated by the titles, these procedures are intended to simulate fires in buildings from ignition through to full maturity. Tests can represent specific fire situations and materials to determine potential for flashover conditions, development of fire growth, and behaviour of various products in a continuous fire environment. In addition to visual interpretations of flame spread both vertically and horizontally, measurements can also include temperature, air flow and gas emissions.

▨ ■ ■ BUILDING REGULATIONS: PART B

One of the major aims of this part of the Building Regulations is to limit the spread of fire, and this is achieved by considering the use of a building, the fire resistance of structural elements and surface finishes, the size of the building or parts of a building, and the degree of isolation between buildings or parts of buildings.

Part B to the Building Regulations contains three regulations that are concerned with fire spread under the headings of:

- Internal fire spread (linings) – Regulation B2;
- Internal fire spread (structure) – Regulation B3;
- External fire spread – Regulation B4.

Approved Document B, which supports these regulations, is comprehensive and gives recommendations for and guidance on meeting the performance requirements set out in the actual regulations. It is not proposed to analyse each recommendation fully but rather to use the recommendations as illustrations as to how the objectives can be achieved.

A full understanding of the terminology used is important to comprehend the recommendations being made in the Approved Document, and this is given in Appendix E under the heading of definitions. Figure 7.2.1 illustrates some of the general definitions given in this appendix. Other interpretations that must be clearly understood are elements of structure, fire stops, relevant and notional boundaries, and unprotected areas. These are illustrated in Figs 7.2.2, 7.2.3, 7.2.4 and 7.2.5, respectively.

The use of a building enables it to be classified into one of the seven purpose groups given in Approved Document B, Appendix D, Table D1. The purpose groups can apply to a whole building, or to a separated part or compartment of a building. All buildings covered by Part B of the Building Regulations should be included in one of these purpose groups. Each purpose group can be summarised:

1. **Residential (dwellings)**   Subdivides into three groups:
   (a) Flat.
   (b) Dwelling-house with habitable accommodation containing a floor level over 4.500 m above the ground.
   (c) Dwelling-house with habitable accommodation containing floor levels less than 4.500 m above ground.
   Note: Where a flat also functions as a workplace for its occupants and others, provisions for escape in event of a fire are given in Approved Document B1 (Vol. 2), *Means of escape from flats*, Section 2.52, Live/work units.
2. (a) **Residential (institutional)**   Hospital, care home, school or other similar establishment where persons sleep on the premises.
   (b) **Other**   Hotel, boarding house, hostel and any other residential purpose not described above.
3. **Office**   Any premises used for office and administration work.
4. **Shop and commercial**   Includes business premises used for any form of retail trade, such as a restaurant or hairdressers, or where members of the public can enter to deliver goods for repair or treatment.
5. **Assembly and recreation**   A public building or place of assembly of people for functions, which could be social, recreational, entertainment, exhibition, education, etc.
6. **Industrial**   Premises used for manufacture, production, processing, repairs, etc.

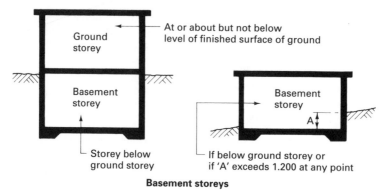

Basement storeys

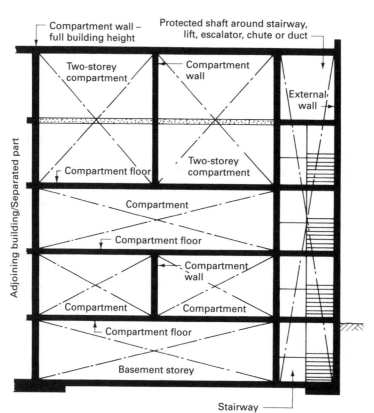

Figure 7.2.1   Approved Document B: general definitions

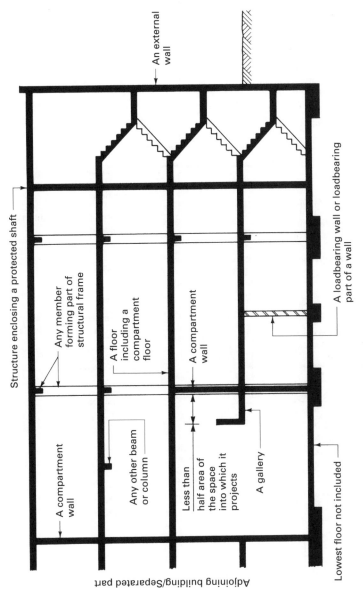

Figure 7.2.2 Approved Document B: elements of structure

Fire-stop – a seal which closes an imperfection of fit or design tolerance between elements or components, to prevent the passage of smoke or flame.

Cavity barrier – construction which closes a concealed space against smoke or flame penetration.

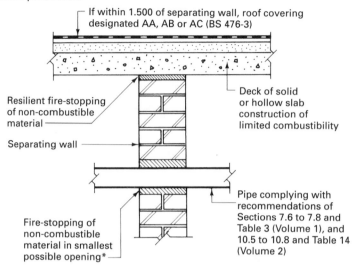

* *Alternatively*: 1. Proprietary sealing system or collar.
2. Sleeving for pipes of low melting point, e.g. uPVC, extending ∢1 m beyond separating element. Sleeve stopped as shown.

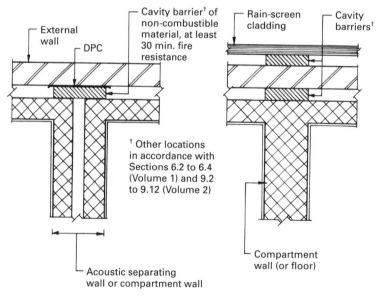

Figure 7.2.3 Approved Document B: fire-stopping and cavity barriers

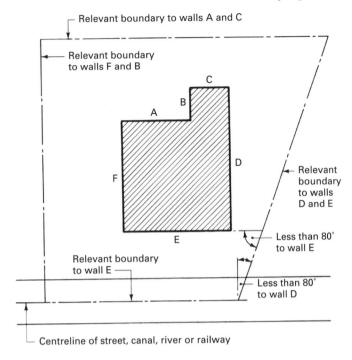

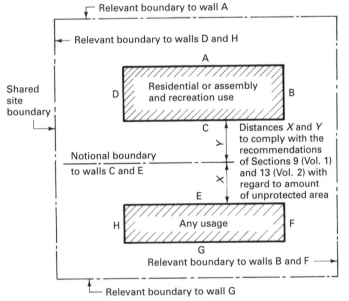

Figure 7.2.4   Approved Document B: boundaries and space separation

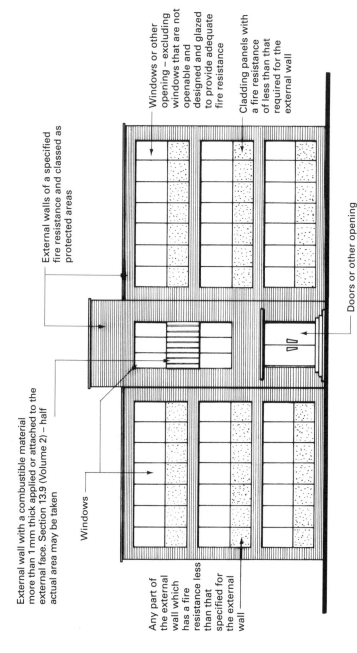

External wall with a combustible material more than 1 mm thick applied or attached to the external face. Section 13.9 (Volume 2) – half actual area may be taken

Windows

Any part of the external wall which has a fire resistance less than that specified for the external wall

External walls of a specified fire resistance and classed as protected areas

Windows or other opening – excluding windows that are not openable and designed and glazed to provide adequate fire resistance

Cladding panels with a fire resistance of less than that required for the external wall

Doors or other opening

Figure 7.2.5   Approved Document B: unprotected areas – fire resistance less than that given in Appendix A, Table A2

7. **Storage and other non-residential**   Subdivides into two categories:

   (a) Place for storage, deposit or placement of goods and materials (vehicles excluded), and any other non-residential purpose not described in 1–6, except for detached garages and carports not exceeding 40 m^2, which are included in the dwelling-house purpose group.

   (b) Car parks solely for the accommodation of cars, motor cycles, passenger and light goods vehicles of less than 2,500 kg gross weight.

The use of fire-resistant cells or compartments within a building (Fig. 7.2.1) is a means of confining an outbreak of fire to the site of origin for a reasonable time to allow the occupants a chance to escape and the firefighters time to tackle, control and extinguish the fire. The Approved Document B3, Sections 5 (Volume 1) and 8 (Volume 2), provides guidance on maximum recommended dimensions for buildings or compartments. These are expressed in terms relating to height of building, floor area and cubic capacity for the various purpose groups. Table A2 in Appendix A gives the recommended minimum periods of fire resistance for all elements of structure (see Fig. 7.2.2). These minimum periods of fire resistance are stated in minutes according to:

- purpose group;
- ground or upper storey (height of top floor above ground or of separated part of building);
- basement storey including floor over (depth of lowest basement).

The recommended minimum periods of fire resistance given in the tables are of prime concern to the designer and contractor. These tables do not state how these minimum periods are to be achieved, but reference is made to the current edition of the BRE Report BR 128: *Guidelines for the construction of fire-resisting structural elements*, which gives appropriate and common methods of construction for the various notional periods of fire resistance for walls, beams, columns and floors. The guidelines are written and presented in a tabulated format, which needs to be translated into working details. Figures 7.2.6 to 7.2.13 show typical examples taken from this document and other sources of reference such as manufacturers' data. Another useful resource is Loss Prevention Standards produced by the Loss Prevention Certification Board. These provide details of minimum standards to satisfy building insurers, which are often higher than building regulations. It is therefore most important for building designers and detailers to ensure that their specifications satisfy all authorities, as post-construction changes are very expensive!

The examples in Figures 7.2.9 to 7.2.11 for fire protection to structural sections assume a **section factor** greater than 140 (section factor = H_p/A (m^{-1})). Where this number is specified as less than 140, a reduced amount of fire insulation can apply and less fire resistance can be expected. However, this will be adequate for many situations. Calculation of the section factor is based on:

H_p = Perimeter of section exposed to fire (m) [The heated perimeter].
A = Cross-sectional area of the steel section (m^2) – data available in BS 4: *Structural steel sections* and other tables of standard steel sections.

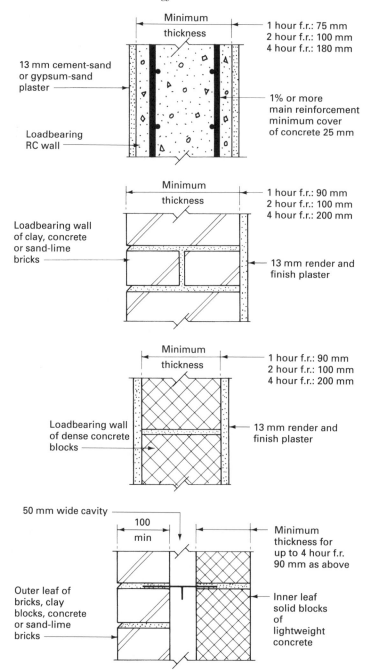

Figure 7.2.6   Fire resistance: walls of masonry construction

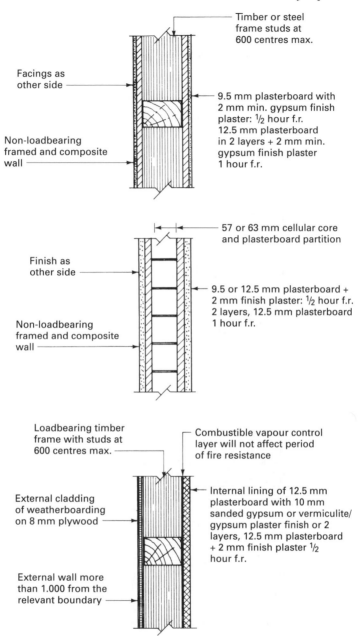

Timber or steel
frame studs at
600 centres max.

Facings as
other side

9.5 mm plasterboard with
2 mm min. gypsum finish
plaster: 1/2 hour f.r.
12.5 mm plasterboard
in 2 layers + 2 mm min.
gypsum finish plaster
1 hour f.r.

Non-loadbearing
framed and composite
wall

57 or 63 mm cellular core
and plasterboard partition

Finish as
other side

9.5 or 12.5 mm plasterboard +
2 mm finish plaster: 1/2 hour f.r.
2 layers, 12.5 mm plasterboard
1 hour f.r.

Non-loadbearing
framed and composite
wall

Loadbearing timber
frame with studs at
600 centres max.

Combustible vapour control
layer will not affect period
of fire resistance

External cladding
of weatherboarding
on 8 mm plywood

Internal lining of 12.5 mm
plasterboard with 10 mm
sanded gypsum or vermiculite/
gypsum plaster finish or 2
layers, 12.5 mm plasterboard
+ 2 mm finish plaster 1/2
hour f.r.

External wall more
than 1.000 from the
relevant boundary

Figure 7.2.7   Fire resistance: framed and composite walls

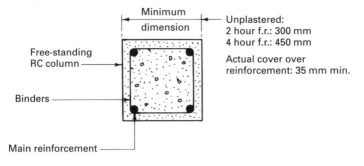

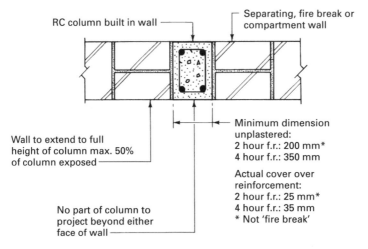

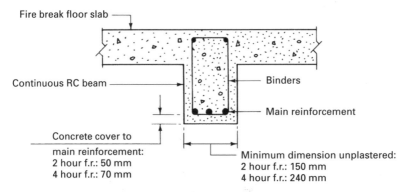

Notes: 1. A fire-break wall is a separating wall between compartments, having a f.r. of at least 240 mins.
2. A fire-break floor is constructed of non-combustible materials, having a f.r. of at least 120 mins.

Figure 7.2.8   Fire resistance: RC columns and beams

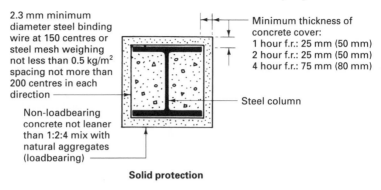

2.3 mm minimum diameter steel binding wire at 150 centres or steel mesh weighing not less than 0.5 kg/m² spacing not more than 200 centres in each direction

Minimum thickness of concrete cover:
1 hour f.r.: 25 mm (50 mm)
2 hour f.r.: 25 mm (50 mm)
4 hour f.r.: 75 mm (80 mm)

Steel column

Non-loadbearing concrete not leaner than 1:2:4 mix with natural aggregates (loadbearing)

Solid protection

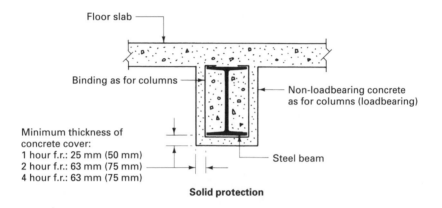

Floor slab

Binding as for columns

Non-loadbearing concrete as for columns (loadbearing)

Minimum thickness of concrete cover:
1 hour f.r.: 25 mm (50 mm)
2 hour f.r.: 63 mm (75 mm)
4 hour f.r.: 63 mm (75 mm)

Steel beam

Solid protection

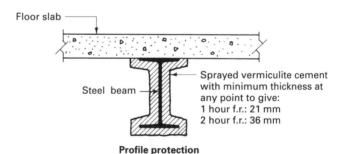

Floor slab

Steel beam

Sprayed vermiculite cement with minimum thickness at any point to give:
1 hour f.r.: 21 mm
2 hour f.r.: 36 mm

Profile protection

Figure 7.2.9 Fire resistance: steel columns and beams (section factor > 140)

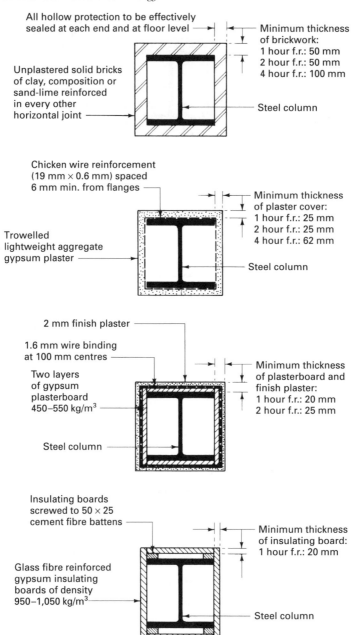

Figure 7.2.10   Fire resistance: hollow protection to steel columns (section factor > 140)

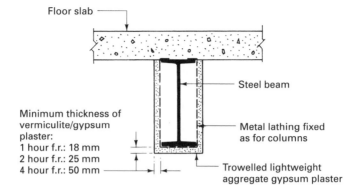

Floor slab

Steel beam

Minimum thickness of
vermiculite/gypsum
plaster:
1 hour f.r.: 18 mm
2 hour f.r.: 25 mm
4 hour f.r.: 50 mm

Metal lathing fixed
as for columns

Trowelled lightweight
aggregate gypsum plaster

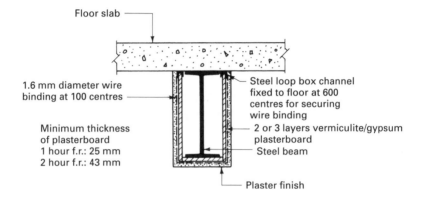

Floor slab

1.6 mm diameter wire
binding at 100 centres

Steel loop box channel
fixed to floor at 600
centres for securing
wire binding

Minimum thickness
of plasterboard
1 hour f.r.: 25 mm
2 hour f.r.: 43 mm

2 or 3 layers vermiculite/gypsum
plasterboard

Steel beam

Plaster finish

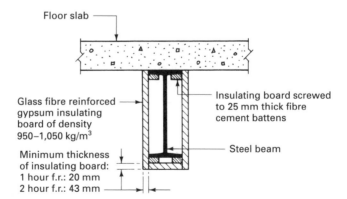

Floor slab

Glass fibre reinforced
gypsum insulating
board of density
950–1,050 kg/m³

Insulating board screwed
to 25 mm thick fibre
cement battens

Steel beam

Minimum thickness
of insulating board:
1 hour f.r.: 20 mm
2 hour f.r.: 43 mm

Figure 7.2.11   Fire resistance: hollow protection to steel beams (section factor > 140)

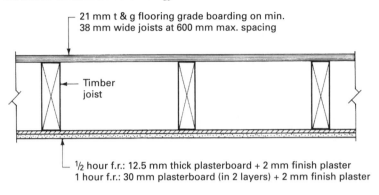

21 mm t & g flooring grade boarding on min.
38 mm wide joists at 600 mm max. spacing

Timber
joist

½ hour f.r.: 12.5 mm thick plasterboard + 2 mm finish plaster
1 hour f.r.: 30 mm plasterboard (in 2 layers) + 2 mm finish plaster

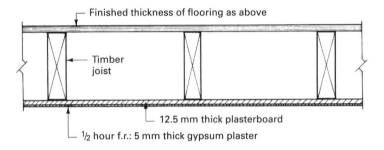

Finished thickness of flooring as above

Timber
joist

12.5 mm thick plasterboard

½ hour f.r.: 5 mm thick gypsum plaster

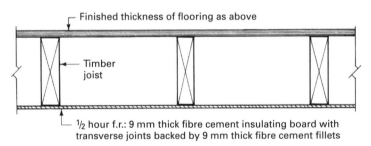

Finished thickness of flooring as above

Timber
joist

½ hour f.r.: 9 mm thick fibre cement insulating board with
transverse joints backed by 9 mm thick fibre cement fillets

Note: Plasterboard nailed at max. 150 mm spacing, 20 mm min. nail length
beyond board.

Figure 7.2.12   Fire resistance: timber floors

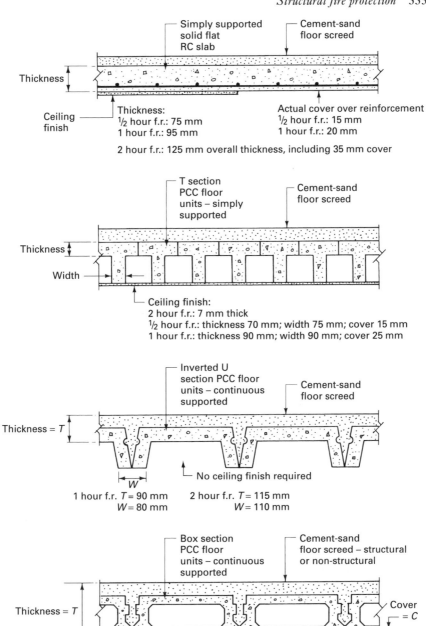

Figure 7.2.13   Fire resistance: concrete floors

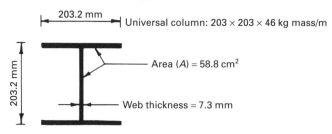

Universal column: 203 × 203 × 46 kg mass/m

203.2 mm

203.2 mm

Area (A) = 58.8 cm^2

Web thickness = 7.3 mm

Isolated column with profiled fire protection:

$H_p = (2 \times 203.2) + (2 \times 203.2) + 2(203.2 - 7.3)$
$H_p = 1,204.6$ mm   or   1.2046 m
$A = 58.8$ cm^2 or   0.00588 m^2
$H_p/A = 1.2046/0.00588 = 205$

Isolated column with solid or hollow fire protection:

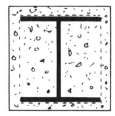

$H_p = (2 \times 203.2) + (2 \times 203.2) + 2 \ (203.2 - 7.3)$
$H_p = 1204.6$ mm   or   1.2046 m
$A = 58.8$ cm^2 or   0.00588 m^2
$H_p/A = 1.2046/0.00588 = 205$

Column exposed on three sides:

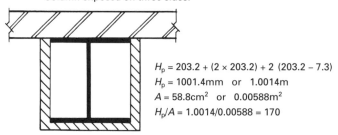

$H_p = 203.2 + (2 \times 203.2) + 2 \ (203.2 - 7.3)$
$H_p = 1001.4$ mm   or   1.0014 m
$A = 58.8$ cm^2 or   0.00588 m^2
$H_p/A = 1.0014/0.00588 = 170$

Figure 7.2.14   Typical section factor calculations

Note: The section factor is defined as the ratio of heated perimeter to section area. Heated perimeter is the inside perimeter of fire protection. For protection of box sections, take the perimeter of the smallest rectangle enclosing the section. Some sample section factor calculations are shown in Fig. 7.2.14.

Readers are also encouraged to study manufacturers' literature on the many patent ready-cut, easy-to-fix fire protection systems for standard structural members, and other methods such as the use of intumescent paints and materials, which expand to form a thick insulating coating or strip on being heated by fire.

The degree of space separation needed between buildings to limit the spread of fire by radiation is covered in Part B4: Sections 8 to 10 (Volume 1) and 12 to 14 (Volume 2) of Approved Document B. This deals with external walls and roofs and, in particular, unprotected areas permitted in relationship to the distance of the building or compartment within the building from the relevant boundary. It differentiates between buildings with external walls occurring within or over 1 metre from the relevant boundary. External wall surfaces must satisfy the fire-resistance requirements given in Tables A1 and A2 of Appendix A, or be constructed with a limited amount of combustible material (see Diagram 40, Approved Document B4, Section 12 (Vol. 2)).

The Approved Document B4 (External Fire Spread) provides four methods to determine the acceptable amount of unprotected external wall area. The first is concerned with small residential buildings not included in the Institutional group, over 1 metre from the relevant boundary. A simple table and diagram given in the Approved Document shows the relationship between boundary distance, maximum length of building side and the maximum total of unprotected area permitted (Sections 9.16, Diagram 22 (Volume 1) and 13.19, Diagram 46 (Volume 2)).

The second method can be used for other buildings or compartments that are not less than 1 metre from the relevant boundary and not over 10 metres high (excludes open-sided car parks). Data are interpolated from Approved Document B4, Sections 9, Table 4 (Volume 1) and 13, Table 15 (Volume 2), to obtain an acceptable percentage of unprotected area relative to the various purpose grouping. Alternatively, other methods described in the BRE Report BR 187: *External fire spread: building separation and boundary distances* may be used. These are known as the **enclosing rectangle method** and the **aggregate notional area method**. The enclosing rectangle method can be used to ascertain the maximum unprotected area for a given boundary position or to find the nearest position of the boundary for a given building design. The method consists in placing an enclosing rectangle around the unprotected areas, noting the width and height of the enclosing rectangle, and calculating the unprotected percentage in terms of the enclosing rectangle. This information will enable the distance from the boundary to be read direct from the tables given in the BRE report. Note that in compartmented buildings the enclosing rectangle is taken for each compartment and not for the whole facade. Typical examples are shown in Fig. 7.2.15.

The aggregate notional area method is an alternative to the previous method described above. The principle of separation is still followed, but by a more precise method, which will involve more investigation than reference to an enclosing rectangle. Reference is made to an aggregate notional area, which is calculated by taking the sum of each relevant unprotected area and multiplying by a factor given in the BRE report. Which unprotected areas are relevant is given in the appendix.

The method entails dividing the relevant boundary into a series of 3.000 m spacings called vertical data and projecting from each point a datum line to the nearest point on the building. A baseline is drawn through each vertical datum at right angles to the datum line and a series of semicircles of various radii drawn from this point represent the distance and hence the factors for calculating the aggregate notional area (see Fig. 7.2.16). For buildings or compartments with a residential, assembly or office use the result should not exceed 210 m^2; for other purpose groups the result should not exceed 90 m^2. In practical terms this method would not normally be used unless the situation was critical or the outline of the building was irregular in shape.

The specific recommendations for compartment walls, openings in compartment walls, compartment floors, protected shafts, doors, stairways and protecting structure are set out in Approved Document B. Each purpose group is considered separately, although many of the recommendations are similar for a number of purpose groups.

For roof coverings the Building Regulations are concerned mainly with the performance of roofs when exposed externally to fire, which questions the application and use of plastics and other combustibles, particularly in roof lights. Thermoplastic materials divide broadly into two categories: TP(a) (rigid) or TP(b). TP(a) considers rigid rooflight products having a class 1 rating to BS 476-7, or satisfying the test requirements of BS 2782-5, *Methods of testing plastics*. TP(b) applies to rigid polycarbonates less than 3 mm thick and other products satisfying tests to BS 2782-5. Diagram 23 and Tables 6 and7 in Section 10 (Volume 1) and Diagram 47 and Tables 17 and 18 in Section 14 (Volume 2) of Approved Document B4 indicate minimum acceptable spacings for rooflights, distances from boundaries and maximum areas applicable to these categories of thermoplastics.

Note: There is also a TP(a) flexible category, which is limited to products less than 1 mm thickness as used for curtains and drapes.

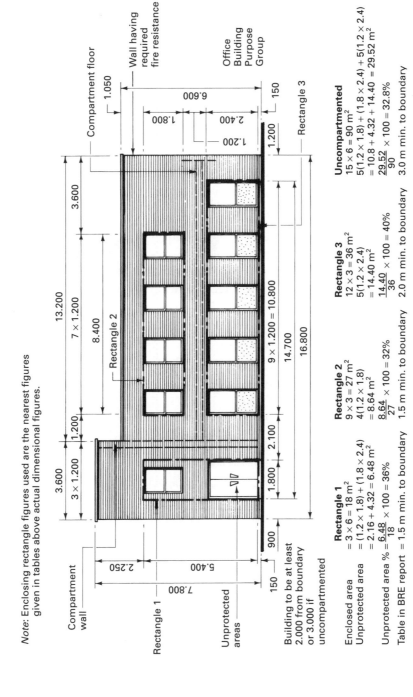

Figure 7.2.15   *Unprotected areas: enclosing rectangle method (reference: BRE Report BR 187)*

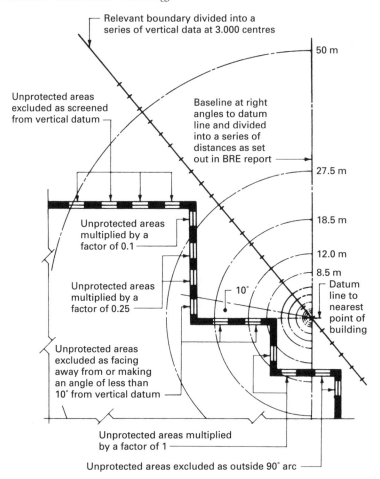

Relevant boundary divided into a series of vertical data at 3.000 centres

50 m

Unprotected areas excluded as screened from vertical datum

Baseline at right angles to datum line and divided into a series of distances as set out in BRE report

27.5 m

18.5 m

Unprotected areas multiplied by a factor of 0.1

12.0 m

8.5 m

Unprotected areas multiplied by a factor of 0.25

10°

Datum line to nearest point of building

Unprotected areas excluded as facing away from or making an angle of less than 10° from vertical datum

Unprotected areas multiplied by a factor of 1

Unprotected areas excluded as outside 90° arc

Notes: 1. Procedure repeated for each vertical datum position.
2. Calculations carried out for any side of a building or compartment.
3. For each position aggregate notional area of the unprotected areas must not exceed:
   210 m² for residential, assembly or office use.
   90 m² for all other purpose groups.

Figure 7.2.16   Unprotected areas: aggregate notional area method (*reference*: BRE Report BR 187)

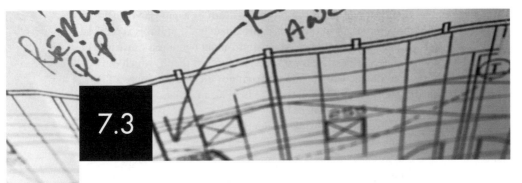

7.3

MEANS OF ESCAPE IN CASE OF FIRE

Means of escape from within a building is concerned with the safe discharge of personnel if an outbreak of fire should occur. It is designed to provide the occupants with the opportunity to reach an area of designated safety. This takes into account factors such as the risks to human life, unfamiliarity with building layout, problems of smoke, and the short space of time available to evacuate the premises before the problems become almost insurmountable.

Fear is a natural human response when confronted with uncontrolled fire and in particular fear of **smoke**, which is justified by the fact that more deaths are caused by smoke and heated gases than by burns. Statistics show that on average approximately 54% of deaths in fires are caused by smoke, 40% by burns and scalds and 6% by other causes. Smoke may be defined as visible suspension in atmosphere of solid and/or liquid particles resulting from combustion or pyrolysis. Although the above statement is true, smoke can also be caused by the release into the air of a variety of chemical compounds. The main dangers, contained in smoke, to human life are the carbon dioxide and carbon monoxide gases that are normal products of combustion.

The presence of these gases does not always cause the greatest hazard in human terms, because the density of smoke is more likely to create fear than the undetectable gases. Buoyant and mobile dense smoke will spread rapidly within a building or compartment during a fire, masking or even obliterating exit signs and directions. Gases, other than those mentioned previously, are generally irritants that can affect the eyes, causing watering which further impairs the vision, and can also affect the respiratory organs, causing reactions to slow and a loss of directional sense. It is worth remembering that smoke, being less dense than air, rises, and that taking up a position as near to the floor as possible will increase the chances of escape.

Carbon dioxide has no smell and is always present in the atmosphere, but because it is a product of combustion its volume increases at the expense of oxygen during a fire. The gas is not poisonous but can cause death by asphyxia. The normal amount of oxygen present in the air is approximately 21%; if this is reduced to 12% abnormal fatigue can be experienced; down to about 6% it can cause nausea, vomiting and loss of consciousness; below 6% respiration is difficult, which can result in death. Carbon dioxide will not support combustion, and can cause a fire to be extinguished if the content by volume exceeds 14%, a fact used by firefighters in their efforts to deal with an outbreak of fire.

Carbon monoxide, like carbon dioxide, is odourless, but it is very poisonous and, having approximately the same density as air, will spread rapidly. A very small concentration (0.2% by volume) of this colourless gas can cause death in about 40 minutes. The first effects are dizziness and headaches, followed in 5 to 10 minutes by loss of consciousness leading to death. As the concentration increases, so the time lapse from the initial dizziness to death decreases, so that by the time the concentration has reached about 1.3% by volume death can take place within a minute or two.

The **heat** that is associated with fire and smoke can also be injurious and even fatal. Temperatures in excess of 100 °C can cause damage to the windpipe and lungs, resulting in death within 30 minutes or sooner as the temperature rises. An interesting fact that emerges from statistics is that females generally have longer survival periods than males, and as would be expected the survival time decreases with age. Injuries caused by heat are generally in the form of burns, followed very often by shock, which can be fatal in many cases.

The above has been written not to frighten, but to emphasise the necessity for an adequate means of escape to give occupants and visitors in buildings a reasonable chance to reach an area of safety should a fire occur. To this end a maze of legislation and advisory documentation exists to guide the designer in planning escape routes without being too restrictive on the overall design concept. In the UK this has included regional by-laws, the Building Regulations, the Fire Precautions Act 1971, the Health and Safety at Work etc. Act 1974, the Fire Safety and Safety of Places of Sports Act 1987, the Management of Health and Safety at Work Regulations 1999, the Fire Precautions (Workplace) Regulations 1997, and references to BS 5588: *Fire precautions in the design, construction and use of buildings.* More recently, the Regulatory Reform (Fire Safety) Order 2005 (see page 363).

■■■ BUILDING REGULATIONS

Building Regulation B1 requires that, in case of fire, a means of escape leading from the building to a place of safety outside the building must be capable of being safely and effectively used at all times. The regulation covers all building types with the exception of prisons. Design sections that satisfy the requirements of Regulation B1 are divided as follows:

- dwelling-houses, Volume 1 only, the following sections are in Volume 2:
- flats;
- general provisions for the common parts of flats;
- design for horizontal escape – buildings other than dwellings;
- design for vertical escape – buildings other than dwellings;
- general provisions common to buildings other than dwelling-houses.

PLANNING ESCAPE ROUTES

When escape routes are being planned the occupancy must be considered. Occupants of flats will be familiar with the layout of the premises, whereas customers in a shop may be completely unfamiliar with their surroundings. In schools the fundamental principle is the provision of an alternative means of escape, and in hospitals the main concern is with the adequacy of the means of escape from all parts of the building.

In the context of means of escape in case of fire the building and its contents are of secondary importance. The provision of a safe escape route should, however, allow at the same time an easy access for the fire service using the same routes, and because these routes are protected the risk of fire spread is minimised. In practice the provision of an adequate means of escape and structural fire protection of the building and its contents are virtually inseparable. Each building has to be considered as an individual exercise, but certain common factors prevail in all cases:

- An outbreak of fire does not necessarily imply the evacuation of the entire building.
- Rescue facilities of the local fire service should not be considered as part of the planning of means of escape.
- Persons should be able to reach safety without assistance when using the protected escape routes.
- All possible sources of an outbreak and the course the fire is likely to take should be examined and the escape routes planned accordingly.

■■■■ FIRE PRECAUTIONS ACT 1971 (See page 363)

This Act is designed to make further provisions for the protection of persons from fire risks. The provisions of this Act do not apply to Northern Ireland, and Section 11, which deals with means of escape, does not apply to the old Greater London Council area, which has its own by-laws, or to Scotland, which has its own regulations. It has since been incorporated with and amended by the provisions of the Health and Safety at Work etc. Act 1974.

A fire certificate is required on any premises designated by the Minister of State at the Home Office within the following classes of use:

- sleeping accommodation;
- institutions providing treatment or care;
- entertainment, recreation or instruction;
- teaching, training or research;
- any purpose involving access to the premises by members of the public by payment or otherwise;
- a place of work.

Certain types of buildings and premises can be exempted from the provisions of this Act and different provision made relative to their designated use. These are:

- premises with existing fire certificates covered by the Offices, Shops and Railway Premises Act 1963, as incorporated in the Health and Safety at Work etc. Act 1974;
- premises with existing fire certificates covered by the Factories Act 1961, as incorporated in the Health and Safety at Work etc. Act 1974;
- a prison, remand centre, detention centre or borstal institution;
- churches, chapels and places of worship;
- houses occupied as a single dwelling;
- special hospitals (Mental Health Act 1983);
- premises occupied solely for purposes of the armed forces.

In dwellings where circumstances indicate a serious fire risk the fire authority may make it compulsory to have a fire certificate in particular where the premises have a room used as living accommodation that is:

- below the ground floor of a building;
- two or more floors above the ground;
- a room of which the floor is 6.000 m or more above the surface of the ground on any side of the building.

The Secretary of State for the Environment has power, by virtue of Sections 4 and 6 of the Public Health Act 1961, Section 11 of the Fire Precautions Act 1971 and the Building Act 1984, to make building regulations with regard to means of escape in case of fire.

Applications for a fire certificate are made to the chief executive of the fire authority on a form in the Fire Precautions (Application for Certificate) Regulations 1989. The relevant fire authority can be obtained from the local authority or council offices. It is generally the occupier's responsibility to make the application and submit any plans, specifications and details required. Penalties for offences under this Act incur substantial fines and/or two years' imprisonment.

The basic principles in the Act relating to means of escape in case of fire are also found in Approved Document B1 to the Building Regulations. These affect:

1. Limitation of travel distances.
2. Escape routes:
    - travel distance within rooms;
    - travel distance from rooms to a stairway or final exit;
    - travel within stairways and to final exit;
    - travel along corridors;
    - external means (stairs, routes other than stairs and flat roofs).
3. Provision of protected corridors, lobbies, shafts and stairways. These are a means for escape separated from the remainder of the building with fire-resisting construction.

The basic design principles relating to escape routes are shown in Figs 7.3.1, 7.3.2 and 7.3.3. The examples shown apply to hotels of Building Regulations, Purpose Group 2(b). Approved Document B1, Volume 2 Section 2 should be consulted for flats and Section 3 other applications.

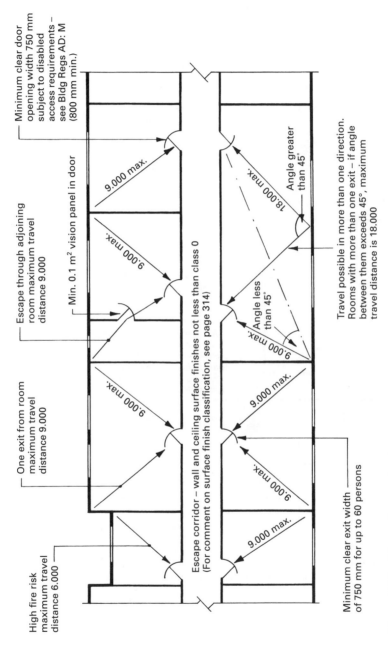

Figure 7.3.1 Means of escape: hotels

Minimum clear door opening width 750 mm subject to disabled access requirements – see Bldg Regs AD: M (800 mm min.)

Escape through adjoining room maximum travel distance 9.000

Min. 0.1 m² vision panel in door

One exit from room maximum travel distance 9.000

High fire risk maximum travel distance 6.000

Escape corridor – wall and ceiling surface finishes not less than class 0 (For comment on surface finish classification, see page 314)

Minimum clear exit width of 750 mm for up to 60 persons

Travel possible in more than one direction. Rooms with more than one exit – if angle between them exceeds 45°, maximum travel distance is 18.000

Angle greater than 45°

Angle less than 45°

9.000 max.

9.000 max.

9.000 max.

9.000 max.

9.000 max.

9.000 max.

18.000 max.

9.000 max.

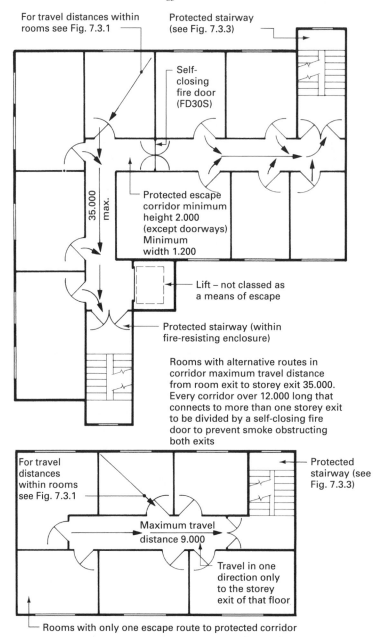

For travel distances within rooms see Fig. 7.3.1

Protected stairway (see Fig. 7.3.3)

Self-closing fire door (FD30S)

35.000 max.

Protected escape corridor minimum height 2.000 (except doorways) Minimum width 1.200

Lift – not classed as a means of escape

Protected stairway (within fire-resisting enclosure)

Rooms with alternative routes in corridor maximum travel distance from room exit to storey exit 35.000. Every corridor over 12.000 long that connects to more than one storey exit to be divided by a self-closing fire door to prevent smoke obstructing both exits

For travel distances within rooms see Fig. 7.3.1

Protected stairway (see Fig. 7.3.3)

Maximum travel distance 9.000

Travel in one direction only to the storey exit of that floor

Rooms with only one escape route to protected corridor

Figure 7.3.2   Means of escape: hotels

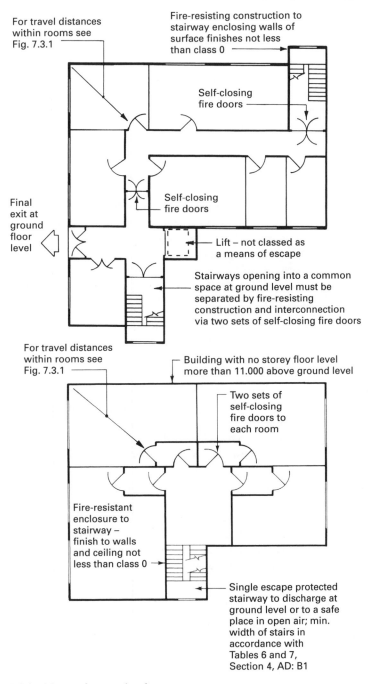

For travel distances within rooms see Fig. 7.3.1

Fire-resisting construction to stairway enclosing walls of surface finishes not less than class 0

Self-closing fire doors

Self-closing fire doors

Final exit at ground floor level

Lift – not classed as a means of escape

Stairways opening into a common space at ground level must be separated by fire-resisting construction and interconnection via two sets of self-closing fire doors

For travel distances within rooms see Fig. 7.3.1

Building with no storey floor level more than 11.000 above ground level

Two sets of self-closing fire doors to each room

Fire-resistant enclosure to stairway – finish to walls and ceiling not less than class 0

Single escape protected stairway to discharge at ground level or to a safe place in open air; min. width of stairs in accordance with Tables 6 and 7, Section 4, AD: B1

Figure 7.3.3   Means of escape: hotels

THE FIRE PRECAUTIONS (WORKPLACE) REGULATIONS 1997 (See page 363)

These regulations made significant changes to responsibilities for fire safety in buildings. The Workplace Regulations, coupled with revisions to Section 12 of the Fire Precautions Act 1971, moved responsibility from the local fire authority (who had previously commented on design proposals, conducted inspections and issued fire certificates) to the employer of people in both existing and new buildings. The role of the fire authority under these regulations is to advise and 'police' self-compliance and assessment of fire risk.

Employers need to ensure adequacy of firefighting measures such as automatic fire control and detection equipment, e.g. sprinklers, smoke alarms, etc., and ensure that non-automatic equipment such as hoses and portable extinguishers is accessible and appropriately indicated. Indication is endorsed by the Health and Safety (Safety Signs and Signals) Regulations 1996. Employees must be nominated with special responsibilities and trained in procedures in the event of fire.

Under Part II, Section 5, of the regulations there are specific requirements for emergency escape routes and exits. These must remain clear at all times and the following be complied with:

1. Emergency escape routes must lead directly to a nominated place of safety.
2. Clear access must be maintained for efficient evacuation of employees in an emergency.
3. Number, location and capacity of escape routes must be adequate, with regard to the use of the workplace and number of personnel employed within it (see Building Regulations AD, B1: *Means of warning and escape* and BS 5588: *Fire precautions in the design, construction and use of buildings*).
4. Emergency escape doors to open with the direction of escape.
5. Emergency escape doors must not have any facility for locking or fastening that would impede their use.
6. Sliding and revolving doors are not to be deployed in exits specifically designated as escape routes.
7. Designated emergency routes and exits must be suitably indicated by signs (see Health and Safety (Safety Signs and Signals) Regulations 1996).
8. Emergency routes and exits must be illuminated and provided with emergency lighting facilities if normal lighting fails.

DWELLING-HOUSES

Provision for escape from fire in dwellings can be categorised under three headings:

1. Dwelling-houses of one or two storeys, with floors not more than 4.500 m above ground level.
2. Dwelling-houses with one floor more than 4.500 m above ground level.
3. Dwelling-houses of two or more storeys with more than one floor over 4.500 m above ground level.

1. With the exception of kitchens, each habitable room should open directly onto a hallway or protected stairway, which in turn connects to the property entrance door or a window through which escape can be made. Such windows should have minimum opening dimensions of 0.450 m high × 0.450 m wide and an opening area of at least 0.330 m². The lowest level of opening in the window is 0.800 m (see *Note*) and the maximum, 1.100 m above floor level. With skylights and dormers the minimum opening height above floor level can reduce to 0.600 m (see *Note*).

Note: Building Regulations, Approved Document K: *Protection from falling, collision and impact.*

Inner rooms, i.e. rooms off a room, are discouraged unless they are a kitchen, laundry or utility room, dressing room or bathroom, or they have their own openable window that could be used as a means of escape.

Balconies and flat roofs may be considered a means of escape if the roof is part of the same building and the route leads to a storey exit or external escape route. The escape route should be constructed to at least a 30 minute fire-resisting standard, including any opening within 3.000 m of that escape route.

A basement stairway could be blocked by fire or smoke. Therefore, if a basement contains habitable accommodation, an alternative means for egress should be provided. This may be:

- an external door; or
- a window of dimensions specified above; or
- a protected stairway to a final exit.

Loft conversions from a two-storey house currently attract certain provisions as defined in Section 2 of Approved Document B to the Building Regulations. These are generally as described in the next category for houses with one floor more than 4.500 m above ground level. New access stairs should be in an enclosure continuing from the existing stairs or accessing the existing stairs. Depending on the age and type of construction of the house, the new floor joists will probably need structurally upgrading from their former function as ceiling ties, and the floor construction will need to achieve a 30 minute fire resistance. Escape windows from within the roof space are as previously described.

2. Dwelling-houses with a floor more than 4.500 m above ground level should have at least two internal stairways. Each of these stairways is to provide an effective means of escape. If it is impractical to install two stairways in a dwelling, all upper storeys should be served by a protected stairway enclosed by 30 minute fire-resisting construction. This stairway should extend directly to a final exit. Alternatively, the stairway can give access to at least two escape routes to final exits at ground level, each route separated from the other by 30 minute fire-resisting construction and fire doors.

A different approach is to separate the upper storey from the lower storeys with 30 minute fire-resisting construction, and provide the upper storey with a designated fire escape route to its own final exit.

Note: Fire-resisting construction includes the need for fire doors. Traditionally these have self-closing devices, but current thinking indicates that, where used in dwellings, these can pose a hazard to children. Also, many occupants remove door closers as they regard them as an inconvenience. Therefore, guidance promotes the benefits of manually closing these doors, particularly at night.

3. Dwelling-houses of two or more storeys with more than one floor over 4.500 m above ground level have the same requirements as the previous category. The following additional provisions apply:

- Each storey over 7.500 m above ground level should have an alternative escape route with access via the protected stairway to an upper storey, or from a landing within the protected stairway to an alternative escape route on the same storey. At or about 7.500 m above ground level the protected stairway should be separated from the lower storeys by fire-resisting construction.
- Alternatively, the whole house should be fitted with a domestic sprinkler system. See BS 9251: *Sprinkler systems for residential and domestic occupancies. Code of practice.*

COMPARTMENTATION AND SEPARATION

Every wall separating semi-detached and terraced houses should be constructed as a fire-resisting compartment wall. The fire resistance period is usually 60 minutes: see Table A2 in the Approved Document to Part B of the Building Regulations where some variation may be seen to apply depending on the building height.

An attached or integral garage should be separated, where in contact with the rest of the house, by 30 minute fire-resisting construction. Any opening for access from the house into the garage should be at least 100 mm above the garage floor level (petrol leakage) and be fitted with a self-closing FD30S fire door.

▪▪▪▪ FLATS

At one time only accommodation over 24.000 m above ground level would have been considered, this being the height over which external rescue by the fire service was impracticable. It has become apparent that even with dwellings within reach of firefighters' ladders external rescue is not always possible. This is because present-day traffic conditions and congestion may prevent the appliance from approaching close to the building or may delay the arrival of the fire service. Section 2 of Approved Document B1, Volume 2 to the Building Regulations specifically relates to this class of buildings and has special provision for accommodation with floors over 4.500 m above ground level.

As with other forms of buildings the only sound basis for planning means of escape from apartments is to identify the positions of all possible sources of any outbreak of fire and to predict the likely course the fire, smoke and gases would follow. The planning should be considered in three stages:

1. Risk to occupants of the dwelling in which the fire originates.
2. Risk to occupants of adjoining dwellings should the fire or smoke penetrate the horizontal escape route or common corridor.
3. Risk to occupants above the level of the outbreak, particularly on the floor immediately above the source of the fire.

■ **Stage 1**   Most of the serious accidents and deaths occur in the room in which the outbreak originates. Fires in bedrooms have increased in the last decade, mainly because of the increased use of electrical appliances such as heated blankets that have been improperly maintained or wired. Social habits, such as watching television, have enabled fires starting in other rooms to develop to a greater extent before detection. Fires occurring in the circulation spaces such as halls and corridors are a serious hazard, and the use of naked flame heaters and smoking in these areas should be discouraged. Risks to occupants in maisonettes are higher than those incurred in flats because fire and smoke will spread more rapidly in the vertical direction than in the horizontal direction.

■ **Stage 2**   This is concerned mainly with the safety of occupants using the horizontal escape route, where the major aim is to ensure that, should a fire start in any one dwelling, it will not adversely affect or obstruct the escape of occupants of any other dwelling on the same or adjoining floor.

■ **Stage 3**   This is concerned with occupants using a vertical escape route, which in fact means a stairway, for in this context lifts are not considered as a means of escape because of:
  ■ time delay in lift answering call;
  ■ limited capacity of lift;
  ■ possible failure of the electricity supply in the event of a fire.

The main objective is to remove the risk of fire or smoke entering a stairway and rendering it impassable above that point. This objective can be achieved by taking the following protective measures:

  ■ Where there is more than one stairway serving a ventilated common corridor there should be a smoke-stop door across the corridor between the doors in the enclosing walls of the stairways to ensure that both stairs are not put at risk in the event of a fire (see Fig. 7.3.4).
  ■ Where there is only one stairway serving a ventilated common corridor there must be a smoke-stop door between the door in the enclosing walls of the stairway and any door covering a potential source of fire, and the lobby so formed should be permanently ventilated through an adjoining external wall (see Fig. 7.3.4).
  ■ Means for ventilating common corridors are necessary because some smoke will inevitably penetrate the corridor as occupants escape through the entrance door to an affected apartment. Ventilation may be by natural or mechanical means.

The basic means of escape requirements for flats can be listed as follows:

■ Every flat should have a protected entrance hall of 30 minute fire resistance.
■ Every living room should have an exit into the protected entrance hall.

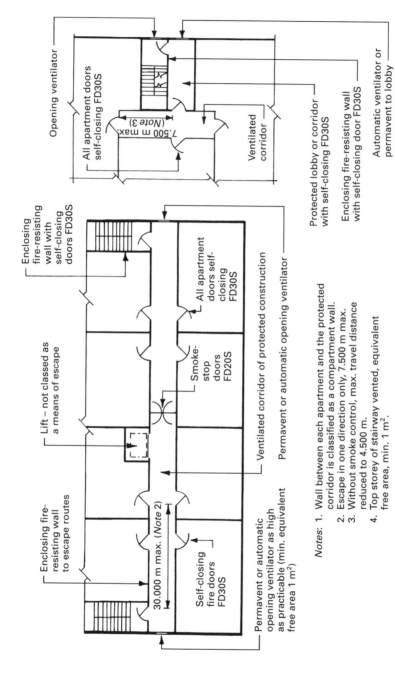

Figure 7.3.4 Main stairway protection: flats

Opening ventilator

All apartment doors self-closing FD30S

7.500 m max. (Note 3)

Ventilated corridor

Enclosing fire-resisting wall with self-closing doors FD30S

Lift – not classed as a means of escape

All apartment doors self-closing FD30S

Smoke-stop doors FD20S

Ventilated corridor of protected construction

Permavent or automatic opening ventilator

Enclosing fire-resisting wall to escape routes

30.000 m max. (Note 2)

Self-closing fire doors FD30S

Permavent or automatic opening ventilator as high as practicable (min. equivalent free area 1 m²)

Protected lobby or corridor with self-closing FD30S

Enclosing fire-resisting wall with self-closing door FD30S

Automatic ventilator or permavent to lobby

Notes: 1. Wall between each apartment and the protected corridor is classified as a compartment wall.
2. Escape in one direction only, 7.500 m max.
3. Without smoke control, max. travel distance reduced to 4.500 m.
4. Top storey of stairway vented, equivalent free area, min. 1 m².

- Bedrooms should be nearer to the entrance door than the living rooms or kitchen, unless an alternative exit is available from the bedroom areas.
- Doors opening onto a protected entrance hall to be self-closing FD30 fire doors (see section on fire doors in this chapter and in Chapter 7.1 of *Construction Technology*).
- Maximum travel distance from any habitable room exit to the entrance door to be 9.000 m; if exceeded an alternative route is to be provided.
- Locate cooking facilities remote from entrance doors.

A typical flat example is shown in Fig. 7.3.5.

The basic means of escape requirements for maisonettes can be similarly listed:

- All maisonettes to have a protected entrance hall and stairway.
- All habitable rooms to have direct access to this hall or stairway.
- Doors opening on to hall or stairway to be self-closing (FD30).
- Hall and stairway to have 30 minute fire-resistant walls.
- Alternatively, means of escape from each habitable room must be provided except that at entrance level, e.g. exit through windows (dimensions as described for dwelling-houses).

A typical maisonette example is shown in Fig. 7.3.6.

Readers are encouraged to study the many examples of means of escape planning for flats given in Section 2 of Approved Document B1, Volume 2 to the Building Regulations and BS 5588-1: *Fire precautions in the design, construction and use of buildings. Code of practice for residential buildings.*

▧▦▩ OFFICE BUILDINGS

BS 5588-11: *Fire precautions in the design, construction and use of buildings. Code of practice for shops, offices, industrial, storage and other similar buildings* covers office buildings of all sizes and heights in the context of means of escape in case of fire, and complements Building Regulation B1.

The planning procedures for means of escape set out in the code are based on attempting to identify the positions of all possible sources of outbreak of fire and to predict the courses that might follow such an outbreak and in particular the passage of the fire and the passage of the smoke and gases ensuing from the fire. The basic planning objectives are to provide protected escape routes in the horizontal and vertical directions that will enable persons within the building confronted by an outbreak of fire to turn away and make a safe escape without outside assistance. To achieve this planning objective there must be sufficient exits to allow the occupants to reach an area of safety without delay. It is not always necessary to plan for the complete evacuation of the building because compartmentation of the building will restrict the fire initially to an area within that compartment.

Two important factors to be considered in planning escape routes are width and travel distance. The **width** is based upon an evacuation time of 2.5 minutes through a storey exit on the assumption that a unit exit width of 500 mm will allow the flow of 40 persons per minute. See Tables 2 and 3 in the code and Table 4 in Section 3

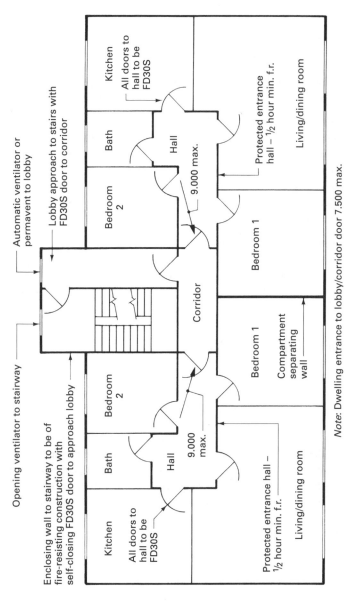

Figure 7.3.5   Means of escape: example of flats with only one stairway

Note: Dwelling entrance to lobby/corridor door 7.500 max.

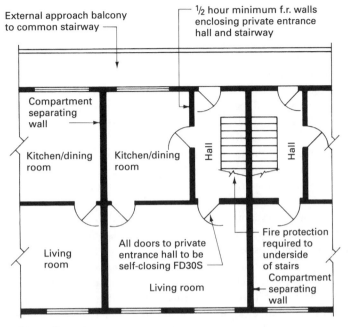

External approach balcony to common stairway

½ hour minimum f.r. walls enclosing private entrance hall and stairway

Compartment separating wall

Kitchen/dining room

Kitchen/dining room

Hall

Hall

Living room

All doors to private entrance hall to be self-closing FD30S

Living room

Fire protection required to underside of stairs

Compartment separating wall

Lower floor plan

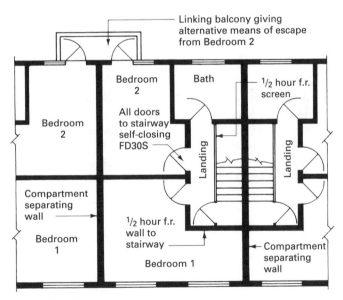

Linking balcony giving alternative means of escape from Bedroom 2

Bedroom 2

Bath

½ hour f.r. screen

Bedroom 2

All doors to stairway self-closing FD30S

Landing

Landing

Compartment separating wall

Bedroom 1

½ hour f.r. wall to stairway

Bedroom 1

Compartment separating wall

Upper floor plan

Figure 7.3.6   Means of escape: maisonette example

of Approved Document B1, Volume 2 of the Building Regulations for guidance on assessing the likely population density and clear widths of escape routes related to the maximum number of persons.

Travel distances are considered under two headings: direct distance and travel distance. A direct distance is the shortest distance from any point within the floor area to the nearest storey exit measured so as to ignore walls, partitions and fittings. Generally the maximum direct distance is 12.000 m for escape in one direction or 30.000 m for escape in more than one direction. The travel distance is defined as the actual distance travelled from any point within the floor area to the nearest storey exit having regard to the layout of walls, partitions and fittings. Generally the maximum travel distances are 18.000 m for escape in one direction and 45.000 m for travel in more than one direction. Note that both maximum distances must not be exceeded, and that if alternative exits subtend an angle of less than 45° with one another and are not separated by fire-resisting construction they are classified as exits in one direction only. See also Sections 3.32 to 3.37 in the Approved Document for small premises.

The horizontal escape routes should be clear gangways with non-slippery even surfaces, and if ramps are included these should have an easy gradient of a slope of not more than 1 in 12. (See also Approved Document M: *Access to and use of buildings.*) The minimum clear headroom should be 2.000 m with no projections from walls except for normal handrails. The minimum fire resistance should be not less than 30 minutes unless a higher resistance is required by the Building Regulations. Limitations on glazing for both escape corridors and doors are given in the text and accompanying tables within Approved Document B and the code of practice.

The code recommends that there should not be less than two protected stairways available from any storey unless the premises are small or a basement storey in a localised fire-risk situation, with direct distance no more than that given in Table 1 of the code. Additional protected stairways should be provided as necessary to meet the requirements for travel distance. Where more than one stairway is provided it is usual to assume that one stairway would be blocked by the fire and therefore the remaining stairways must have sufficient width to cater for the total persons to be evacuated. A typical example of means of escape from an office building is shown in Fig. 7.3.7.

▨▧▨■ SHOPS AND SIMILAR PREMISES

BS 5588-11 covers all classes of shops and other similar types of premises such as cafés, restaurants, public houses and premises where goods are received for treatment, examples being dry cleaners and shoe repairers. The main objectives of the code's requirements are to provide safety from fire by means of planning and providing protection of both horizontal and vertical escape routes for any area threatened by fire, thus enabling any person confronted by an outbreak of fire to make an unassisted escape. Under the Fire Precautions Act 1971 as amended by the Health and Safety at Work etc. Act 1974 an adequate means of escape is required for all shops, and for particular shops a fire certificate will be required.

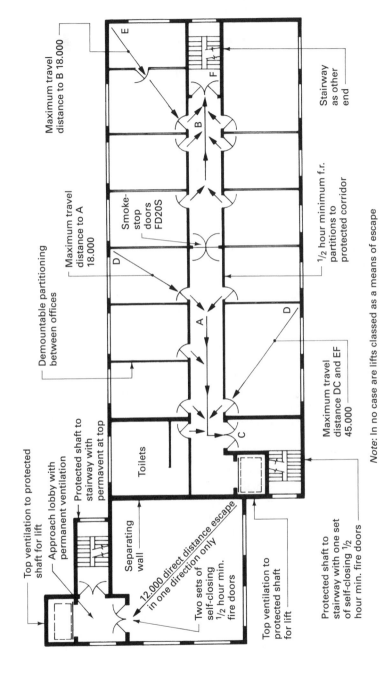

Figure 7.3.7 Means of escape: office buildings – upper floor level < 5.000 m above ground

Top ventilation to protected shaft for lift

Approach lobby with permanent ventilation

Protected shaft to stairway with permavent at top

Toilets

Separating wall

12.000 direct distance escape in one direction only

Two sets of self-closing ½ hour min. fire doors

Top ventilation to protected shaft for lift

Protected shaft to stairway with one set of self-closing ½ hour min. fire doors

Maximum travel distance DC and EF 45.000

Demountable partitioning between offices

Maximum travel distance to A 18.000

Smoke-stop doors FD20S

Maximum travel distance to B 18.000

Stairway as other end

½ hour minimum f.r. partitions to protected corridor

Note: In no case are lifts classed as a means of escape

In planning escape routes there must be sufficient exit facilities, such as protected routes and stairways, to allow the public and staff to reach areas of safety without undue delay. It is not always necessary to plan for complete evacuation immediately an outbreak of fire occurs. Compartmentation of multi-storey shops will limit the rate of fire spread, and in such cases a reasonable assumption would be to design for the immediate evacuation of the floor on which the outbreak occurs plus the floor above, which is the next portion of the building at risk. Escape widths are based upon an evacuation time of 2.5 minutes and on the assumption that a unit exit width of 500 mm permits the passage of 40 persons per minute. Guidance as to possible population densities is given in the code by means of a table that shows the floor space per person for types of rooms or storeys. Similarly, another table gives clear widths for exits related to the total number of persons in a room or storey.

The maximum permitted travel distance is another important factor, and can be taken as a maximum distance measured around obstructions such as fixed counters or a direct distance measured over the obstructions. The distances given in a table take into account the number of exits, the relationship of alternative exits and whether the floor under consideration is a ground floor or an upper floor. Generally the maximum travel distances in one direction are 18.000 m and 45.000 m where exits subtend an angle of more than 45°. The maximum direct travel distances are 12.000 m and 30.000 m for similar conditions.

The vertical escape route via the stairs leading to the final exit leading direct to the open air must also cater for the anticipated number of persons likely to use the stairs for evacuation purposes. A table giving guidance on minimum stair widths related to population density and height of building is given in the code. Alternatively, Tables 6, 7 and 8 in Approved Document B1, Volume 2 may be used. It is recommended that approach lobbies are used for all stairs in buildings over 18.000 m high because of the higher risks to occupants in high buildings (see Fig. 7.3.8). In calculating the number of stairways required it should be assumed that one stairway would be inaccessible in the event of a fire if two or more stairways are actually required.

Small shops with floor areas of not more than 280 m² in one occupancy and of not more than two storeys plus a basement are treated separately. The maximum travel or direct distance is governed by the floor's position within the building. Basement and upper floor distances are 18.000 m and 12.000 m for single exits, with ground-floor distances being 27.000 m and 18.000 m respectively (see Fig. 7.3.8). Where more than one exit exists the distances are 45.000 m and 30.000 m respectively.

STAIRS

These should be of non-combustible materials and continuous, leading ultimately to the final exit door to the place of safety. The recommended dimensions of going, rise, handrails and maximum number of risers per flight are shown in Fig. 7.3.9. Fire-resistant glazing is permitted but should be restricted to the portion of the wall above the handrails and preferably designed in accordance with the recommendations of BS 476-22.

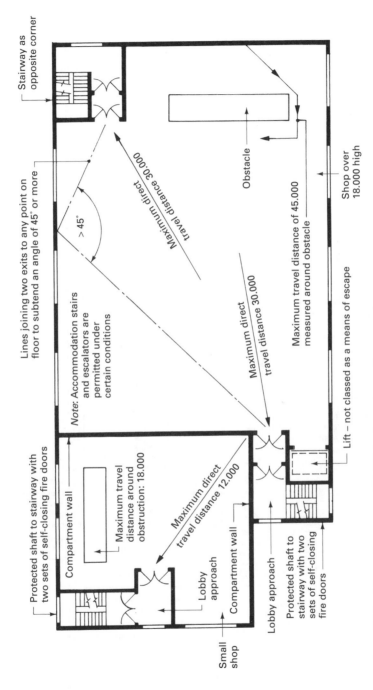

Figure 7.3.8   Means of escape: small and large shops – upper floor example

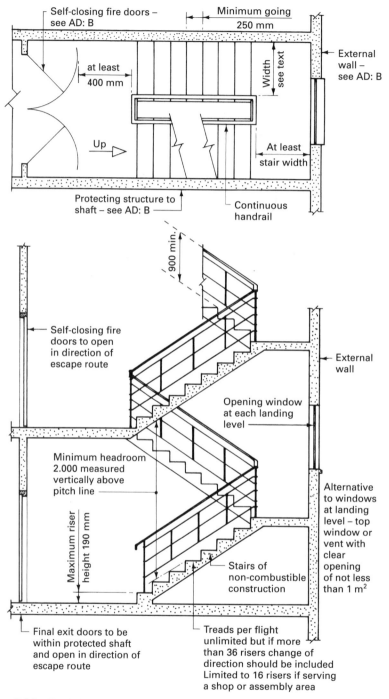

Figure 7.3.9　Typical escape stairway details

Minimum widths of escape stairs depend on the number provided, the type of building served, the anticipated number of persons likely to use it, and the number of storeys served. Tables 4 and 5 in BS 5588-11 can be used for design guidance and Table 6 from Section 4 of Approved Document B1, Volume 2 to the Building Regulations. Also, a formula method can be applied:

$$P = 200w + 50(w - 0.3)(n - 1)$$

where P = no. of people to be accommodated (except ground floor)
w = width of stair
n = number of storeys

For example, in a 15-storey building with 1,000 people, $P = 1,000$ and $n = 15$. Therefore, using the formula:

$$
\begin{aligned}
1,000 &= 200w + 50(w - 0.3)(15 - 1) \\
&= 200w + (50w - 15)(14) \\
&= 200w + 700w - 210 \\
&= 900w - 210 \\
\therefore \quad w &= 1,210/900 = 1.345 \text{ m}
\end{aligned}
$$

■■■ DOORS

Factors relating to the integrity, insulation and stability of doors in the presence of fire are considered in some detail in Chapter 7.1 of the accompanying volume, *Construction Technology*, and in Chapter 7.2 of this volume. Specific reference is also made to BS 476-22: *Methods for determination of fire resistance of non-loadbearing elements of construction*. See also Fig. 7.3.10 for a typical example of a fire door with variations to achieve FD30 and FD60 ratings.

SMOKE-STOP DOORS

The function of this form of door is obvious from its title, and no special requirements are recommended as to door thickness or area of glazing when using 6 mm wired glass. The most vulnerable point for the passage of smoke is around the edges, and therefore a maximum gap of 3 mm is usually specified, together with a brush draught-excluder seal.

AUTOMATIC FIRE DOORS

In compartmented industrial buildings it is not always convenient to keep the fire doors in the closed position as recommended. A closed door in such a situation may impede the flow and circulation of people and materials, thus slowing down or interrupting the production process. To seal the openings in the compartment walls in the event of a fire automatic fire doors or shutters can be used. These can be held in the open position in normal circumstances by counterbalance weights or electromagnetic devices, and should a fire occur they will close automatically, usually by gravitational forces. The controlling device can be a simple fusible link,

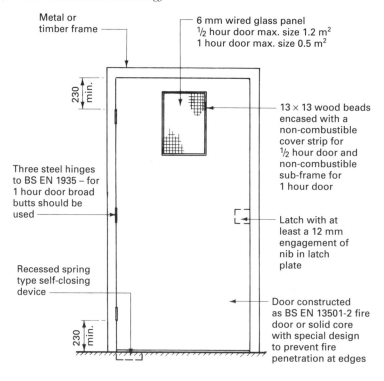

Metal or timber frame

6 mm wired glass panel
$\frac{1}{2}$ hour door max. size 1.2 m^2
1 hour door max. size 0.5 m^2

230 min.

13 × 13 wood beads encased with a non-combustible cover strip for $\frac{1}{2}$ hour door and non-combustible sub-frame for 1 hour door

Three steel hinges to BS EN 1935 – for 1 hour door broad butts should be used

Latch with at least a 12 mm engagement of nib in latch plate

Recessed spring type self-closing device

230 min.

Door constructed as BS EN 13501-2 fire door or solid core with special design to prevent fire penetration at edges

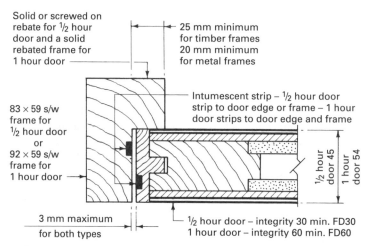

Solid or screwed on rebate for $\frac{1}{2}$ hour door and a solid rebated frame for 1 hour door

25 mm minimum for timber frames
20 mm minimum for metal frames

Intumescent strip – $\frac{1}{2}$ hour door strip to door edge or frame – 1 hour door strips to door edge and frame

83 × 59 s/w frame for $\frac{1}{2}$ hour door or 92 × 59 s/w frame for 1 hour door

$\frac{1}{2}$ hour door 45
1 hour door 54

3 mm maximum for both types

$\frac{1}{2}$ hour door – integrity 30 min. FD30
1 hour door – integrity 60 min. FD60

References:
BS EN 1935: *Building hardware. Single axis hinges.*
BS EN 13501-2: *Fire classification of construction products and building elements.*

Figure 7.3.10   Fire doors

a smoke or heat detector, which could be linked to the fire alarm system. A typical detail of an automatic sliding fire door is shown in Fig. 7.3.11.

■ ■ ■ PRESSURISED STAIRWAYS

The traditional methods of keeping escape stairways clear of smoke during a fire by having access lobbies and/or natural ventilations are not always acceptable in modern designs where, for example, the stairways are situated in the core of the building or where the building has a full air conditioning and ventilation design. Pressurisation is a method devised to prevent smoke logging in an unfenestrated stairway by a system of continuous pressurisation, which will keep the stairway clear of smoke if the doors remain closed. Only a small fan installation is required to maintain a nominal pressure of not less than 50 Pa. In addition to controlling smoke, pressurisation also increases the fire resistance of doors opening into the protected area. It is essential with these installations that the doors are well sealed to prevent pressure loss.

■ ■ ■ EXTERNAL ESCAPE STAIRS

Having now completed a brief study of fire and the way it affects the design of buildings choice of materials and circulation patterns it must be obvious to the reader that this is an important topic that, if ignored or treated lightly, can have disastrous consequences not only for a structure but also for human life.

■ ■ ■ THE REGULATORY REFORM (Fire Safety) ORDER 2005

This introduced changes to fire safety legislation in England and Wales. In Scotland these have been introduced through the Fire (Scotland) Act and in Northern Ireland new legislation is being processed. The order supersedes many existing fire safety procedures and repeals many established statutes, not least the Fire Precautions Act and the Fire Precautions (Workplaces) Regulations. Whilst the established legislation provides a sound principle, emphasis is now on a duty of fire safety care and fire risk assessment by an appointed *Responsible Person* and others in control of a building. The traditional procedure of obtaining fire certificates no longer applies.

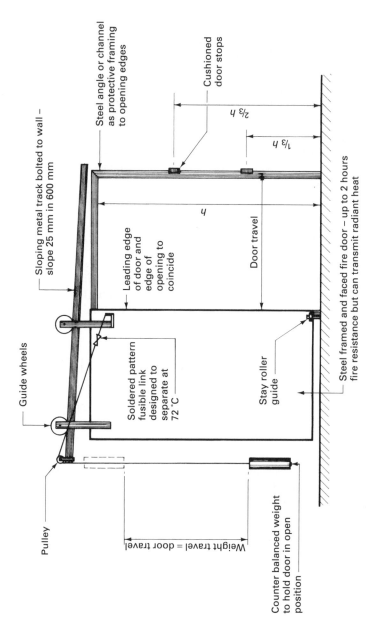

Figure 7.3.11   Automatic sliding fire door details

Note: Protective roof and sides omitted for clarity

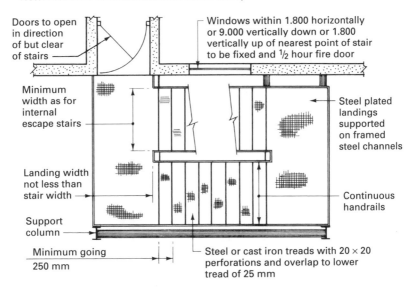

Doors to open in direction of but clear of stairs

Windows within 1.800 horizontally or 9.000 vertically down or 1.800 vertically up of nearest point of stair to be fixed and ½ hour fire door

Minimum width as for internal escape stairs

Steel plated landings supported on framed steel channels

Landing width not less than stair width

Continuous handrails

Support column

Minimum going 250 mm

Steel or cast iron treads with 20 × 20 perforations and overlap to lower tread of 25 mm

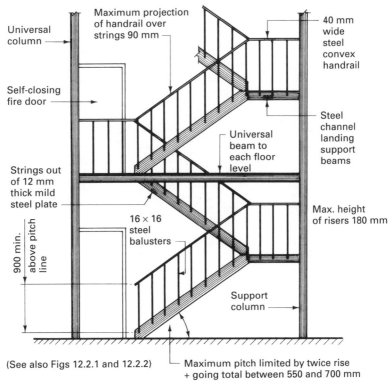

Maximum projection of handrail over strings 90 mm

Universal column

40 mm wide steel convex handrail

Self-closing fire door

Universal beam to each floor level

Steel channel landing support beams

Strings out of 12 mm thick mild steel plate

16 × 16 steel balusters

Max. height of risers 180 mm

900 min. above pitch line

Support column

(See also Figs 12.2.1 and 12.2.2)

Maximum pitch limited by twice rise + going total between 550 and 700 mm

Figure 7.3.12 Typical external steel escape stairway details

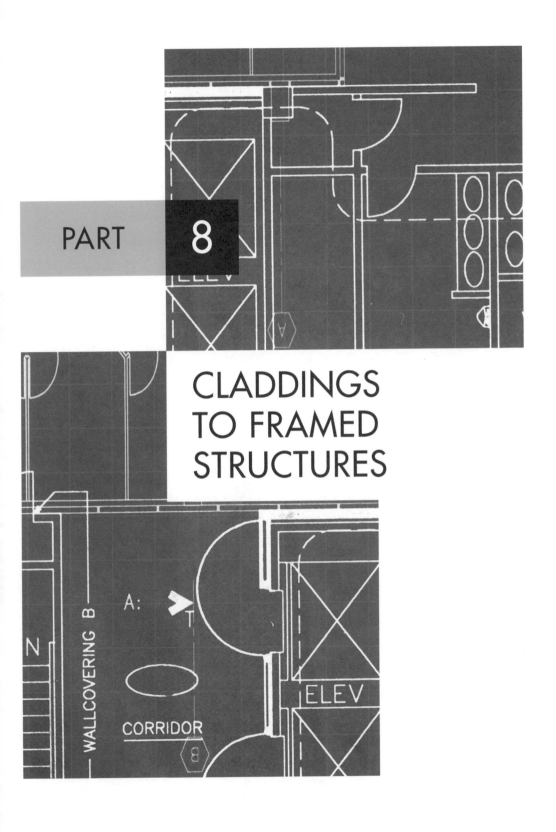

PART 8

CLADDINGS
TO FRAMED
STRUCTURES

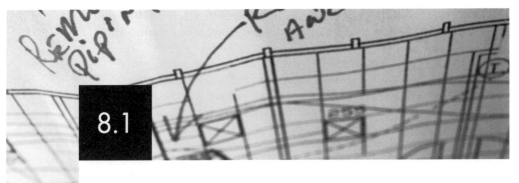

CLADDING PANELS

Claddings are a form of masking or infilling a structural frame, and can be considered under the following headings:

- Panel walls with or without attached facings (see Chapter 10.6 of *Construction Technology*).
- Concrete and similar cladding panels.
- Light infill panels.
- Curtain walling, which can be defined as a sheath cladding that encloses the entire structure, such as fully glazed systems.

All forms of cladding should fulfil the following functions:

- self-support between the framing members;
- resistance to rain penetration;
- resistance to both positive and negative wind pressures;
- resistance to wind penetration;
- thermal insulation;
- sound insulation;
- fire resistance;
- admittance of natural daylight and ventilation;
- constructed to a suitable size for handling and placing;
- durability – limited maintenance;
- aesthetically acceptable for the situation.

■■■ CONCRETE CLADDING PANELS

These are usually made of precast concrete with a textured face in a storey height or undersill panel format. The storey height panel is designed to span vertically from beam to beam and if constructed to a narrow module will give the illusion of a tall building. Undersill panels span horizontally from column to column and are used where a high wall/window ratio is required. Combinations of both formats are also possible.

Concrete cladding panels should be constructed of a dense concrete mix and suitably reinforced with bar reinforcement or steel-welded fabric. The reinforcement should provide the necessary tensile resistance to the stresses induced in the final position and to the stresses set up during transportation and hoisting into position. Lifting lugs, positions or holes should be incorporated into the design to ensure that the panels are hoisted in the correct manner so that unwanted stresses are not induced. The usual specification for cover of concrete over reinforcement is 25 mm minimum. If thin panels are being used the use of galvanised or stainless steel reinforcement should be considered to reduce the risk of corrosion.

When designing or selecting a panel the following must be taken into account:

- column or beam spacing;
- lifting capacities of plant available;
- jointing method;
- exposure conditions;
- any special planning requirements as to finish or texture.

The greatest problem facing the designer and installer of concrete panels is one of jointing to allow for structural and thermal movements and at the same time provide an adequate long-term joint (see Chapters 8.3 and 8.4). Typical examples of storey height and undersill panels are shown in Figs 8.1.1 and 8.1.2.

Where a stone facing is required to a framed structure, possibly to comply with planning requirements, it may be advantageous to use a composite panel. These panels have the strength and reliability of precast concrete panel design and manufacture but the appearance of traditional stonework. This is achieved by casting a concrete backing to a suitably keyed natural or reconstructed stone facing and fixing it to the frame by traditional masonry fixing cramps or by conventional fixings (see Fig. 8.1.3).

Thermal insulation can be achieved when using precast concrete panels by creating a cavity, as shown in Figs 8.1.1 and 8.1.2. However, continuity of insulation will require an insulative material lining with supplementary blockwork extending over the face of the edge beams to coincide with mineral wool (or similar) insulation slabs to the roof/floor slab soffit. These slabs can be faced with plasterboard containing an integral vapour control layer. It may also be possible to incorporate insulation in a sandwich cladding panel, as shown in Fig. 8.3.4.

Concrete cladding panels can be large and consequently heavy. To reduce the weight they are often designed to be relatively thin (50 to 75 mm) across the centre portion and stiffened around the edges with suitably reinforced ribs, which usually occur on the back face but can be positioned on the front face as a feature, which can also limit the amount of water which can enter the joint.

An alternative lightweight cladding material is glass-fibre-reinforced plastic (GRP). This consists of glass fibre reinforcement impregnated with resin, incorporating fillers, pigments and a suitable catalyst as a hardener. The resultant panels are lightweight, durable, non-corrosive, have good weather resistance, can be

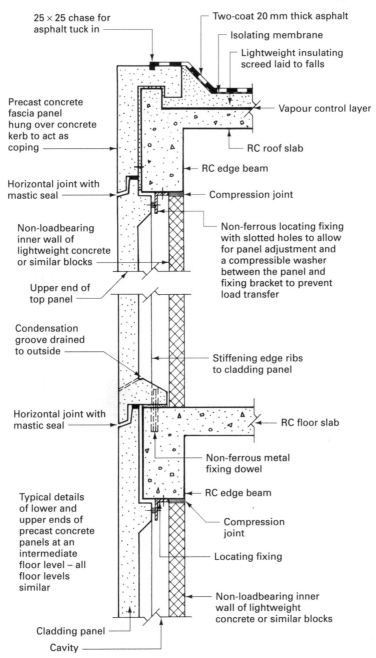

25 × 25 chase for asphalt tuck in

Two-coat 20 mm thick asphalt

Isolating membrane

Lightweight insulating screed laid to falls

Precast concrete fascia panel hung over concrete kerb to act as coping

Vapour control layer

RC roof slab

RC edge beam

Horizontal joint with mastic seal

Compression joint

Non-loadbearing inner wall of lightweight concrete or similar blocks

Non-ferrous locating fixing with slotted holes to allow for panel adjustment and a compressible washer between the panel and fixing bracket to prevent load transfer

Upper end of top panel

Condensation groove drained to outside

Stiffening edge ribs to cladding panel

Horizontal joint with mastic seal

RC floor slab

Non-ferrous metal fixing dowel

RC edge beam

Typical details of lower and upper ends of precast concrete panels at an intermediate floor level – all floor levels similar

Compression joint

Locating fixing

Non-loadbearing inner wall of lightweight concrete or similar blocks

Cladding panel

Cavity

Note: For buildings where energy conservation and thermal bridging prevention measures are required (UK – see Bldg Reg. L), a supplementary thickness of insulation and inner walling can extend over the edge beam. Additional insulation is applied to the soffit of the RC roof/floor slabs. Cavity insulation may also be considered.

Figure 8.1.1 Typical storey height concrete cladding panel

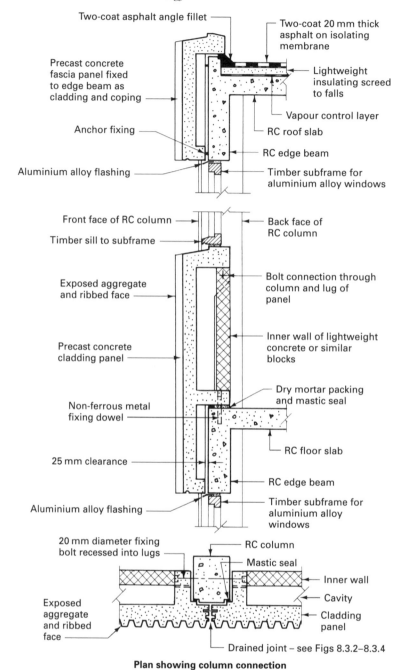

Two-coat asphalt angle fillet

Two-coat 20 mm thick asphalt on isolating membrane

Precast concrete fascia panel fixed to edge beam as cladding and coping

Lightweight insulating screed to falls

Anchor fixing

Vapour control layer

RC roof slab

RC edge beam

Aluminium alloy flashing

Timber subframe for aluminium alloy windows

Front face of RC column

Back face of RC column

Timber sill to subframe

Exposed aggregate and ribbed face

Bolt connection through column and lug of panel

Precast concrete cladding panel

Inner wall of lightweight concrete or similar blocks

Non-ferrous metal fixing dowel

Dry mortar packing and mastic seal

25 mm clearance

RC floor slab

RC edge beam

Aluminium alloy flashing

Timber subframe for aluminium alloy windows

20 mm diameter fixing bolt recessed into lugs

RC column

Mastic seal

Inner wall

Cavity

Exposed aggregate and ribbed face

Cladding panel

Drained joint – see Figs 8.3.2–8.3.4

Plan showing column connection

Figure 8.1.2   Typical undersill concrete cladding panel (see note to Fig. 8.1.1)

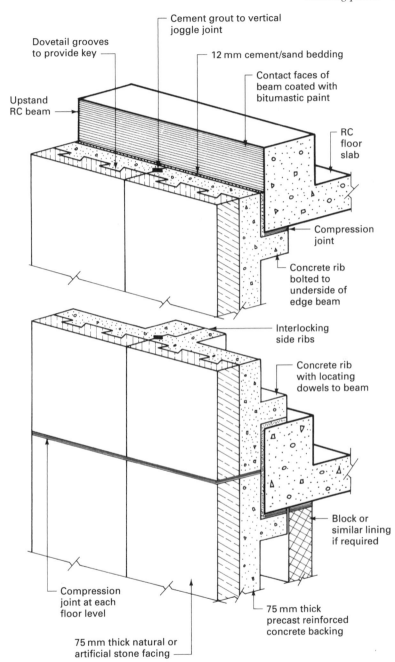

Figure 8.1.3 Typical storey height composite panel

moulded to almost any profile, and have good aesthetic properties. As they age, their colour can have a tendency to fade.

Further relevant reading includes the Building Research Establishment's (BRE) *Cladding Pack (2005)*. This is a collection taken from the Establishment's BRE Digests, Information Papers and Good Building Guides. Pack contents:

- Insulated external cladding systems (GG 31).
- Metal cladding: assessing the thermal performance of built-up systems which use 'Z' spacers (IP 10/02).
- Reinforced plastics cladding panels (DG 161).
- Wall cladding defects and their diagnosis (DG 217).
- Wall cladding: Designing to minimise defects due to inaccuracies and movements (DG 223).

The BRE also publishes *A cladding bibliography of selected publications*: BRE Report 194.

8.2

INFILL PANELS

The functions of an infill panel are as listed previously for cladding panels in general. Infill panels are lightweight and usually glazed to give good internal natural daylighting conditions. The panel layout can be so arranged to expose some or all of the structural members, creating various optical impressions. For example, if horizontal panels are used, leaving only the beams exposed, an illusion of extra length and/or reduced height can be created (see Fig. 8.2.1). These practices are typical of many existing structures and continue to be appropriate where thermal insulation and thermal bridging are not a design priority, e.g. car parks. Current energy conservation requirements provide an opportunity to over-clad exposed structural members with a superficial cladding when refurbishing and updating these buildings: see Chapter 8.6 on rainscreen cladding.

A wide variety of materials or combinations of materials can be employed, such as timber, steel, aluminium or plastic. Double-glazing techniques can be used to achieve the desired sound or thermal insulation. The glazing module should be such that a reasonable thickness of glass can be specified.

The design of the 'solid' panel is of great importance, because this panel must provide the necessary resistance to fire, heat loss, sound penetration and interstitial condensation. Most of these panels are of composite or sandwich construction, as shown in Figs 8.2.2 and 8.2.3. The extent and type of glazing, plus the thickness of insulating quilt in the lower panels, will depend on local climatic conditions and associated legislation (see note to Fig. 8.2.2).

The jointing problem with infill panels occurs mainly at its junction with the structural frame, and allowance for moisture or thermal movement is usually achieved by using a suitable mastic or sealant (see Chapter 8.4). Masonry infill panels of contemporary cavity construction can be used to preserve traditional features and to attain high standards of sound and thermal insulation and fire protection. Masonry walls are tied to the structural frame with wall ties cast into

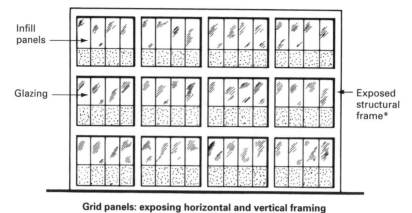

Infill panels

Glazing

Exposed structural frame*

Grid panels: exposing horizontal and vertical framing

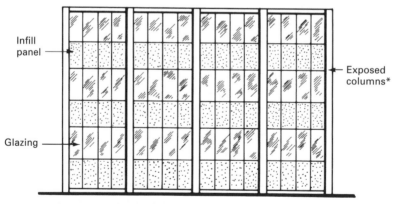

Infill panel

Glazing

Exposed beams*

Horizontal panels: beams exposed to create illusion of length

Infill panel

Glazing

Exposed columns*

Vertical panels: columns exposed to create illusion of height

*See introductory text on previous page.

Figure 8.2.1   Typical infill panel arrangements

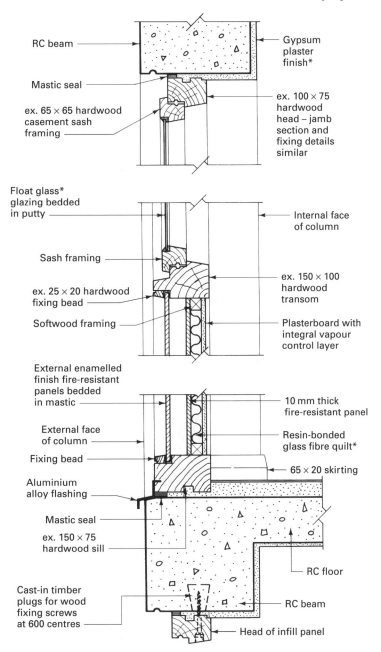

RC beam

Mastic seal

ex. 65 × 65 hardwood casement sash framing

Gypsum plaster finish*

ex. 100 × 75 hardwood head – jamb section and fixing details similar

Float glass* glazing bedded in putty

Sash framing

ex. 25 × 20 hardwood fixing bead

Softwood framing

Internal face of column

ex. 150 × 100 hardwood transom

Plasterboard with integral vapour control layer

External enamelled finish fire-resistant panels bedded in mastic

External face of column

Fixing bead

Aluminium alloy flashing

Mastic seal

ex. 150 × 75 hardwood sill

Cast-in timber plugs for wood fixing screws at 600 centres

10 mm thick fire-resistant panel

Resin-bonded glass fibre quilt*

65 × 20 skirting

RC floor

RC beam

Head of infill panel

*Note: Construction shown can be upgraded to improve thermal insulation standards by applying thermal laminate gypsum plasterboard to interior of exposed RC beams and underside of the RC floor, installing double glazing (see *Construction Technology*) and an insulated screed, and increasing the thickness of the lower panel to contain more glass fibre insulant.

Figure 8.2.2 Typical timber infill panel details

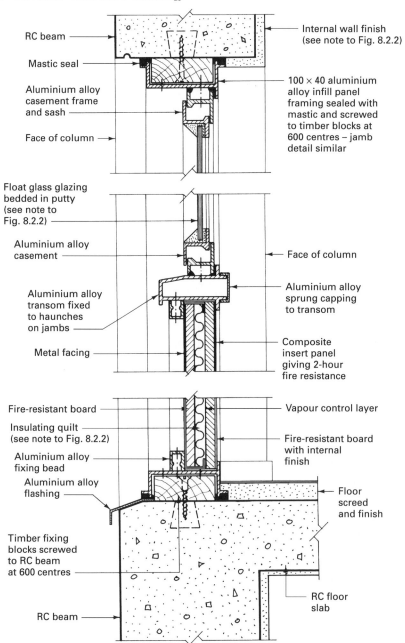

RC beam

Mastic seal

Aluminium alloy
casement frame
and sash

Face of column

Internal wall finish
(see note to Fig. 8.2.2)

100 × 40 aluminium
alloy infill panel
framing sealed with
mastic and screwed
to timber blocks at
600 centres – jamb
detail similar

Float glass glazing
bedded in putty
(see note to
Fig. 8.2.2)

Aluminium alloy
casement

Aluminium alloy
transom fixed
to haunches
on jambs

Metal facing

Face of column

Aluminium alloy
sprung capping
to transom

Composite
insert panel
giving 2-hour
fire resistance

Fire-resistant board

Insulating quilt
(see note to Fig. 8.2.2)

Aluminium alloy
fixing bead

Aluminium alloy
flashing

Timber fixing
blocks screwed
to RC beam
at 600 centres

RC beam

Vapour control layer

Fire-resistant board
with internal
finish

Floor
screed
and finish

RC floor
slab

Figure 8.2.3   Typical metal infill panel details

the concrete columns or with a purpose-made cladding support system.
Figures 8.2.4 and 8.2.5 show the principles of application.

Most infill panels are supplied as a manufacturer's modular system, because
purpose-made panels can be uneconomic, but whichever method is chosen
the design aims remain constant: that is, to provide a panel that fulfils all the
required functions and has a low long-term maintenance factor. Note that many
of the essentially curtain-walling systems are adaptable as infill panels, which
gives the designer a wide range of systems from which to select the most suitable.

One of the maintenance problems encountered with infill panels, and probably
to a lesser extent with the concrete claddings, is the cleaning of the facade and
in particular the glazing. All buildings collect dirt, the effects of which can vary
with the material: concrete and masonry tend to accept dirt and weather naturally,
whereas impervious materials such as metals and glass do not accept dirt and can
corrode or become less efficient.

If glass is allowed to become coated with dirt its visual appearance is less
acceptable, its optical performance lessens because clarity of vision is reduced,
and the useful penetration of natural daylight diminishes. The number of times
that cleaning will be necessary depends largely upon the area, ranging from three-
monthly intervals in non-industrial areas to six-weekly intervals in areas with a high
pollution factor.

Access for cleaning glazed areas can be external or internal. Windows at ground
level present no access problems and present only the question of choice of method
such as hand cloths or telescopic poles with squeegee heads. Low- and medium-rise
structures can be reached by ladders or a mobile scaffold tower and usually present
very few problems. High-rise structures need careful consideration. External access
to windows is gained by using a cradle suspended from roof level; this can be in the
form of a temporary system consisting of counterweighted cantilevered beams from
which the cradle is suspended. Permanent systems, which are incorporated as part
of the building design, are more efficient and consist of a track on which a mobile
trolley is mounted and from which davit arms can be projected beyond the roof
edge to support the cradle. A single track fixed in front of the roof edge could also
be considered; these are simple and reasonably efficient but the rail is always visible
and can therefore mar the building's appearance.

Internal access for cleaning the external glass face can be achieved by using
windows such as reversible sashes, horizontal and vertical sliding sashes, but the
designer is restricted in his choice to the reach possible by the average person.
It cannot be overemphasised that such windows can be a very dangerous hazard
unless carefully designed so that all parts of the glazed area can be reached by the
person cleaning the windows while standing firmly on the floor.

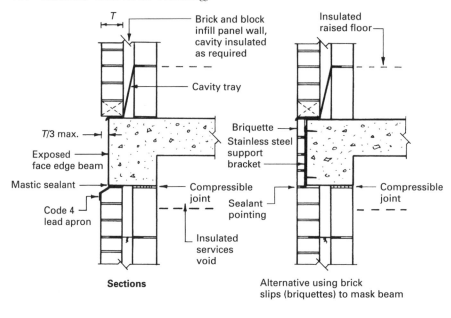

Sections

Alternative using brick
slips (briquettes) to mask beam

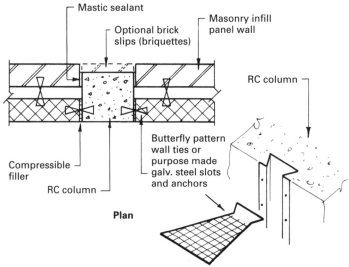

Plan

Figure 8.2.4   Non-loadbearing masonry infill panel

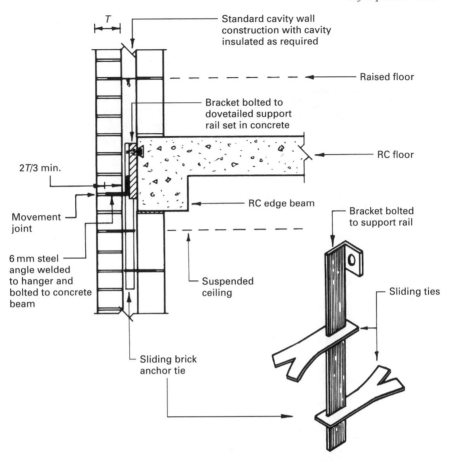

Figure 8.2.5 Non-loadbearing masonry cladding panel

8.3

JOINTING

When incorporating precast concrete cladding panels in a framed structure the problem of making the joints waterproof is of paramount importance. Joints should be designed so that they fulfil the following requirements:

- exclude wind, rain and snow;
- allow for structural, thermal and moisture movement;
- good durability;
- easily maintained;
- maintain the thermal and sound insulation properties of the surrounding cladding;
- easily made or assembled.

Experience has shown that bad design, poor workmanship or lack of understanding of the function of a joint has led to water penetration through the joints between cladding panels. To overcome this problem it is essential that both the designer and the site operative fully appreciate the design principles and the need for accurate installation. Suitable joints can be classified under two headings:

- Filled joints.
- Drained joints.

Filled joints are generally satisfactory if the cladding panel module is small because, if incorporated in large module panels, filled joints can crack and allow water to penetrate. This failure is due either to the filling materials being incapable of accommodating movement or to a breakdown of adhesion between the filling material and the panel. Research has shown that if the above failures are possible the most effective alternative is the drained joint.

▨▪◾ FILLED JOINTS

These joints are not easy to construct and rely mainly upon mortars, sealants, mastics or preformed gaskets to provide the barrier against the infiltration of wind and rain. They are limited in their performance by the amount to which the sealing material(s) can accommodate movement and to a certain extent by their weathering properties, such as their resistance to ultraviolet rays. The disadvantages of filled joints can be listed as:

- difficulty in making and placing the joints accurately, particularly with combinations of materials;
- providing for structural, thermal and moisture movement;
- suitable only for small module claddings.

For typical detail see Fig. 8.3.1.

▨▪◾ DRAINED JOINTS

These joints have been designed and developed to overcome the disadvantages of the filled joint by:

- designing the joint to have a drainage zone;
- providing an air-tight seal at the rear of the joint.

Drained joints have two components that must be considered: the vertical joint and the horizontal joint.

VERTICAL JOINTS

These consist basically of a deep narrow gap between adjacent panels where the rear of the joint is adequately sealed to prevent the passage of air and moisture. The width of the joint does not significantly affect the amount of water reaching the rear seal, for the following reasons:

- Most of the water entering the joint (approximately 80%) will do so by following over the face of the panel; the remainder (approximately 20%) will enter the joint directly, and most of this water entering the joint will drain down within the first 50 mm of the joint depth. Usually the deciding factor for determining the joint width is the type of mastic or sealant being used and its ability to accommodate movement.
- There are checks on the amount of water entering the drainage zone such as ribs to joint edges, exposed aggregate external surfaces and the use of baffles.

Baffles are loose strips of material such as neopreen, butyl rubber or plasticised PVC, which are unaffected by direct sunlight and act as a first line of defence to water penetration. The baffles are inserted, after the panels have been positioned and fixed, either by pulling them through prepared grooves or by direct insertion into the locating grooves from the face or back of the panel according to the joint design. Care must be taken when inserting baffles by the pulling method because

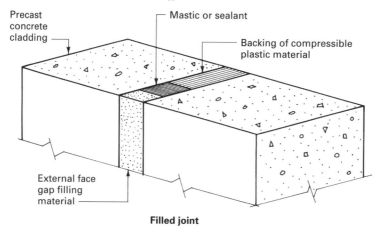

Filled joint

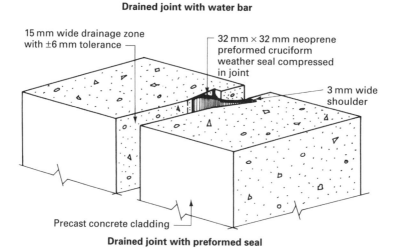

Drained joint with water bar

Drained joint with preformed seal

Figure 8.3.1 Typical filled and drained joints

they invariably stretch during insertion and they must be allowed to return to their original length before trimming off the surplus to ensure adequate cover at the intersection of the vertical and horizontal joint.

Adequate sealing at the back of the joint is of utmost importance, because some water will usually penetrate past the open drainage zone or the baffle, and any air movement through the joint seal will also assist the passage of water or moisture. Drained joints that have only a back seal or a baffle and seal can have a cold bridge effect on the internal face giving rise to local condensation: therefore consideration must be given to maintaining the continuity of the thermal insulation value of the cladding (see typical details in Figs 8.3.1 and 8.3.2).

HORIZONTAL JOINTS

These are usually in the form of a rebated lap joint, the upper panel being lapped over the top edge of the lower panel. As with the vertical joints, the provision of an adequate back seal to prevent air movement through the joint is of paramount importance. The seal must also perform the function of a compression joint: therefore the sealing strip is of a compressible material such as bituminised foamed polyurethane or a preformed cellular rubber strip.

The profile of the joint is such that any water entering the gap by flowing from the panel face or by being blown in by the wind is encouraged to drain back onto the face of the lower panel. The depth of joint overlap is usually determined by the degree of exposure, and ranges from 50 mm for normal exposure to 100 mm for severe exposure (see Fig. 8.3.3 for typical details). Note that the effective overlap of a horizontal joint is measured from the bottom edge of the baffle in the vertical joint to the seal, and not from the rebated edge of the lower panel (see Fig. 8.3.4).

■■■ INTERSECTION OF JOINTS

This is an important feature of drained joint design and detail, because it is necessary to shed any water draining down the vertical joint onto the face of the lower panels where the vertical and horizontal joints intersect, as the joints are designed to cater only for the entry of water from any one panel connection at a time. The usual method is to use a flashing starting at the back of the panel, dressed over and stuck to the upper edge of the lower panel, as shown in Fig. 8.3.4. The choice of material for the flashing must be carefully considered, because it must accept the load of the upper panel and any movements made while the panel is positioned and secured. Also, it should be a material that is durable but will not give rise to staining of the panel surface. Experience has shown that suitable materials are bitumen–coated woven glasscloth and synthetic rubber sheet.

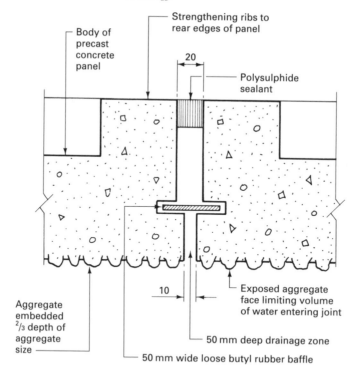

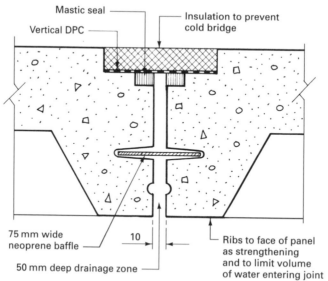

Figure 8.3.2   Typical drained joints using baffles

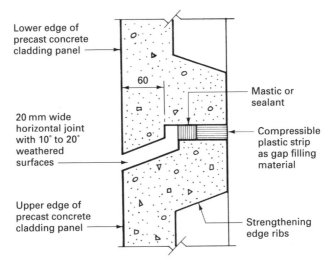

Lower edge of
precast concrete
cladding panel

60

Mastic or
sealant

20 mm wide
horizontal joint
with 10° to 20°
weathered
surfaces

Compressible
plastic strip
as gap filling
material

Upper edge of
precast concrete
cladding panel

Strengthening
edge ribs

Joint for low to moderate exposure

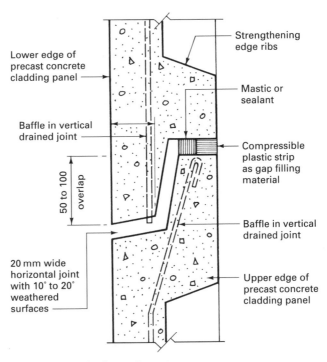

Strengthening
edge ribs

Lower edge of
precast concrete
cladding panel

Mastic or
sealant

Baffle in vertical
drained joint

50 to 100 overlap

Compressible
plastic strip
as gap filling
material

Baffle in vertical
drained joint

20 mm wide
horizontal joint
with 10° to 20°
weathered
surfaces

Upper edge of
precast concrete
cladding panel

Joint for moderate to severe exposure

Figure 8.3.3 Typical horizontal joints

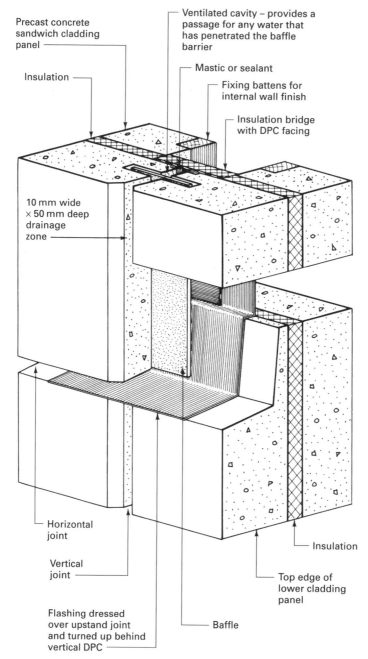

Precast concrete sandwich cladding panel

Insulation

Ventilated cavity – provides a passage for any water that has penetrated the baffle barrier

Mastic or sealant

Fixing battens for internal wall finish

Insulation bridge with DPC facing

10 mm wide × 50 mm deep drainage zone

Horizontal joint

Vertical joint

Flashing dressed over upstand joint and turned up behind vertical DPC

Baffle

Insulation

Top edge of lower cladding panel

Figure 8.3.4 Typical drained joint intersection detail

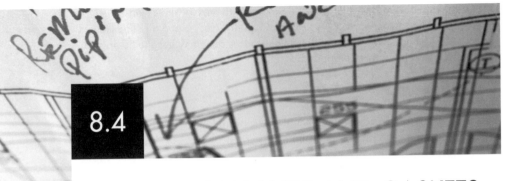

8.4

MASTICS, SEALANTS AND GASKETS

Materials that are to be used for sealing joints, whether in the context of claddings or for sealing the gap between a simple frame and the reveals, have to fulfil the following requirements:

■ provide a weathertight seal;
■ accommodate movement due to thermal expansion, wind loadings, structural movement and/or moisture movement;
■ accommodate and mask tolerance variations;
■ stability: that is, to remain in position without slumping;
■ should not give rise to the staining of adjacent materials.

Mastics and sealants have a limited life ranging from 10 to 25 years, which in most cases is less than the initial design life of the structure: therefore all joints must be designed in such a way that the seals can be renewed with reasonable ease, efficiency and cost. The common form of sealing material, called **putty**, is unsuitable for many applications because it hardens soon after application and cannot therefore accept the movements that can be expected in claddings and similar situations. It was this inability to accommodate movement that led to the development of mastics and sealants.

A wide variety of mastics and sealants are available to the designer and builder, giving a variety of properties and applications. A text of this nature does not lend itself to a complete analysis of all the types available but only to a comparison of some of the most frequently used.

Mastics and sealants can be applied in a variety of ways such as gun or knife applications. The most common method is by hand-held gun using disposable cartridges fitted into the body of the applicator. Most guns can be fitted with various nozzles to produce a neat bead of the required shape and size. Mastics are also available for knife application, the material being supplied in tins or kegs up to 25 kg capacity.

The joint should be carefully prepared to receive the mastic or sealant by ensuring that the contact surfaces are free from all dirt, grease and oil. The contact surfaces must be perfectly dry, and in some instances they may have to be primed before applying the jointing compound.

■■■ MASTICS

These are materials that are applied in a plastic state and form a surface skin over the core, which remains pliable for a number of years.

BUTYL MASTIC

These are basic mastics with the addition of butyl rubber or related polymers, making them suitable for glazing. They have a durability of up to 10 years and can accommodate both negative and positive movements up to 5% using a maximum joint width of 20 mm and a minimum joint depth of 6–10 mm.

OIL-BOUND MASTICS

These are the most widely used mastics, made with non-drying oils; they have similar properties to the butyl mastics but in some cases a better durability of up to 15 years. They have a joint width of 25 mm maximum although some special grades can be up to 50 mm, but in all cases the minimum joint depth is 12 mm.

■■■ SEALANTS

These are capable of accommodating greater movement than mastics; they are more durable, but dearer in both material and installation costs. They are applied in a plastic state and are converted by chemical reactions into an elastomer or synthetic rubber.

TWO-PART POLYSULPHIDE AND POLYURETHANE SEALANTS

These are well-established elastomer sealants having an excellent durability of 25 years or more, with a movement accommodation of approximately 15%. They are supplied as two components: a polysulphide or polyurethane base and a curing or vulcanising agent. Dense impervious surfaces such as glass may need priming, particularly before polyurethane is applied. The two parts are mixed shortly before use, the mixture having a pot life of approximately 4 hours. The two parts will chemically react to form a synthetic rubber, and should be used with a maximum joint width of 25 mm and a minimum joint depth of 6 mm when used in conjunction with metal or glass or a 10 mm minimum joint depth when used in connection with concrete. Two-part polysulphide sealants should conform to BS 4254: *Specification for two-part polysulphide-based sealants.*

ONE-PART POLYSULPHIDE, SILICONE OR URETHANE SEALANTS

These do not require premixing on site before application. They convert to a synthetic rubber or an elastomer by chemical reaction with absorbed moisture from the atmosphere. The joint sizing is similar to that described for a two-part polysulphide but the final movement accommodation is less at approximately plus or minus $12\frac{1}{2}\%$. Silicones and urethanes cure quite quickly whereas polysulphides can take up to 2 months, and during this period the movement accommodation is very low. One-part polysulphides are generally specified for pointing and where early or excessive movements are unlikely to occur. They should conform to BS 5215: *Specification for one-part gun grade polysulphide-based sealants.*

ONE-PART SOLVENT-RELEASE SEALANT

This is a butyl- or acrylic-based material that does not undergo chemical reaction with the atmosphere. It remains soft and pliable, adhering to most surfaces, but may break down in persistently damp exposures.

SILICONE RUBBER SEALANT

This is a one-part sealing compound that converts to an elastomer by absorbing moisture from the atmosphere, and has similar properties to the one-part polysulphide. It is available as pure white, translucent or in a wide range of colours, which makes it suitable for a variety of internal and external applications, e.g. sealing tiles to bath fittings and window/door frames to masonry. It should conform to BS 5889: *Specification for one-part gun grade silicone-based sealants.*

See also: BS EN 26927: *Building construction. Jointing products. Sealants. Vocabulary.*

■■■ GASKETS

As an alternative to using mastics and sealants for jointing, pre-formed gaskets of various designs and shapes are available (see Fig. 8.3.1). They are classified as **structural** or **non-structural**. Structural gaskets are produced from a compound of vulcanised polychloroprene rubber and specified to BS 4255-1: *Specification for non-cellular gaskets.* Non-structural gaskets can be made from synthetic rubber and various plastics in solid or tubular profiles. They too should conform to BS 4255-1. These can have better durability than mastics and sealants, but the design and manufacture of the joint profile requires a high degree of accuracy if a successful joint is to be obtained.

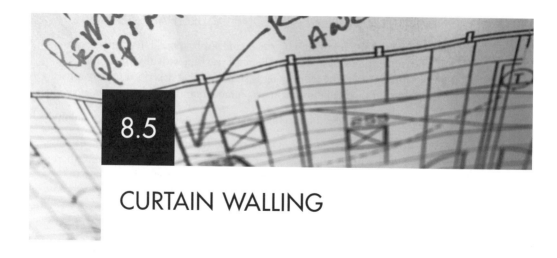

CURTAIN WALLING

Curtain walls are a form of external lightweight cladding attached to a framed structure forming a complete envelope or sheath around the structural frame. They are non-loadbearing claddings, which have to support only their own deadweight and any imposed wind loadings that are transferred to the structural frame through connectors, which are usually positioned at floor levels. For clarification, BS EN 13830: *Curtain walling – product standard* defines curtain walling as 'an external vertical building enclosure produced by elements mainly of metal, timber or plastic'. The basic conception of most curtain walls is a series of vertical mullions spanning from floor to floor interconnected by horizontal transoms forming openings into which can be fixed panels of glass or infill panels of opaque materials. Most curtain walls are constructed by using a patent or proprietary system produced by profile metal fabricators.

The primary objectives of using curtain-walling systems are to:

- provide an enclosure to the structure that will give the necessary protection against the elements;
- make use of dry construction methods;
- impose onto the structural frame the minimum load in the form of claddings;
- express an architectural feature.

To fulfil its primary functions a curtain wall must meet the following requirements:

- **Resistance to the elements** The materials used in curtain walls are usually impervious and therefore in themselves present no problem, but by virtue of the way in which they are fabricated a large number of joints occur. These joints must be made as impervious as the surrounding materials or designed as a drained joint. The jointing materials must also allow for any local thermal, structural or moisture movement, and generally consist of mastics, sealants and/or preformed gaskets of synthetic rubber or PVC.

■ **Assistance in maintaining the designed internal temperatures**   As curtain walls usually include a large percentage of glass the overall resistance to the transfer of heat is low, and therefore preventive measures may have to be incorporated into the design. Another problem with large glazed areas is solar heat gain, because glass will allow the short-wave radiations from the sun to pass through and consequently warm up the surfaces of internal walls, equipment and furniture. These surfaces will in turn radiate this acquired heat in the form of long-wave radiations that cannot pass back through the glazing, thus creating an internal heat build-up. Louvres fixed within a curtain-walling system will have little effect upon this heat build-up but they will reduce solar glare. A system of non-transparent external louvres will reduce the heat gain slightly by absorbing heat and radiating it back to the external air. The usual methods employed to solve the problem of internal heat gain are:
  ■ Deep recessed windows, which could be used in conjunction with external vertical fins.
  ■ Balanced internal heating and ventilation systems.
  ■ Use of special solar control glass such as reflective glasses that are modified during manufacture by depositing a metallic or dielectric reflective layer on the surface of the glass. The efficiency of this form of glazing can be increased if the glass is tilted by 5° to 15° to increase the angle of incidence.

■ **Adequate strength**   Although curtain walls are classified as non-loadbearing they must be able to carry their own weight and resist both positive and negative wind loadings. The magnitude of this latter loading will depend upon three basic factors:
  ■ height of building;
  ■ degree of exposure;
  ■ location of building.
The strength of curtain walling relies mainly upon the stiffness of the vertical component or mullion together with its anchorage or fixing to the structural frame. Glazing beads and the use of compressible materials also add to the resistance of possible wind damage of the glazed and infill panel areas by enabling these units to move independently of the curtain-wall framing.

■ **Required degree of fire resistance**   This is probably one of the greatest restrictions encountered when using curtain-walling techniques because of the large proportion of unprotected areas as defined in the Building Regulations, Approved Document B4: Section 12.7. See also Part 7 of this volume for matters relating to external fire spread. By using suitable materials or combinations of materials the opaque infill panels can normally achieve the required fire resistance to enable them to be classified as protected areas.

■ **Ease of assembly and fixing**   The principal member of a curtain-walling system is usually the mullion, which can be a solid or box section that is fixed to the structural frame at floor levels by means of adjustable anchorages or connectors. The infill framing and panels may be obtained as a series of individual components or as a single prefabricated unit. The main problems are ease of handling, amount of site assembly required and mode of access to the fixing position.

- **Required degree of sound insulation** Sound originating from within the structure may be transmitted vertically through the curtain-walling members. The chief source of this form of structure-borne sound is machinery, and this may be reduced by isolating the offending machines by mounting them on resilient pads and/or using resilient connectors in the joints between mullion lengths. Airborne sound can be troublesome with curtain-walling systems because the lightweight cladding has little mass to offer in the form of a sound barrier, the weakest point being the glazed areas. A reduction in the amount of sound transmitted can be achieved by:
    - reducing the areas of glazing;
    - using sealed windows of increased glass thickness;
    - double glazing in the form of inner and outer panes of glass with an air space of 150 to 200 mm between them.
- **Provision for thermal and structural movements** As curtain walling is situated on an external face of the structure it will be more exposed than the structural frame and will therefore be subject to greater amounts of temperature change, resulting in high thermal movement. The main frame may also be subjected to greater settlement than the claddings attached to its outer face. These differential movements mean that the curtain-walling systems should be so designed, fabricated and fixed that the attached cladding can move independently of the structure. The usual methods of providing for this required movement are to have slotted bolt connections and, to allow for movement within the curtain walling itself, to have spigot connections and/or mastic-sealed joints. Figures 8.5.1 and 8.5.2 show typical established curtain-walling examples to illustrate these principles. To reduce the risk of condensation on the interior of support framing and thermal bridging, a thermal break is incorporated as shown in Fig. 8.5.3.

▨ ▧ ▨ INFILL PANELS

The panels used to form the opaque areas in a curtain-walling system should have the following properties:

- lightweight;
- rigidity;
- impermeability;
- suitable fire resistance;
- suitable resistance to heat transfer;
- good durability requiring little or no maintenance.

No one material has all the above-listed properties, and therefore infill panels are usually manufactured in the form of a sandwich or combination panel. One of the main problems encountered with any form of external sandwich panel is interstitial condensation, which is usually overcome by including a vapour control layer of suitable material situated near the inner face of the panel. A vapour control layer can be defined as a membrane with a vapour resistance greater than 200 MN/g. Suitable materials include adequately lapped sheeting such as aluminium foil,

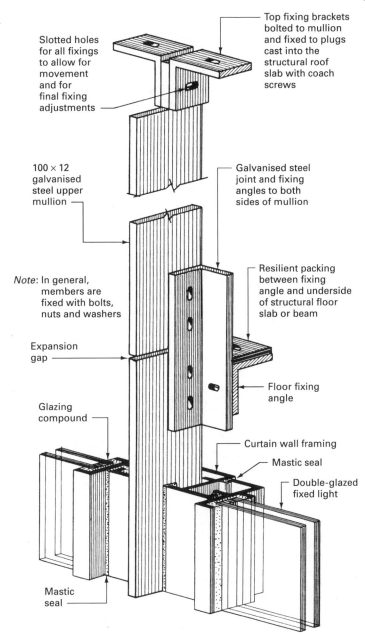

Slotted holes for all fixings to allow for movement and for final fixing adjustments

Top fixing brackets bolted to mullion and fixed to plugs cast into the structural roof slab with coach screws

100 × 12 galvanised steel upper mullion

Galvanised steel joint and fixing angles to both sides of mullion

Note: In general, members are fixed with bolts, nuts and washers

Resilient packing between fixing angle and underside of structural floor slab or beam

Expansion gap

Floor fixing angle

Glazing compound

Curtain wall framing

Mastic seal

Double-glazed fixed light

Mastic seal

Figure 8.5.1 Typical curtain–walling details: 1

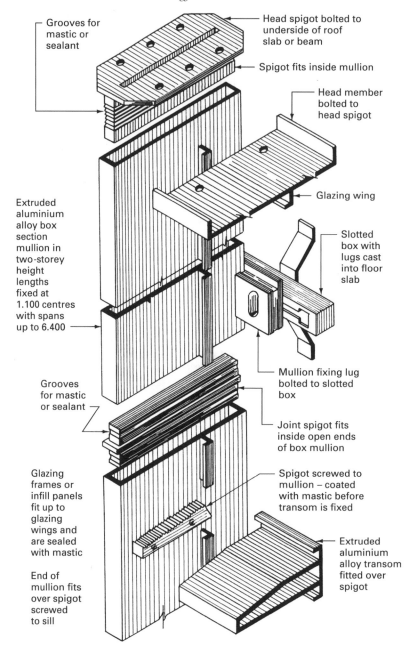

Grooves for
mastic or
sealant

Head spigot bolted to
underside of roof
slab or beam

Spigot fits inside mullion

Head member
bolted to
head spigot

Glazing wing

Extruded
aluminium
alloy box
section
mullion in
two-storey
height
lengths
fixed at
1.100 centres
with spans
up to 6.400

Slotted
box with
lugs cast
into floor
slab

Mullion fixing lug
bolted to slotted
box

Grooves
for mastic
or sealant

Joint spigot fits
inside open ends
of box mullion

Glazing
frames or
infill panels
fit up to
glazing
wings and
are sealed
with mastic

End of
mullion fits
over spigot
screwed
to sill

Spigot screwed to
mullion – coated
with mastic before
transom is fixed

Extruded
aluminium
alloy transom
fitted over
spigot

Figure 8.5.2   Typical curtain-walling details: 2

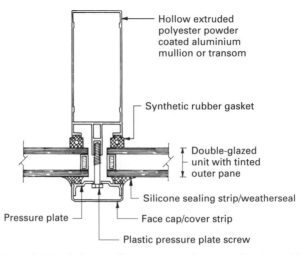

Hollow extruded
polyester powder
coated aluminium
mullion or transom

Synthetic rubber gasket

Double-glazed
unit with tinted
outer pane

Silicone sealing strip/weatherseal

Pressure plate

Face cap/cover strip

Plastic pressure plate screw

Figure 8.5.3   Extruded aluminium mullion or transom incorporating thermal breaks

waterproof building papers, polythene sheet and applied materials such as two coats
of bitumen paint or two coats of chlorinated rubber paint. Care must be taken when
positioning vapour control layers to ensure that an interaction is not set up between
adjacent materials, such as the alkali attack of aluminium if placed next to concrete
or fibre cement.

The choice of external facing for these infill cladding materials is very important
because of their direct exposure to the elements. Plastic and plastic-coated materials
are obvious choices provided they meet the minimum fire regulations as set out in
Approved Document B. One of the most popular materials for the external facing
of infill panels is vitreous enamelled steel or aluminium sheets of 0.7 and 0.8 mm
thickness respectively. In the preparation process a thin coating of glass is fused
onto the metal surface at a temperature of between 800 and 860 °C, resulting in an
extremely hard, impervious, acid and corrosion-resistant panel that will withstand
severe abrasive action; also, the finish will not be subject to crazing or cracking,
resulting in an attractive finish with the strength of the base metal. Aluminium
sheeting can be finished with a silicone polyester coating. When used in
combination with materials such as expanded polystyrene (EPS), rockfibre,
polyurethane, phenolic or polyisocyanurate insulants a lightweight infill panel
giving U-values less than 0.35 W/m^2 K can be achieved. Typical infill panel
examples are shown in Figs 8.5.4 and 8.5.5. Fire performance for both internal
and external surfaces must satisfy the Building Regulations for class 0, tested in
accordance with BS 476-6 and 7. The insulating core should not ignite when tested
to BS 476-12. Panels can be manufactured up to 3.000 m long by 1.000 m wide.

▨▨■■ GLAZING

The primary function of any material fixed into an opening in the external facade
of a building is to provide a weather seal. Glass also provides general daylight

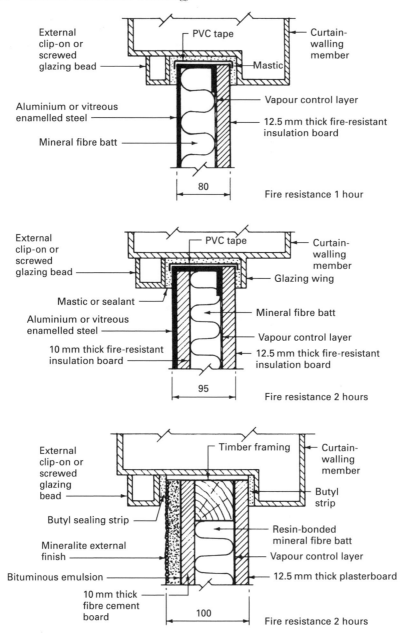

Figure 8.5.4 Typical curtain-walling infill panel details

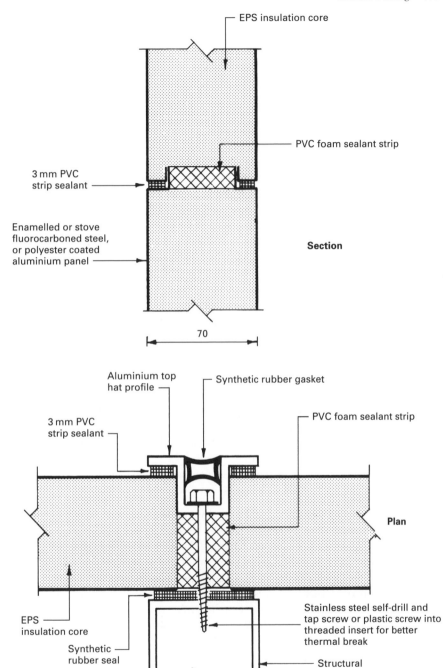

EPS insulation core

PVC foam sealant strip

3 mm PVC
strip sealant

Section

Enamelled or stove
fluorocarboned steel,
or polyester coated
aluminium panel

70

Aluminium top
hat profile

Synthetic rubber gasket

3 mm PVC
strip sealant

PVC foam sealant strip

Plan

EPS
insulation core

Synthetic
rubber seal

Stainless steel self-drill and
tap screw or plastic screw into
threaded insert for better
thermal break

Structural
box frame

Figure 8.5.5 Typical insulated horizontal wall panel cladding

illumination of the interior of the building and daylight for carrying out specific tasks; at the same time it can provide a view out or visual contact with the outside world. It is not really necessary to have the internal space of a building illuminated by natural daylight because this can be adequately covered by a well-designed and installed artificial lighting system, but in psychological and energy conservation terms it is usually considered desirable to have a reasonable proportion of glazed areas.

The nature of work to be carried out and the position of the working plane will largely determine whether daylight from glazed areas alone can provide sufficient illumination for specific work tasks. Many of these tasks are carried out on a horizontal surface, which is best illuminated by vertical light: therefore careful design of window size, window position and possible daylight factors need to be assessed if glazed areas are to provide the main source of work task illumination. The need for visual contact with the outside world, like the need for daylight illumination of the interior, is largely a case of psychological well-being rather than dire necessity, but to provide an acceptable view out the areas of glazing need to be planned bearing in mind the size, orientation and view obtained. The problems of solar heat gain, solar glare, thermal and sound insulation have already been considered at the beginning of this chapter, and therefore no further comment seems necessary under this heading. The major problem remaining when using glass in the facades of high-rise structures and in particular for curtain walling is providing a means of access for cleaning and maintenance.

The cleaning of windows in this context can be a dangerous and costly process, but such glazed areas do need cleaning, for the following reasons:

- prevention of dirt accumulation resulting in a distortion of the original visual appearance;
- maintaining the designed daylight transmission;
- maintaining the clarity of vision out;
- prevention of deterioration of the glazing materials due to chemical and/or dirt attack.

The usual method of cleaning windows is by washing with water using swabs, chamois leather, scrims and squeegee cleaners, all of which are hand held, requiring close access to the glass to be cleaned. Cleaning the internal surfaces does not normally present a problem, but unless pivot or tilt-and-turn windows are used the cleaning of the outside surface will require a means of external access. In low- to medium-rise structures access can be by means of trestles, stepladders or straight ladders, the latter being possible up to 11.000 m, after which they become dangerous because of flexing and the lack of overall control. Tower scaffolds are seldom used for window cleaning because of the cost and time involved in assembly, but lightweight, quickly erected scaffold systems could be considered for heights up to 6.000 m.

Access for external cleaning of curtain-wall facades in high-rise structures is generally by the use of suspended cradles, which can be of a temporary nature as shown in Fig. 2.6.2, or can be of a permanent system designed and constructed as an integral part of the structure. The simplest form is to install a universal beam section at roof level positioned about 450 mm in front of the general facade line and continuous around the perimeter of the roof. A conventional cradle is attached by means of castors located on the bottom flange of the ring or edge beam. Control is by means of ropes from the cradle, which must be lowered to ground level for access purposes.

A better form is where the transversing track is concealed by placing a pair of rails to a 750 mm gauge to which is fixed a trolley having projecting beams or davits that can be retracted and/or luffed, the latter being particularly useful for negotiating projections in the general facade. The trolley may be manually operated from the cradle for transversing, or it may be an electric-powered trolley giving vertical movement of between 5 and 15 m per minute or horizontal movement of 5 to 12 m per minute. This form of trolley is usually considered essential for heights over 45.000 m. In all cases the roof must have been structurally designed to accept the load of the apparatus.

Some form of cable restraint must be incorporated in the design to overcome the problem of unacceptable cradle movement due to the action of wind around the structure. Winds moving up the face of the building will cause the cradle to swing at right angles to the face, giving rise to possible impact damage to the face of the structure as well as placing the operatives in a hazardous situation.

Crosswinds can cause the cradle to move horizontally along the face of the building, with equally disastrous results. Methods of cradle and cable restraint available are the use of suction grips, eyebolts fixed to the facade through which the suspension ropes can be threaded, suspension ropes suitably tensioned at ground level, electrical cutouts at intervals of 15.000 m, which will prevent further cradle movement until a special plug through which the hoist line passes has been inserted, using the mullions as a guide for rollers that are either in contact with the mullion face to prevent lateral movement, or using castors located behind a mullion flange to prevent outward movement.

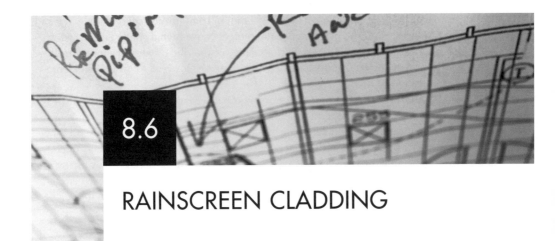

8.6

RAINSCREEN CLADDING

Rainscreen cladding is a facing or facade applied as part of the construction of new buildings or as a protective or decorative overcladding to existing structures. The latter application provides an opportunity to modernise the exterior appearance of a dated building and improve its sound and thermal insulation.

The 'loose-fit' nature of rainscreen cladding with screw, rivet or clip-on fixings allows for future changes, as variations in corporate image or changing owner requirements may determine. The key features of this type of exterior treatment are the provision of:

- an outer weather-resistant layer;
- an outer decorative layer;
- a vertical support frame secured to the structural wall;
- a drained and ventilated cavity between cladding and wall;
- a facility for integral insulation.

The concept is shown in Fig. 8.6.1.

▨ ▨ ▨ NEW CONSTRUCTION

The use of rainscreen cladding allows an opportunity for the designer to use an inexpensive single skin of lightweight concrete blockwork as infilling between the structural frame. This provides the background support for superimposed water-resistant insulation such as resin-bonded rock fibre batts in thicknesses up to 150 mm. Traditional details such as cavity trays and damp-proof membranes/courses are not required. Also, the continuity of insulation considerably reduces any possibility for cold bridging or the occurrence of interstitial condensation.

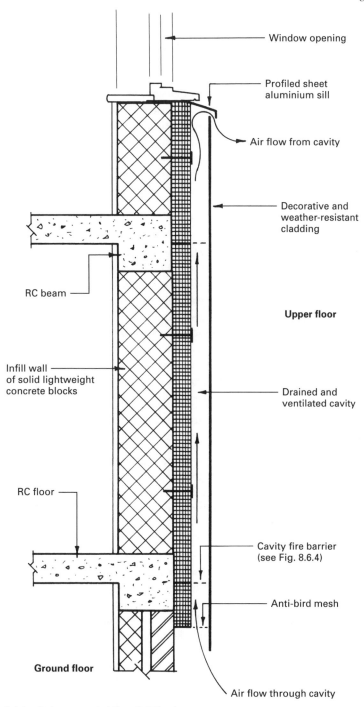

Figure 8.6.1 Rainscreen cladding (RSC): the concept

▩▩■ EXISTING CONSTRUCTION

Weather-damaged, dated and generally unattractive building facades can be considerably improved and protected by supplementary cladding. However, local planning restrictions with particular regard to conservation areas must be consulted before proceeding with the obliteration of an existing facade. Significant change of appearance, modernisation and change of use will necessitate a formal planning application. Nevertheless, the alternative interior approach to updating to comply with current sound and thermal regulations and periods of fire resistance may prove difficult and expensive. The practicalities of application to the interior of walls can be more difficult to effect and more disruptive for the building occupants, and can impose on valuable floor area.

▩▩■ APPLICATION

Vertical battens of timber or metal-profiled framing are secured to the background wall at spacings to suit the cladding. Treated timber battens of 50 mm × 38 mm and 100 mm × 38 mm at panel junctions are normally secured at between 400 and 600 mm vertical spacing to allow continuity of ventilation and drainage. There should be no horizontal battens, noggins or struts. Figure 8.6.2 shows the principles of application, which must be supplemented with anti-bird mesh at upper and lower ventilation openings to prevent entry by birds, vermin or insects. Ventilation openings must also be provided at interruptions such as windows and projecting ring beams. The size of ventilation opening can vary with height of building:

- Up to 5 storeys: 10 mm continuous or equivalent.
- 5 to 15 storeys: 15 mm continuous or equivalent.
- Over 15 storeys: 20 mm continuous or equivalent.

Incorporating insulation behind timber-supported cladding is unlikely to be economical or practical when considering the space required. Several manufacturers produce purpose-made aluminium sections with extendable brackets that can suspend the cladding up to 200 mm from the wall, as shown in Fig. 8.6.3.

▩▩■ FIRE SPREAD

As for any other void or cavity, the cavity created behind the cladding is treated by the Building Regulations as a potential passage for conveyance of fire. Therefore fire barriers must be located in accordance with the spacings specified in Approved Document B3, Section 9 (Volume 2). The vertical support battens or framing are normally adequate, but these will need to be supplemented horizontally. In order to preserve the continuity of ventilation and drainage, horizontal barriers are manufactured from pressed steel, perforated, and coated with an intumescent material (see Fig. 8.6.4).

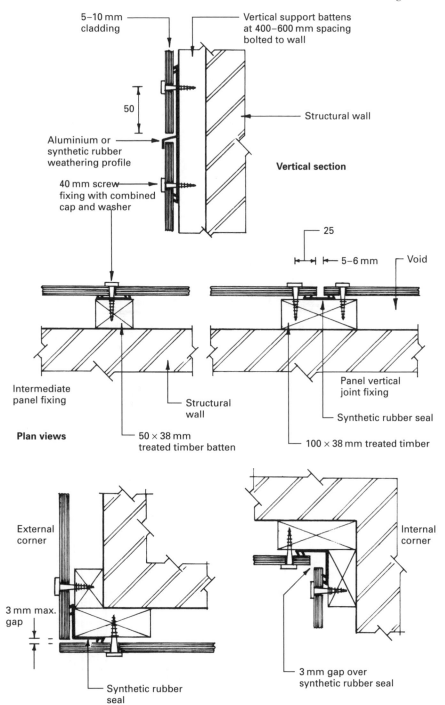

Figure 8.6.2 RSC with timber support battens

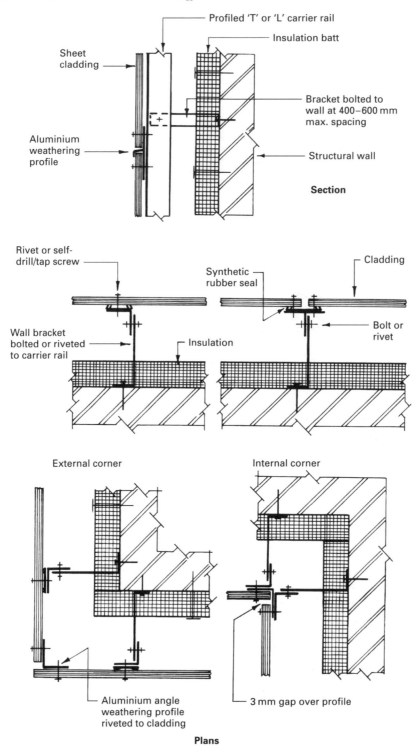

Figure 8.6.3   RSC with profiled aluminium carriers and brackets

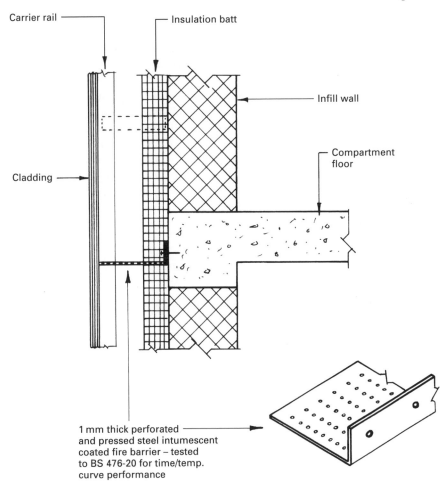

Figure 8.6.4   Cavity fire barrier

■■■ CLADDING MATERIALS

Rainscreen claddings are fire-tested to BS 476-6 and 7 for fire propagation and surface spread of flame respectively, and are also classified for fire resistance depending on purpose grouping. Standards apply to both sides of the cladding, with criteria for assessment depending on building height, purpose group and distance from the boundary. These are determined in the Building Regulations, Approved Document B4, Section 11.6, with reference to class 0 materials of limited combustibility and an Index (I) < 20 (see reference to BS 476-6 in Chapter 7.2).

Sheet claddings are available in thicknesses ranging from 5 to 10 mm in a wide variety of colours. Popular sheet material applications include fibre cement, glass-fibre-reinforced plastics and polyester resins, some with stone finishing, polyester-coated aluminium, enamelled or stove fluorocarboned steel, and resin-impregnated cellulose with melamine facing.

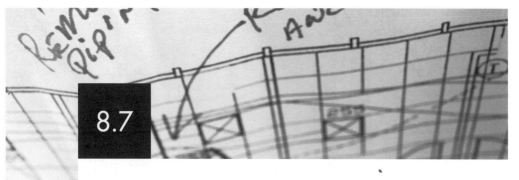

8.7

STRUCTURAL GLASS CLADDING

Toughened glass sheeting of 10 or 12 mm thickness can be considered under two headings:

- structural sealant (silicone) glazing;
- mechanical fixing or suspended glazing.

■■■ STRUCTURAL SEALANT (SILICONE) GLAZING

This system relies on the bonding characteristics of silicone or other elastomeric sealants between the glass and the support frame. Sealants are also selected with regard to their long-term resilience when exposed to ultraviolet light, heat, rain, varying temperatures, thermal movement, atmospheric pollutants, etc. Two types of glazing are used: two-edge and four-edge. The two-edge system uses conventional head and sill rebates with glass retention beads, and the sides are silicone-bonded to mullions or styles. The four-edge system uses a structural adhesive to bond all four sides of the glass pane to a frame, with the sealant functioning as an integral structural link between glazing and frame. The lower edge of each glass pane is supported on resilient setting blocks or spacers of precured silicone to ensure adequate provision of sealant under the lower edge. Total control of production quality (particularly for four-edge systems) will require factory assembly of glazing modules and close supervision of the bonding process. Standard application is to factory-bond the glass to an aluminium frame with *in-situ* silicone finishing/pointing provided between the pane and frame, as shown in Fig. 8.7.1. Some coated glass, and absorbent backgrounds such as timber or masonry, must be carefully primed before application of sealant.

Vertical sections

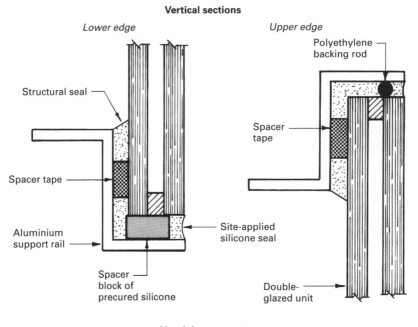

Lower edge

Structural seal

Spacer tape

Aluminium
support rail

Spacer
block of
precured silicone

Site-applied
silicone seal

Upper edge

Polyethylene
backing rod

Spacer
tape

Double-
glazed unit

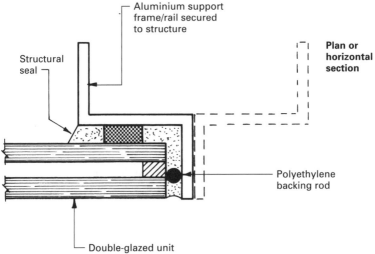

Aluminium support
frame/rail secured
to structure

Plan or
horizontal
section

Structural
seal

Polyethylene
backing rod

Double-glazed unit

Figure 8.7.1   Structural sealant glazing

▨▩■ MECHANICAL FIXING

This method promotes the use of uninterrupted glazing, limiting the visual impact of framing members. A frameless glazing facade is achieved by suspending toughened glass from the building structure, with supports for each pane at its four corners. Some support is also provided from the pane below. Intermediate fixing may also be necessary if panes are sufficiently large and if justified by high wind loading. Fixings comprise stainless steel nuts and bolts, plastic washers, plastic bushes, fibre washers and stainless steel support brackets that attach to the building structural frame. Bolts can be countersunk to fit neatly into preformed holes in the glass. Edges of glass are butt-jointed and sealed with elastomeric (silicone) sealant. Some applications use metal patch plates of about 165 mm square to create a clamping effect at significant stress concentrations around the bolt hole and corners of the glass pane (see Fig. 8.7.2). Heights in excess of 20.000 m are possible based on modules of 1.200 and 1.500 m, but the system is limited to single glazing.

▨▩■ 'PLANAR' GLAZING SYSTEM

This variation of frameless glazing reduces the visual impact of fixing plates. It has been developed by Pilkington UK Ltd on the principle of direct bearing of the glass via nylatron polyamide bushes onto the fixing bolts. Each pane is separately fixed independent of the pane below, with bolts having minimal clamping effect owing to attachment to a spring plate by predetermined torque settings (see Fig. 8.7.3). The spider bracket shown is purpose-made to accommodate all four corner fixings in one cast aluminium unit. The spring plate system creates flexibility by reducing the concentration of stresses around the 'planar' fitting, relative to the patch plate system. A further advantage is application to either single or double glazing, the latter providing opportunity for variations in the selection of glass. A typical specification could be 10 or 12 mm toughened glass outer layer, tinted to deflect solar gains, with a 6 mm clear toughened glass inner layer and a 16 mm air space, as shown in Fig. 8.7.4.

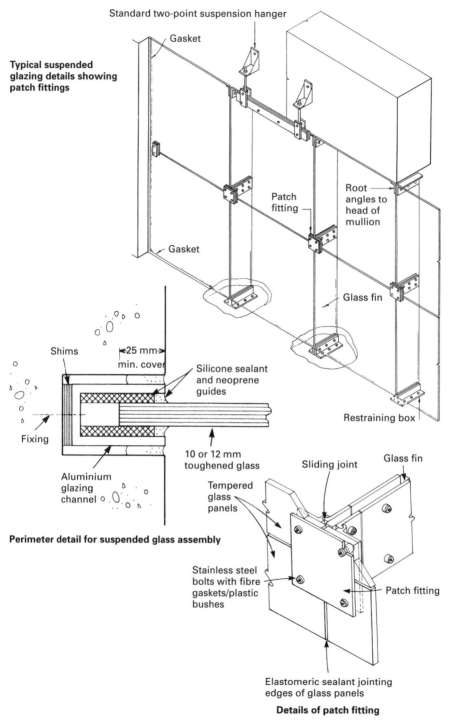

Typical suspended glazing details showing patch fittings

Standard two-point suspension hanger

Gasket

Patch fitting

Root angles to head of mullion

Glass fin

Gasket

Shims

25 mm min. cover

Silicone sealant and neoprene guides

Fixing

10 or 12 mm toughened glass

Aluminium glazing channel

Restraining box

Perimeter detail for suspended glass assembly

Tempered glass panels

Sliding joint

Glass fin

Stainless steel bolts with fibre gaskets/plastic bushes

Patch fitting

Elastomeric sealant jointing edges of glass panels

Details of patch fitting

Figure 8.7.2   Mechanical fixing or suspended glazing (*courtesy*: Pilkington Glass)

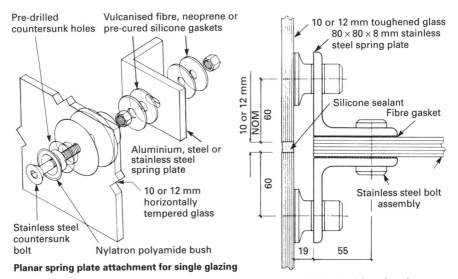

Pre-drilled countersunk holes

Vulcanised fibre, neoprene or pre-cured silicone gaskets

Aluminium, steel or stainless steel spring plate

10 or 12 mm horizontally tempered glass

Stainless steel countersunk bolt

Nylatron polyamide bush

Planar spring plate attachment for single glazing

10 or 12 mm toughened glass 80 × 80 × 8 mm stainless steel spring plate

Silicone sealant
Fibre gasket

10 or 12 mm NOM
60
60

Stainless steel bolt assembly

19 55

Planar type 902 bolt with spring plate attachment to glass fin

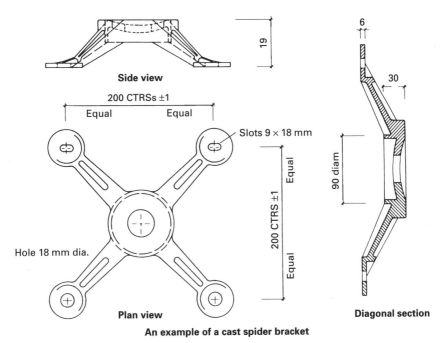

Side view

19

200 CTRSs ±1
Equal   Equal

Slots 9 × 18 mm

Hole 18 mm dia.

200 CTRS ±1
Equal   Equal

Plan view

6

30

90 diam

Diagonal section

An example of a cast spider bracket

Figure 8.7.3 'Planar' glazing system (*courtesy*: Pilkington Glass)

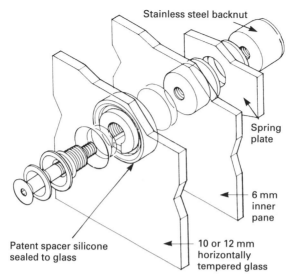

Stainless steel backnut

Spring plate

6 mm inner pane

10 or 12 mm horizontally tempered glass

Patent spacer silicone sealed to glass

Planar spring plate attachment for double glazing

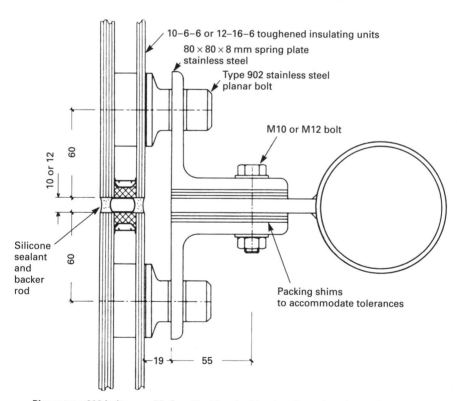

10–6–6 or 12–16–6 toughened insulating units

80 × 80 × 8 mm spring plate stainless steel

Type 902 stainless steel planar bolt

M10 or M12 bolt

10 or 12

60

60

Silicone sealant and backer rod

Packing shims to accommodate tolerances

19   55

Planar type 902 bolt assembly for attaching double-glazed panels to internal structure

Figure 8.7.4   'Planar' double glazing (*courtesy*: Pilkington Glass)

8.8

SUSTAINABLE CONSTRUCTION

Existing buildings originally designed specifically for commercial, industrial or assembly purposes are too often demolished, as it is not financially or practically viable to adapt them for alternative use. Modern premises require vast areas for cabling and other services, but much of our building heritage was designed solely for its initial function, without consideration for the high-technology facilities needed today. The result is a relatively short-term cycle of demolition and new build. When designing and building framed structures and their cladding, the economic and environmental benefits of creating sustainable, versatile and adaptable construction are now recognised. The concept evolved gradually over the latter part of the last century. In 1968 the designer Robert Propst published *The office: A facility based on change.* This reference presented the idea of buildings being adaptable to differing uses, and functions as a strategic life-cycle design plan. Later, in the early 1970s, the architect Alex Gordon developed a design formula that became known as 'Long life, Loose fit, Low energy'.

■■■■ LONG LIFE

The idea is to ensure that the design of a building is not just based on the limited life of its initial function. Planned obsolescence may be appropriate for inexpensive commodities, which can include temporary buildings, site huts, etc. However, commercial buildings represent a huge financial investment, considerable construction time, and significant use of diminishing material resources. Designers have a responsibility to society to ensure maximum benefit from a facility, and to their clients to provide maximum financial return. Long life takes into consideration that a building is not only for tomorrow; it should have an in-built design to adapt and upgrade to future changes of use and occupation. The structure should be a shell or skeletal frame with a number of solutions for division of space and provision for services.

▪▪▪ LOOSE FIT

Loose fit develops the concept further, with provision for accommodating numerous uses over the design life of a building.

The accommodation must be flexible in layout, so that successive occupants can configure and reconfigure the interior to suit their changing needs, working layouts and processes. This requires demountable partitioning, generally of modular format, suspended ceilings, and raised flooring to facilitate state-of-the-art services installations.

The outward appearance or aesthetic perspective can be considered with regard to architectural fashion changes and the corporate image of the building occupants. Loose fit in the form of replacement curtain-walling and over-cladding systems permits a complete exterior change to an existing frame, thereby considerably enhancing a structure without the cost of demolition and rebuilding. Such changes can run parallel with upgrading to current legislative standards, particularly thermal insulation and airtightness.

▪▪▪ LOW ENERGY

Low energy refers to the last consideration under loose fit, which ensures that refurbishment associated with ongoing changes of use is subject to building control and in accordance with current energy conservation measures, e.g. high-efficiency fuel-burning appliances, solar controls, double glazing, etc. Low energy also refers to the time, labour and material cost or wastage that would otherwise be associated with the repetitive cycle of building and demolition that has been needed to meet ongoing architectural design trends and client needs for internal spatial requirements.

Notwithstanding the aforementioned, all structural investments have an economic life: that is, because of deterioration and limited adaptability they reach a point at which the value of the site and the building becomes no more than the value of the land on which the building stands. The objective of sustainable construction is to maximise the investment value in a building by ensuring that it has a long and useful life.

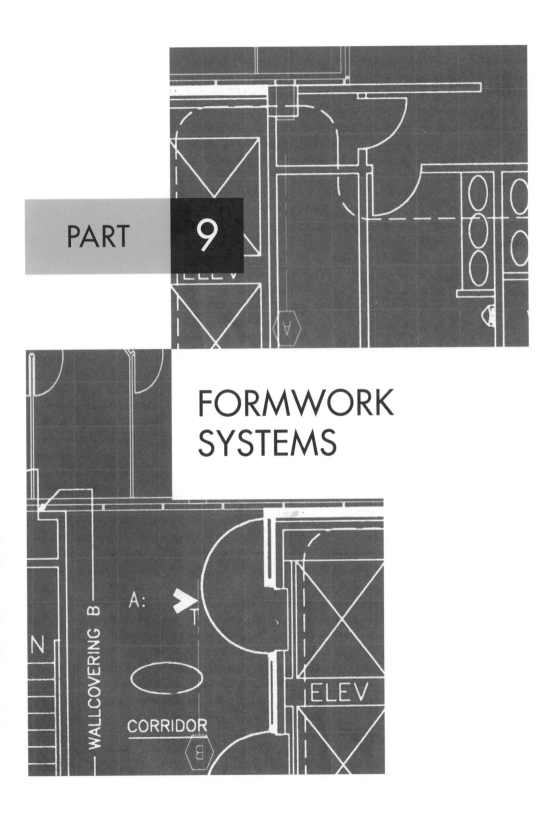

PART 9

FORMWORK
SYSTEMS

WALL FORMWORK

Before considering different applications of formwork, it is worth defining some aspects of this work. Formwork is used generally to describe the temporary work associated with the formation of concrete structures and components. However, it can be subdivided into several areas:

- **Centering** A type of falsework used as a temporary support to masonry arches. See Chapter 3.6 in *Construction Technology*.
- **Falsework** The term applied to timber, scaffolding or other temporary structural framing to support the forms, shutters or moulds for concrete work or masonry arches. Falsework can be of either sawn timber or steel sections, scaffold tube and couplings being an option for the latter. The structure must have sufficient strength to prevent deformation, deflection and collapse under the full load of wet concrete, operatives and equipment. Falsework must remain in position until the permanent structure is sufficiently strong to support itself and ongoing work.
- **Moulds** These are generally associated with the factory manufacture of precast concrete units such as blocks, lintels and feature work of artificial or reconstructed stonework, e.g. sills. Moulds can be of planed timber, steel or plastics. Plastic materials are often used as a smooth lining or facing to timber.
- **Shuttering** A term generally restricted to the use of planed timber, plywood sheeting, steel and other materials used for supporting and shaping the finished concrete on site. Timber products will require mould oil treatment to the surface exposed to concrete, to prevent cement adhering to the timber and subsequently spalling when the shuttering is struck.

Formwork provides the facility for wet concrete and reinforcement to form a particular desired shape with a predetermined strength. Depending upon the complexity of the form, the relative cost of formwork to concrete can be as high as 75% of the total cost to produce the required member. A typical breakdown of percentage costs could be as follows:

- **Concrete**   (materials 28%; labour 12%) = 40%.
- **Reinforcement**   (materials 18%; labour 7%) = 25%.
- **Formwork**   (materials 15%; labour 20%) = 35%.

The above breakdown shows that a building contractor will have to use an economic method of providing the necessary formwork to be competitive in tendering, because this is the factor over which that company has most control.

The economic essentials of formwork can be listed thus:

- **Low cost**   Only the amount of money necessary that will produce the required form to be expended.
- **Strength**   Careful selection of formwork materials and falsework to obtain the most economic balance in terms of quantity used and continuing site activity around the assembled formwork.
- **Finish**   Selection of method, materials and, if necessary, linings to produce the desired result direct from the formwork. Applied finishes are usually specified, and therefore method is the only real factor over which the builder would have any economic control.
- **Assembly**   Consideration must be given to the use of patent systems and mechanical handling plant.
- **Material**   Advantages of using either timber, steel or glass-reinforced plastics should be considered; generally, timber's light weight and adaptability make it the most favoured material, but steel and glass-reinforced plastics will provide more uses than timber, although they cannot be repaired as easily.
- **Design**   Within the confines of the architectural and/or structural design, formwork should be as repetitive and adaptable as possible. Timber forms have limited reuse, with exposed surfaces suffering the most damage. Five or six reuses are usual, whereas steel and plastics can be reused indefinitely if cared for. These are pressed or moulded into fixed shapes, having specific application to repetitive uses such as columns and beams in multi-storey buildings.
- **Joints**   These should be tight enough to prevent grout leakage. Sealing of joints can be achieved with compressible plastic tape between components or with mastic or silicone sealant applied to junctions of adjacent pieces of formwork.

A balance of the above essentials should be achieved, preferably at pre-tender stage, so that an economic and competitive cost can be calculated.

Before proceeding with this section, background reading of the basic principles of formwork explained in Chapter 10.3 of *Construction Technology* is recommended.

In principle the design, fabrication and erection of wall formwork are similar to that previously considered in the context of column formwork in *Construction Technology*. Several basic methods are available that will enable a wall to be cast in large quantities, defined lifts or continuous from start to finish.

▨■■ TRADITIONAL WALL FORMWORK

This usually consists of standard framed panels tied together over their backs with horizontal members called **walings**. The walings fulfil the same function as the yokes or column clamps of providing the resistance to the horizontal force of wet concrete. A 75 mm high concrete kicker is formed at the base of the proposed wall to enable the forms to be accurately positioned and to help prevent the loss of grout by seepage at the base of the form.

The usual assembly is to erect one side of the wall formwork, ensuring that it is correctly aligned, plumbed and strutted. The steel reinforcement cage is inserted and positioned before the formwork for the other side is erected and fixed. Keeping the forms parallel and at the correct distance from one another is most important; this can be achieved by using precast concrete spacer blocks that are cast in, or steel spacer tubes that are removed after casting and curing, the voids created being made good, or alternatively by using one of the many proprietary wall tie spacers available (see Fig. 9.1.1).

To keep the number of ties required within acceptable limits horizontal members or walings are used; these also add to the overall rigidity of the formwork panels and help with alignment. Walings are best if composed of two members, with a space between that will accommodate the shank of the wall tie bolt: this will give complete flexibility in positioning the ties and leave the waling timbers undamaged for eventual reuse. To ensure that the loads are evenly distributed over the pair of walings, plate washers should be specified.

Plywood sheet is the common material used for wall formwork, but this material is vulnerable to edge and corner damage. The usual format is therefore to make up wall forms as framed panels on a timber studwork principle with a plywood facing sheet screwed to the studs so that it can be easily removed and reversed to obtain the maximum number of uses.

Corners and attached piers need special consideration, because the increased pressure at these points could cause the abutments between panels to open up, giving rise to unacceptable grout escape and a poor finish to the cast wall. The walings can be strengthened by including a loose tongue at the abutment position, and extra bracing could be added to internal corners (see Fig. 9.1.2). When considering formwork for attached piers it is usually necessary to have special panels to form the reveals.

▨■■ CLIMBING FORMWORK

This is a method of casting a wall in set vertical lift heights using the same forms in a repetitive fashion, thus obtaining maximum usage from a minimum number of forms. The basic formwork is as shown in Fig. 9.1.3: in the first lift it is positioned against the kicker in the inverted position, the concrete is poured and allowed to cure, after which the forms are removed, reversed and fixed to the newly cast concrete. After each casting and curing of concrete the forms are removed and raised to form the next lift until the required height has been reached.

It is possible to use this method for casting walls against an excavated or sheet–piled back face using formwork to one side only by replacing the through wall tie spacers with loop wall ties (see Fig. 9.1.3).

When using this single–sided method adequate bracing will be required to maintain the correct wall thickness, and when the formwork is reversed, after the

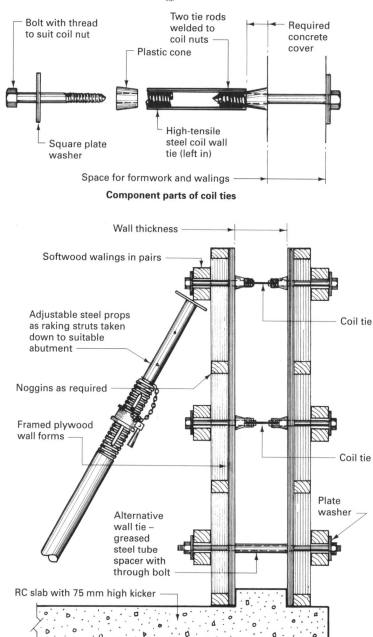

Bolt with thread to suit coil nut

Two tie rods welded to coil nuts

Required concrete cover

Plastic cone

Square plate washer

High-tensile steel coil wall tie (left in)

Space for formwork and walings

Component parts of coil ties

Wall thickness

Softwood walings in pairs

Adjustable steel props as raking struts taken down to suitable abutment

Coil tie

Noggins as required

Framed plywood wall forms

Coil tie

Alternative wall tie – greased steel tube spacer with through bolt

Plate washer

RC slab with 75 mm high kicker

Note: Reinforcement omitted for clarity

Figure 9.1.1   Traditional wall formwork details: 1

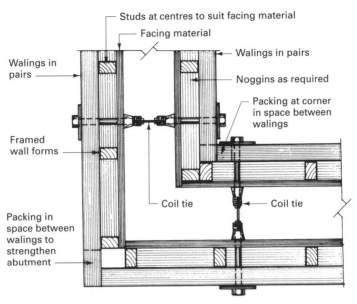

Studs at centres to suit facing material

Facing material

Walings in pairs

Walings in pairs

Noggins as required

Packing at corner in space between walings

Framed wall forms

Coil tie

Coil tie

Packing in space between walings to strengthen abutment

Plan on corner formwork

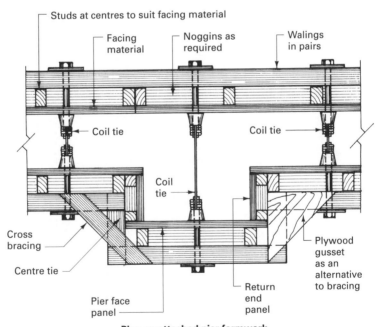

Studs at centres to suit facing material

Facing material

Noggins as required

Walings in pairs

Coil tie

Coil tie

Coil tie

Cross bracing

Centre tie

Plywood gusset as an alternative to bracing

Pier face panel

Return end panel

Plan on attached pier formwork

Figure 9.1.2   Traditional wall formwork details: 2

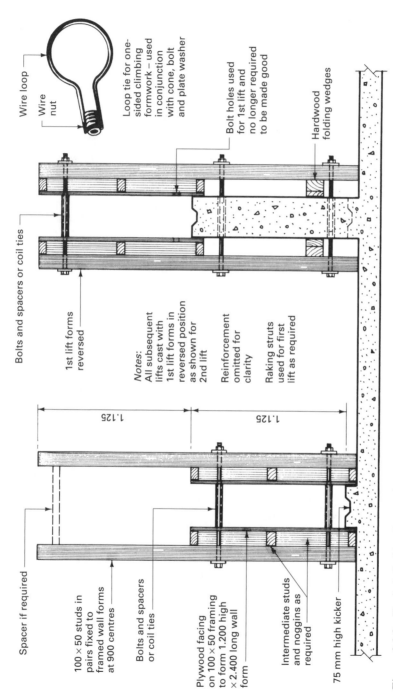

Figure 9.1.3   Typical climbing formwork arrangement

first lift, it must be appreciated that the uprights or soldiers are acting as cantilevers and will therefore need to be stronger than those used in the double-sided version.

▨▨■ SLIDING FORMWORK

This is a system of formwork that slides continuously up the face of the wall being cast, by climbing up and being supported by a series of hydraulic jacks operating on jacking rods. The whole wall is therefore cast as a monolithic and jointless structure, making the method suitable for structures such as water towers, chimneys and the cores of multi-storey buildings that have repetitive floors.

Because the system is a continuous operation, good site planning and organisation are essential, and will involve the following aspects:

- Round-the-clock working, which will involve shift working and artificial lighting to enable work to proceed outside normal daylight hours.
- Careful control of concrete supply to ensure that stoppages of the lifting operation are not encountered. This may mean having standby plant as an insurance against mechanical breakdowns.
- Suitably trained staff accustomed to this method of constructing *in-situ* concrete walls.

The actual architectural and structural design must be suitable for the application of a slipform system; generally the main requirements are a wall of uniform thickness with a minimum number of openings and a height of at least 20.000 m to make the cost of equipment, labour and planning an economic proposition.

The basic components of slip formwork are:

- **Side forms** These need to be strongly braced, and are loadbearing of timber and/or steel construction. Steel forms are heavier than timber, and more difficult to assemble and repair, but they have lower frictional loading, are easier to clean and have better durability. Timber forms are lighter, have better flexibility, are easier to repair and are generally favoured. A typical timber form would consist of a series of 100 mm × 25 mm planed straight-grained staves assembled with a 2 mm wide gap between consecutive boards to allow for swelling, which could give rise to unacceptable friction as the forms rise. The forms are usually made to a height of 1.200 m with an overall sliding clearance of 6 mm by keeping the external panel plumb and the internal panel tapered so that it is 3 mm in at the top and 3 mm out at the bottom, giving the true wall thickness in the centre position of the form. The side forms must be adequately stiffened with horizontal walings and vertical puncheons to resist the lateral pressure of concrete and transfer the loads of working platforms to the supporting yokes.
- **Yokes** These assist in supporting the suspended working platforms and transfer the platform and side form loads to the jacking rods. Yokes are usually made of framed steelwork suitably braced and designed to provide the necessary bearings for the working platforms.

- **Working platforms**   Three working levels are usually provided. The first is situated above the yokes at a height of about 2.000 m above the top of the wall forms for the use of the steel fixers. The second level is a platform over the entire inner floor area at a level coinciding with the top of the wall forms, and is used by the concrete gang for storage of materials and to carry levelling instruments and jacking control equipment. This decking could ultimately be used as the soffit formwork to the roof slab if required. The third platform is in the form of a hanging or suspended scaffold, usually to both sides of the wall, and is to give access to the exposed freshly cast concrete below the slip formwork for the purpose of finishing operations.

- **Hydraulic jacks**   The jacks used are usually specified by their loadbearing capacities, such as 3 tonnes or 6 tonnes, and consist of two clamps operated by a piston. The clamps operate on a jacking rod of 25 to 50 mm diameter according to the design load, and are installed in banks operated from a central control to give an all-round consistent rate of climb. The upper clamp grips the jacking rod, and the lower clamp, being free, rises, pulling the yoke and platforms with it until the jack extension has been closed. The lower clamp now grips the climbing rod while the upper clamp is released and raised to a higher position, when the lifting cycle is recommenced. Factors such as temperature and concrete quality affect the rate of climb, but typical speeds are between 150 and 450 mm per hour.

   The upper end of the jacking rod is usually encased in a tube or sleeve to overcome the problem of adhesion between the rod and the concrete. The jacking rod therefore remains loose in the cast wall and can be recovered at the end of the jacking operation. The 2.500 to 4.000 m lengths of rod are usually joined together with a screw joint arranged so that all such joints do not occur at the same level.

A typical arrangement of sliding formwork is shown diagrammatically in Fig. 9.1.4.
   The site operations commence with the formation of a substantial kicker 300 mm high incorporating the wall and jacking rod starter bars. The wall forms are assembled and fixed together with the yokes, upper working platforms and jacking arrangement, after which the initial concrete lift is poured. The commencing rate of climb must be slow, to allow time for the first batch of concrete to reach a suitable condition before emerging from beneath the sliding formwork. A standard or planned rate of lift is usually reached within about 16 hours after commencing the lifting operation.

   Openings can be formed in the wall by using framed formwork with splayed edges, to reduce friction, tied to the reinforcement. Small openings can be formed using blocks of expanded polystyrene, which should be 75 mm less in width than the wall thickness so that a layer of concrete is always in contact with the sliding forms to eliminate friction. The concrete cover is later broken out and the blocks removed. Chases for floor slabs can be formed with horizontal boxes drilled to allow the continuity reinforcement to be passed through and to be bent back within the thickness of the wall so that when the floor slabs are eventually cast the reinforcement can be pulled out into its correct position.

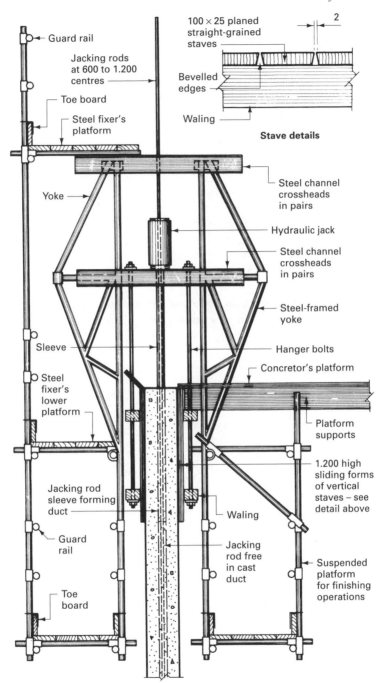

Guard rail

Jacking rods
at 600 to 1.200
centres

Toe board

Steel fixer's
platform

Yoke

Sleeve

Steel
fixer's
lower
platform

Jacking rod
sleeve forming
duct

Guard
rail

Toe
board

100 × 25 planed
straight-grained
staves

2

Bevelled
edges

Waling

Stave details

Steel channel
crossheads
in pairs

Hydraulic jack

Steel channel
crossheads
in pairs

Steel-framed
yoke

Hanger bolts

Concretor's platform

Platform
supports

1.200 high
sliding forms
of vertical
staves – see
detail above

Waling

Jacking
rod free
in cast
duct

Suspended
platform
for finishing
operations

Figure 9.1.4 Diagrammatic arrangement of sliding formwork

■ ■ ■ PERMANENT FORMWORK

In certain circumstances formwork is left permanently in place because of the difficulty and/or cost of removing it once the concrete has been cast. A typical example of these circumstances would be when a beam and slab raft foundation with shallow upstand beams and an *in-situ* slab have been constructed. Apart from the cost aspect, consideration must be given to any nuisance that such an arrangement could cause in the finished structure, such as the likelihood of fungal or insect attack and the possible risk of fire.

Permanent formwork can also be a means of utilising the facing material as both formwork and outer cladding, especially in the construction of *in-situ* reinforced concrete walls. The external face or cladding is supported by the conventional internal face formwork, which can in certain circumstances overcome the external strutting or support problems often encountered with high-rise structures.

This method is, however, generally limited to thin small modular facing materials, the size of which is governed by the supporting capacity of the internal formwork. Figure 9.1.5 shows a typical example of the application of this aspect of permanent formwork.

The methods described above for the construction of *in-situ* reinforced concrete walls can also be carried out by using a patent system of formwork.

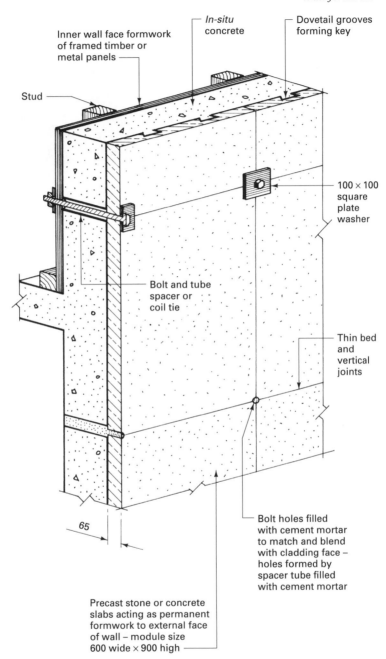

Inner wall face formwork of framed timber or metal panels

Stud

In-situ concrete

Dovetail grooves forming key

100 × 100 square plate washer

Bolt and tube spacer or coil tie

Thin bed and vertical joints

65

Bolt holes filled with cement mortar to match and blend with cladding face – holes formed by spacer tube filled with cement mortar

Precast stone or concrete slabs acting as permanent formwork to external face of wall – module size 600 wide × 900 high

Figure 9.1.5   Example of permanent formwork

9.2

PATENT FORMWORK

Patent formwork is sometimes called system formwork, and is usually identified by the manufacturer's name. All proprietary systems have the same common aim, and most are similar in their general approach to solving some of the problems encountered with formwork for modular designed or repetitive structures. As shown in the previous chapter, formwork is one area where the contractor has most control over the method and materials to be used in forming an *in-situ* reinforced concrete structure. In trying to design or formulate the ideal system for formwork the following must be considered:

- **Strength**   To carry the concrete and working loads.
- **Lightness without strength reduction**   To enable maximum-size units to be employed.
- **Durability without prohibitive costs**   To gain maximum usage of materials.
- **Good and accurate finish straight from the formwork**   To reduce the costly labour element of making good and patching, which in itself is a difficult operation to accomplish without it being obvious that this kind of treatment was found necessary.
- **Erection and dismantling times**.
- **Ability to employ unskilled or semi-skilled labour**.

Patent or formwork systems have been devised to satisfy most of the above requirements by the standardisation of forms and by easy methods of securing and bracing the positioned formwork.

The major component of any formwork system is the unit panel, which should fulfil the following requirements:

- available in a wide variety of sizes based on a standard module, usually multiples and submultiples of 300 mm;
- manufactured from durable materials;

- covered with a facing material that is durable and capable of producing the desired finish;
- interchangeable so that they can be used for beams, columns and slabs;
- formed so that they can be easily connected together to form large unit panels;
- lightweight so that individual unit panels can be handled without mechanical aid;
- designed so that the whole formwork can be assembled and dismantled easily by unskilled or semi-skilled labour;
- capable of being adapted so that non-standard width inserts of traditional formwork materials can be included where lengths or widths are not exact multiples of the unit panels.

Most unit panels consist of a framed tray made from light metal angle or channel sections stiffened across the width as necessary. The edge framing is usually perforated with slots or holes to take the fixing connectors and waling clips or clamps. The facing is of sheet metal or plywood, some manufacturers offering a choice. Longitudinal stiffening and support are given by clamping over the backs of the assembled panels, special walings of hollow section, or in many systems standard scaffold tubes. Vertical support where required can be given by raking standard adjustable steel props, or where heavy loadings are encountered most systems have special vertical stiffening and support arrangements based on designed girder principles, which also provide support for an access and working platform. Spacing between opposite forms is maintained using wall ties or similar devices, as previously described for traditional formwork.

Walls that are curved in plan can be formed in a similar manner to the straight wall techniques described above using a modified tray that has no transverse stiffening members, thus making it flexible. Climbing formwork can also be carried out using system formwork components, but instead of reversing the forms as described for traditional formwork climbing shoes are bolted to the cast section of the wall to act as bearing corbels to support the soldiers for each lift.

When forming beams and columns the unit panels are used as side or soffit forms held together with steel column clamps, or in the case of beams conventional strutting or using wall ties through the beam thickness. Some manufacturers produce special beam box clamping devices that consist of a cross-member surmounted by attached and adjustable triangulated struts to support the side forms.

Many patent systems for the construction of floor slabs that require propping during casting use basic components of unit panels, narrow width (150 mm) filler panels, special drop-head adjustable steel props, joists and standard scaffold tubes for bracing. The steel joists are lightweight, purpose made, and are supported on the secondary head of the prop in the opposite direction to the filler panels, which are also supported by the secondary or drop head of the prop. The unit panels are used to infill between the filler panels and pass over the support joist; the upper head of the prop is at the same level as the slab formwork and indeed is part of the slab soffit formwork.

After casting the slab and allowing it to gain sufficient strength the whole of the slab formwork can be lowered and removed, leaving the undisturbed prop head to give the partially cured slab a degree of support. This method enables the formwork to be removed at a very early stage, releasing the unit panels, filler pieces and joists for reuse, and at the same time accelerating the drying out of the concrete by allowing a free air flow on both sides of the slab.

Slabs of moderate spans (up to 7.500 m) that are to be cast between loadbearing walls or beams without the use of internal props can be formed using unit panels supported by steel telescopic floor centres. These centres are made in a simple lattice form extending in one or two directions according to span, and are light enough to be handled by one operative. They are precambered to compensate any deflections when loaded. Typical examples of system formwork are shown in Figs 9.2.1 to 9.2.4.

■ ■ ■ TABLE FORMWORK

This special class of formwork has been devised for use when casting large repetitive floor slabs in medium- to high-rise structures. The main objective is to reduce the time factor in erecting, striking and re-erecting slab formwork by creating a system of formwork that can be struck as an entire unit, removed, hoisted and repositioned without any dismantling.

The basic requirements for a system of table formwork are:

- a means of adjustment for aligning and levelling the forms;
- adequate means of lowering the forms so that they can be dropped clear of the newly cast slab; generally the provision for lowering the forms can also be used for final levelling purposes;
- means of manoeuvring the forms clear of the structure to a point where they can be attached to the crane for final extraction, lifting and repositioning ready to receive the next concrete pour operation;
- a means of providing a working platform at the external edge of the slab to eliminate the need for an independent scaffold, which would be obstructive to the system.

The basic support members are usually a modified version of inverted adjustable steel props. These props, suitably braced and strutted, carry a framed decking that acts as the soffit formwork. To manoeuvre the forms into a position for attachment to the crane a framed wheeled arrangement can be fixed to the rear end of the tableform so that the whole unit can be moved forward with ease. The tableform is picked up by the crane at its centre of gravity by removing a loose centre board to expose the framework. The unit is then extracted clear of the structure, hoisted in the balanced horizontal position, and lowered onto the recently cast slab for repositioning.

Another method, devised by Kwikform Ltd, uses a special lifting beam that is suspended from the crane at predetermined sling points, which are lettered so that the correct balance for any particular assembly can be quickly identified, each table having been marked with the letter point required. The lifting beam is connected to the working platform attachment of the tableform so that when the unit formwork is lowered and then extracted by the crane the tableform and the lifting beam are in

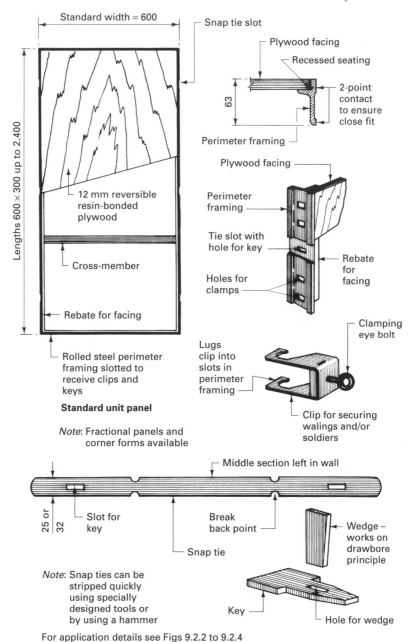

Standard width = 600

Snap tie slot

Plywood facing

Recessed seating

63

2-point contact to ensure close fit

Perimeter framing

Plywood facing

Lengths 600 × 300 up to 2.400

12 mm reversible resin-bonded plywood

Perimeter framing

Tie slot with hole for key

Cross-member

Holes for clamps

Rebate for facing

Rebate for facing

Rolled steel perimeter framing slotted to receive clips and keys

Standard unit panel

Note: Fractional panels and corner forms available

Lugs clip into slots in perimeter framing

Clamping eye bolt

Clip for securing walings and/or soldiers

Middle section left in wall

25 or 32

Slot for key

Break back point

Wedge – works on drawbore principle

Snap tie

Note: Snap ties can be stripped quickly using specially designed tools or by using a hammer

Key

Hole for wedge

For application details see Figs 9.2.2 to 9.2.4

Figure 9.2.1   System formwork: typical components

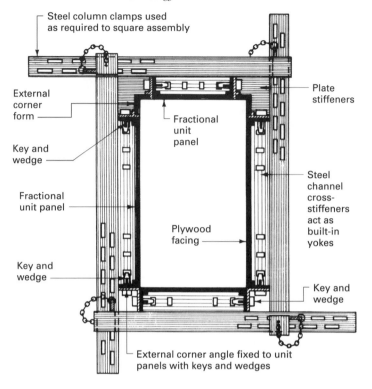

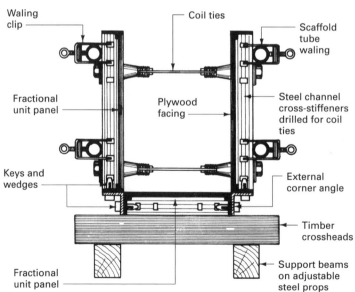

Figure 9.2.2   System formwork: columns and beams

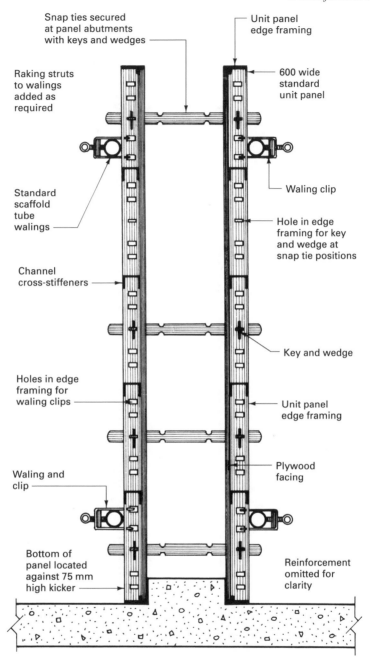

Snap ties secured at panel abutments with keys and wedges

Unit panel edge framing

Raking struts to walings added as required

600 wide standard unit panel

Standard scaffold tube walings

Waling clip

Hole in edge framing for key and wedge at snap tie positions

Channel cross-stiffeners

Key and wedge

Holes in edge framing for waling clips

Unit panel edge framing

Waling and clip

Plywood facing

Bottom of panel located against 75 mm high kicker

Reinforcement omitted for clarity

Figure 9.2.3 System formwork: walls

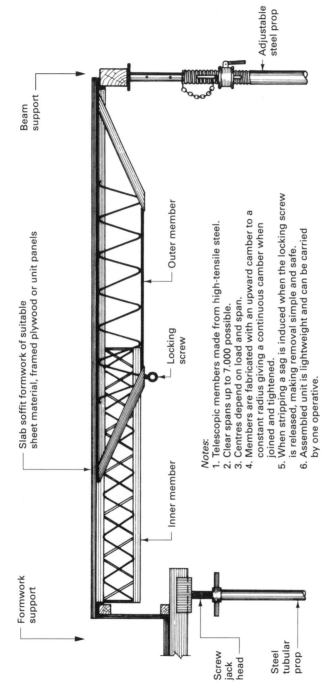

Adjustable steel prop

Beam support

Outer member

Inner member

Slab soffit formwork of suitable sheet material, framed plywood or unit panels

Locking screw

Formwork support

Screw jack head

Steel tubular prop

Notes:
1. Telescopic members made from high-tensile steel.
2. Clear spans up to 7.000 possible.
3. Centres depend on load and span.
4. Members are fabricated with an upward camber to a constant radius giving a continuous camber when joined and tightened.
5. When stripping a sag is induced when the locking screw is released, making removal simple and safe.
6. Assembled unit is lightweight and can be carried by one operative.

Figure 9.2.4   System formwork: slabs using telescopic centres

perfect balance. This method of system formwork is illustrated diagrammatically in Fig. 9.2.5.

The selection of formwork for multi-storey and complex structures is not a simple matter; it requires knowledge and experience in design appreciation, suitability of materials and site operations. Some large building contracting organisations will have their own specialist staff for designing and detailing formwork. Alternatively, as main contractors they may prefer to limit their direct involvement to administration only and seek to subcontract this area of work to a company that provides a combined design and install service. These subcontractors will have experienced and specialist site staff for the fabrication of formwork or if using a patent system to supervise the semi-skilled or unskilled labour being used. Badly designed and/or erected formwork can result in failure of the structure during the construction period or inaccurate and unacceptable members being cast, both of which can be financially disastrous for a company and ruinous to the firm's reputation.

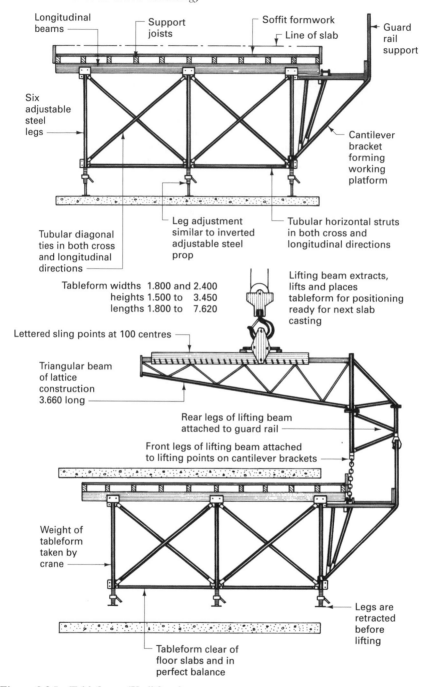

Longitudinal beams — Support joists — Soffit formwork — Line of slab — Guard rail support

Six adjustable steel legs

Cantilever bracket forming working platform

Tubular diagonal ties in both cross and longitudinal directions

Leg adjustment similar to inverted adjustable steel prop

Tubular horizontal struts in both cross and longitudinal directions

Tableform widths 1.800 and 2.400
heights 1.500 to 3.450
lengths 1.800 to 7.620

Lifting beam extracts, lifts and places tableform for positioning ready for next slab casting

Lettered sling points at 100 centres

Triangular beam of lattice construction 3.660 long

Rear legs of lifting beam attached to guard rail

Front legs of lifting beam attached to lifting points on cantilever brackets

Weight of tableform taken by crane

Legs are retracted before lifting

Tableform clear of floor slabs and in perfect balance

Figure 9.2.5   Tableforms: 'Kwikform' system

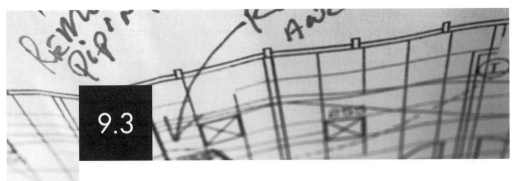

9.3

CONCRETE SURFACE FINISHES

The appearance of concrete members is governed by their surface finish, and this is influenced by three main factors: colour, texture and surface profile. The method used to produce the concrete member will have some degree of influence over the finish obtained, because a greater control of quality is usually possible with precast concrete techniques under factory-controlled conditions. Most precast concrete products can be cast in the horizontal position, which again promotes better control over the resultant casting, whereas *in-situ* casting of concrete walls and columns must be carried out using vertical-casting techniques, which not only have a lesser degree of control but also limit the types of surface treatment that can be successfully obtained direct from the mould or formwork.

■■■ COLOUR

The colour of concrete as produced direct from the mould or formwork depends upon the colour of the cement being used and, to a lesser extent, upon the colour of the fine aggregate. The usual methods of obtaining variations in the colour of finished concrete are:

- **Using a coloured cement** The range of colours available is limited, and most are pastel shades; if a pigment is used to colour cement the cement content of the mix will need to be increased by approximately 10% to counteract the loss of strength due to the colouring additive.
- **Using the colour of the coarse aggregate** The outer matrix or cement paste is removed to expose the aggregate, which imparts not only colour but also texture to the concrete surface.

Concrete can become stained during the construction period, resulting in a mottled appearance. Some of the causes of this form of staining are as follows:

- **Using different-quality timber within the same form**   Generally timber that has a high absorption factor will give a darker concrete than timber with a low absorption factor. The same disfigurement can result from using old and new timber in the same form, because older timber tends to give darker concrete than new timber with its probable higher moisture content and hence lower absorption.
- **Formwork detaching itself from the concrete**   This allows dirt and dust to enter the space and attach itself to the 'green' concrete surface.
- **Type of release agent used**   Generally the thinner the release agent used the better will be the result; even coating over the entire contact surface of the formwork or mould is also of great importance.

Staining occurring on mature concrete, usually after completion and occupation of the building, is very often due to bad design, poor selection of materials or poor workmanship. Large overhangs to parapets and sills without adequate throating can create damp areas that are vulnerable to algae or similar growths and pollution attack. Poor detailing of damp-proof courses can create unsightly stains because they fail to fulfil their primary function of providing a barrier to dampness infiltration. Efflorescence can occur on concrete surfaces, although it is not as common as on brick walls. The major causes of efflorescence on concrete are allowing water to be trapped for long periods between the cast concrete and the formwork, and poorly formed construction or similar joints allowing water to enter the concrete structure. Removal of efflorescence is not easy or always successful, and therefore the emphasis should always be on good design and workmanship in the first instance. Methods of efflorescence removal range from wire brushing through various chemical applications to mechanical methods such as grit-blasting.

■■■ TEXTURE

When concrete first became acceptable as a substitute for natural stone in major building works the tendency was to try to recreate the smooth surface and uniform colour possible with natural stones. This kind of finish is difficult to achieve using the medium of concrete for the following reasons:

- Natural shrinkage of concrete can cause hairline cracks on the surface.
- Texture and colour can be affected by the colour of the cement, water content, degree of compaction and the quality of the formwork.
- Pinholes on the surface can be caused by air being trapped between the concrete and the formwork.
- Rough patches can result from the formwork adhering to the concrete face.
- Grout leakage from the formwork can cause fins or honeycombing on the cast concrete.

Under site conditions it is not an easy task to keep sufficient control over the casting of concrete members to guarantee that these faults will not occur. A greater control is, however, possible with the factory-type conditions prevailing in a well-organised precast concrete works. Note that attempts to patch, mask or make good the above defects are nearly always visible.

Methods that can be used to improve the appearance of a concrete surface include:

- finishes obtained direct from the formwork or mould intended to conceal the natural defects by attracting the eye to a more obvious and visual point;
- special linings placed within the formwork to produce a smooth or profiled surface;
- removal of the surface matrix to expose the aggregate;
- applied surface finishes such as ceramic tiles, renderings and paints.

Formwork can be designed and constructed to highlight certain features, such as the joints between form members or between concrete pours or lifts, by adding to the inside of the form small fillets to form recessed joints, or conversely by recessing the formwork to form raised joints, the axiom being: if you cannot hide or mask a joint, make a feature of it. The use of sawn boards to imprint the grain pattern can give a visually pleasing effect to concrete surfaces, particularly if boards of a narrow width are used.

A wide variety of textured, patterned and profiled surfaces can be obtained by using different linings within the formwork. Typical materials used are thermoplastics, glass fibre mouldings, moulded rubber and PVC sheets; all of these materials can be obtained to form various patterns, giving many uses, and are easily removed from the concrete surface. Glass fibre and thermoplastic linings have the advantage of being mouldable to any reasonable shape or profile. Ribbed and similar profiles can be produced by fixing materials such as troughed or corrugated steel into the form as a complete lining or as a partial lining to give a panelled effect. Coarse fabrics such as hessian can also be used to produce various textures, provided the fabric is first soaked with a light mineral oil and squeezed to remove the excess oil before fixing, as a precaution against the fabric becoming permanently embedded on the concrete surface.

The most common method of imparting colour and texture to a concrete surface is to expose the coarse aggregate, which can be carried out by a variety of methods depending generally upon whether the concrete is 'green', mature, or yet to be cast. Brushing and washing is a common method employed on 'green' concrete, and is carried out using stiff wire or bristle brushes to loosen the surface matrix and clean water to remove the matrix and clean the exposed aggregate. The process should be carried out as soon as practicable after casting, which is usually within 2 to 6 hours of the initial pour but certainly not more than 18 hours after casting. When treating horizontal surfaces brushing should commence at the edges, whereas brushing to vertical surfaces should start at the base and be finished by washing from the top downwards. Skill is required with this method of exposing aggregates to ensure that only the correct amount of matrix is removed, the depth being dependent on the size of aggregate used. The use of retarding agents can be employed, but it is essential that these are used only in strict accordance with the manufacturer's instructions.

Treatments that can be given to mature concrete to expose the aggregate include tooling, bush hammering and blasting techniques. These methods are generally employed when the depth of the matrix to be removed is greater than that usually associated with the brushing method described above. The most suitable age for treating the mature concrete surface depends upon such factors as the type of cement used and the conditions under which the concrete is cured, the usual period being from 2 to 8 weeks after casting.

The usual hand-tooling methods that can be applied to natural stone, such as point tooling and chiselling, can be applied to a mature concrete surface but can be expensive if used for large areas. Tooling methods should not be used on concrete if gravel aggregates have been specified, because these tend to shatter and leave unwanted pits on the surface. Hard point tooling near to sharp arrises should also be avoided because of the tendency to spalling at these edges; to overcome this problem a small plain margin at least 10 mm wide should be specified.

Except for small areas, or where special treatment is required, bush hammering has largely replaced the hand tooling techniques to expose the aggregate. Bush hammers are power-operated, hand-held tools to which two types of head can be used to give a series of light but rapid blows (e.g. 1,750 blows per minute) to remove about 3 mm of matrix for each passing. Circular heads with 21 cutting points are easier to control if working on confined or small areas than the more common and quicker roller head with its 90 cutting points.

Blasting techniques using sand, shot or grit are becoming increasingly popular because these methods do not cause spalling at the edges and generally result in a very uniform textured surface. Blasting can be carried out at almost any time after casting but preferably within 16 hours to 3 days of casting, which gives a certain degree of flexibility to a programme of work. This method is of a specialist nature, and is normally carried out by a subcontractor who can usually expose between 4 and 6 m² of surface per hour. Sand is usually dispersed in a jet of water, whereas grit is directed onto the concrete surface from a range of about 300 mm in a jet of compressed air, and is the usual method employed. Shot is dispersed from a special tool with an enclosed head that collects and recirculates the shot.

Any method of removing the surface skin or matrix will reduce the amount of concrete cover over the reinforcement, and therefore if this type of surface treatment is to be used an adequate cover of concrete should be specified, such as 45 mm minimum, before surface treatment.

When exposing aggregates by the above methods the finished result cannot be assured, because the distribution of the aggregates throughout the concrete is determined by the way in which the fine and coarse aggregates disperse themselves during the mixing, placing and compacting operations. If a particular colour or type of exposed aggregate is required it will be necessary to use the chosen aggregate throughout the mix, which may be both uneconomic and undesirable.

A method that can be used for casting *in-situ* concrete and will ensure the required distribution of aggregate over the surface and enable a selected aggregate to be used on the surface is the aggregate transfer method. This entails sticking the aggregate to the rough side of pegboard sheets with a mixture of water-soluble cellulose compounds and sand fillers. The resultant mixture has a cream-like texture and is spread evenly over the pegboard surface to a depth of one-third the aggregate size. The aggregate may be sprinkled over the surface and lightly tamped; alternatively it may be placed by hand, after which the prepared board is allowed to set and dry for about 36 hours. This prepared pegboard is used as a liner to the formwork with a loose baffle of hardboard or plywood immediately in front of the aggregate face as protection during the pouring of the cement. The baffle is removed as work proceeds. This is not an easy method of obtaining a specific exposed aggregate finish to *in-situ* work, and it is generally recommended that, where possible, simpler and just as effective precast methods should be considered.

Precast concrete casting can be carried out in horizontal or vertical moulds. Horizontal casting can have surface treatments carried out on the upper surface or on the lower surface that is in contact with the mould base. Plain smooth surfaces can be produced by hand trowelling the upper surface, but this is expensive in labour costs and requires a high degree of skill if a level surface of uniform texture and colour is to be obtained, whereas a good plain surface can usually be achieved direct from the mould using the lower face method. Various surface textures can be obtained from upper face casting by using patterned rollers, profiled tamping boards, using a straight-edged tamping board drawn over the surface with a sawing action or by scoring the surface with brooms or rakes. Profiled finishes from lower face casting can be obtained by using a profiled mould base or a suitable base liner without any difficulty.

Exposed aggregate finishes can be obtained from either upper or lower face horizontal casting by spraying and brushing or by the tooling methods previously described. Alternatives for upper face casting are to trowel selected aggregates into the freshly cast surface or to trowel in a fine (10 mm) aggregate dry mortar mix and use a float with a felt pad to pick up the fine particles and cement, leaving a clean exposed aggregate finish.

The sand bed method of covering the bottom of the mould with a layer of sand and placing the selected aggregate into the sand bed and immediately casting the concrete over the layer of aggregate is the usual method employed with lower face casting. Cast-on finishes such as bricks, tiles and mosaic can be fixed in a similar manner to those described above for exposed aggregates. For all practical purposes vertical casting of precast concrete members needs the same considerations as those described previously in the context of *in-situ* casting.

Another surface treatment that could be considered for concrete is grinding and polishing, although this treatment is expensive and time consuming, being usually applied to surfaces such as slate, marble and terrazzo. The finish can be obtained by using a carborundum rotary grinder incorporating a piped water supply to the centre of the stone. The grinding operation should be carried out as soon as the mould or formwork has been removed, by first wetting the surface and then grinding, allowing the stone dust and water to mix, forming an abrasive paste. The operation is completed by washing and brushing to remove the paste residue. Dry grinding is possible, but this creates a great deal of dust, from which the operative's eyes and lungs must be protected by wearing protective masks.

It is very often wrongly assumed that by adding texture or shape to a concrete surface any deficiency in design, formwork fabrication or workmanship can be masked. This is not so; a textured surface, like the plain surface, will only be as good as the design and workmanship involved in producing the finished component, and any defects will be evident with all finishes.

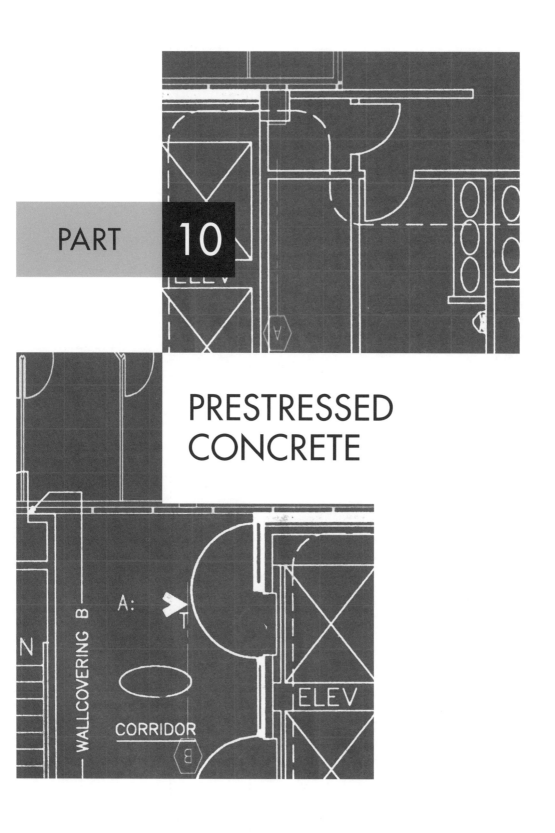

PART 10

PRESTRESSED CONCRETE

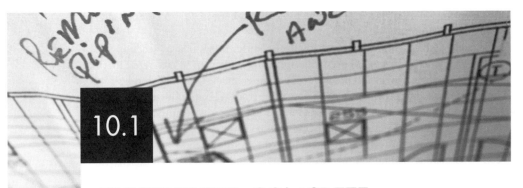

PRESTRESSED CONCRETE: PRINCIPLES AND APPLICATIONS

The basic principle of prestressing concrete is very simple. If a material has little tensile strength it will fracture immediately its own tensile strength is exceeded, but if such a material is given an initial compression then when load-creating tension is applied the material will be able to withstand the force of this load as long as the initial compression is not exceeded. At this stage in the study of construction technology readers will already be familiar with the properties of concrete, which result in a material of high compressive strength with low tensile strength. By combining concrete with steel reinforcing bars of the correct area and pattern, ordinary concrete can be given an acceptable amount of tensile strength. Prestressing techniques are applied to concrete in an endeavour to make full use of the material's high compressive strength.

Attempts were made at the end of the nineteenth century to induce a prestress into concrete, but these were largely unsuccessful because the prestress could not be maintained. A French civil engineer, Marie Eugene Leon Freyssinet (1879–1962), showed in the early 1920s how this problem could be overcome, and demonstrated the type of concrete and prestressing steel that was required. His most significant contributions were the quantitative assessment of creep and shrinkage, and the realisation that only a high-strength steel at a high stress would achieve a permanent prestress in concrete.

In normal reinforced concrete the designer is unable to make full use of the high tensile strength of steel or of the high compressive strength of the concrete. When loaded above a certain limit, tension cracks will occur in a reinforced concrete member, and these should not generally be greater than 0.3 mm in width, as recommended in BS 8110: *Structural use of concrete*. This stage of cracking will normally be reached before the full strength potential of either steel or concrete has been obtained. In prestressed concrete the steel is stretched and securely anchored within its mould box before the concrete is placed. After concreting the steel will then try to regain its original length, but because it is fully restricted it will be subjecting the concrete to a compressive force throughout its life. A comparison of methods is shown in Fig. 10.1.1.

Concrete while curing will shrink; it will also suffer losses in cross-section due to creep when subjected to pressure. Shrinkage and creep in concrete can normally be reduced to an acceptable level by using a material of high strength with a low workability. Mild steel will also suffer from relaxation losses, which is the phenomenon of the stresses in steel under load decreasing towards a minimum value after a period of time. This can be counteracted by increasing the initial stress in the steel. If mild steel is used to induce a compressive force into a concrete member the amount of shrinkage, creep and relaxation that occurs will cancel out any induced stress. The special alloy steels used in prestressing, however, have different properties, enabling the designer to induce extra stress into the concrete member, thus counteracting any losses due to shrinkage and creep and at the same time maintaining the induced compressive stress in the concrete component.

The high-quality strength concrete specified for prestress work should take into account the method of stressing. For pre-tensioned work a minimum 28-day cube strength of 40 N/mm^2 is required, whereas for post-tensioned work a minimum 28-day cube strength of 30 N/mm^2 is required. Steel in the form of wire or bars used for prestressing should conform to the recommendations of BS 5896: *Specification for high tensile steel wire and strand for the prestressing of concrete*, which covers steel wire manufactured from cold-drawn plain carbon steel. The wire can be plain round, crimped or indented with a diameter range of 2 to 7 mm. Crimped and indented bars will develop a greater bond strength than plain round bars, and are available in 4, 5 and 7 mm diameters. Another form of stressing wire or tendon is strand, which consists of a straight core wire around which are helically wound further wires to form a 6 over 1 or 7-wire strand, or a 9 over 9 over 1 giving a 19-wire strand tendon.

Seven-wire strand is the easiest to manufacture and is in general use for tendon diameters up to 15 mm. The wire used to form the strand is cold-drawn from plain carbon steel, as recommended in BS 5896. To ensure close contact of the individual wires in the tendon the straight core wire is usually 2% larger in diameter than the outer wires, which are helically wound around it at a pitch of 12 to 16 times the nominal diameter of the strand.

Tendons of strand can be used singly or in groups to form a multi-strand cable. The two major advantages of using strand are:

■ A large prestressing force can be provided in a restricted area.
■ It can be produced in long flexible lengths and can therefore be stored on drums, thus saving site space and reducing site labour requirements by eliminating the site fabrication activity.

Having now considered the material requirements, the basic principles of prestressing can be considered. It works in the same way that compressing several bricks together stops the middle ones falling out. A prestressing force inducing precompression into a concrete member can be achieved by anchoring a suitable tendon at one end of the member and applying an extension force at the other end,

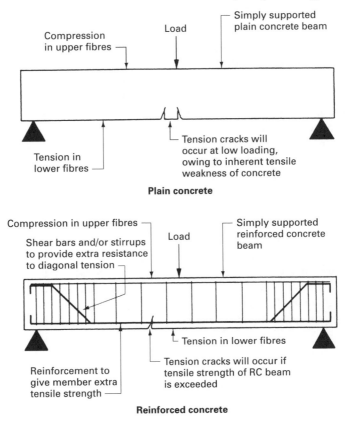

Compression
in upper fibres

Load

Simply supported
plain concrete beam

Tension in
lower fibres

Tension cracks will
occur at low loading,
owing to inherent tensile
weakness of concrete

Plain concrete

Compression in upper fibres

Shear bars and/or stirrups
to provide extra resistance
to diagonal tension

Load

Simply supported
reinforced concrete
beam

Tension in lower fibres

Reinforcement to
give member extra
tensile strength

Tension cracks will occur if
tensile strength of RC beam
is exceeded

Reinforced concrete

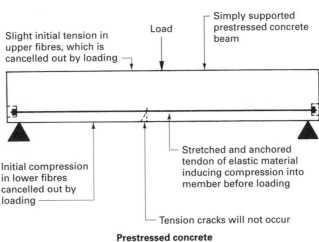

Slight initial tension in
upper fibres, which is
cancelled out by loading

Load

Simply supported
prestressed concrete
beam

Initial compression
in lower fibres
cancelled out by
loading

Stretched and anchored
tendon of elastic material
inducing compression into
member before loading

Tension cracks will not occur

Prestressed concrete

Figure 10.1.1   Structural concrete: comparison of methods

which can be anchored when the desired extension has been reached. Upon release the anchored tendon, in trying to regain its original length, will induce a compressive force into the member. Figure 10.1.2 shows a typical arrangement in which the tendon inducing the compressive force is acting about the neutral axis and is stressed so that it will cancel out the tension induced by the imposed load *W*. The stress diagrams show that the combined or final stress will result in a compressive stress in the upper fibres equal to twice the imposed load. The final stress must not exceed the characteristic strength of the concrete as recommended in BS 8110, and if the arrangement given in Fig. 10.1.2 is adopted the stress induced by the imposed load will be only half its maximum.

To obtain a better economic balance the arrangement shown in Fig. 10.1.3 is normally adopted, where the stressing tendon is placed within the lower third of the section. The basic aim is to select a stress that, when combined with the dead load, will result in a compressive stress in the lower fibres equal to the maximum stresses induced by any live loads, resulting in a final stress diagram having in the upper fibres a compressive stress equal to the characteristic strength of the concrete and a zero stress in the bottom fibres. This is the pure theoretical case, and is almost impossible to achieve in practice, but provided any induced tension occurring in the lower fibres is not in excess of the tensile strength of the concrete used, an acceptable prestressed condition will exist.

▨▨■ PRESTRESSING METHODS

There are two methods of producing prestressed concrete:

- pre-tensioning;
- post-tensioning.

PRE-TENSIONING

In this method the wires or cables are stressed before the concrete is cast around them. The stressing wires are anchored at one end of the mould and are stressed by hydraulic jacks from the other end until the required stress has been obtained. It is common practice to overstress the wires by some 10% to counteract the anticipated losses that will occur because of creep, shrinkage and relaxation. After stressing the wires the side forms of the mould are positioned and the concrete is placed around the tensioned wires; the casting is then usually steam-cured for 24 hours to obtain the desired characteristic strength, a common specification being 28 N/mm^2 in 24 hours. The wires are cut or released, and the bond between the stressed wires and the concrete will prevent the tendons from regaining their original length, thus inducing the prestress.

At the extreme ends of the members the bond between the stressed wires and concrete will not be fully developed, because low frictional resistance will result in a contraction and swelling at the ends of the wires, forming what is in effect a cone-shape anchor. The distance over which this contraction takes place is called the **transfer length** and is equal to 80 to 120 times the wire diameter. Usually

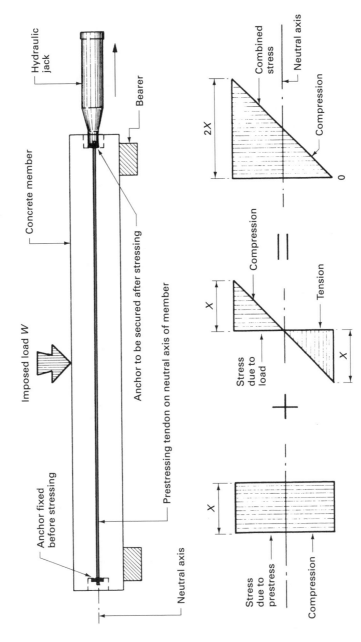

Figure 10.1.2   Prestressing principles: 1

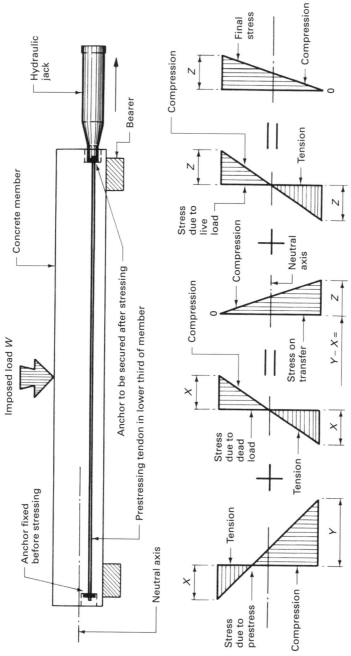

Figure 10.1.3 Prestressing principles: 2

small-diameter wires (2 to 5 mm) are used so that for any given total area of stressing wire a greater surface contact area is obtained. The bond between the stressed wires and concrete can also be improved by using crimped or indented wires.

Pre-tensioning is the prestressing method used mainly by manufacturers of precast components such as floor units and slabs, employing the long line method of casting, where precision metal moulds up to 120.000 m long can be used with spacers or dividing plates positioned along the length to create the various lengths required; a typical arrangement is shown in Fig. 10.1.4.

POST-TENSIONING

In this method the concrete is cast around ducts in which the stressing tendons can be housed, and the stressing is carried out after the concrete has hardened. The tendons are stressed from one or both ends, and when the stress required has been reached the tendons are anchored at their ends to prevent them from returning to their original length, thus inducing the compressive force. The anchors used form part of the finished component. The ducts for housing the stressing tendons can be formed by using flexible steel tubing or inflatable rubber tubes. The void created by the ducting will enable the stressing cables to be threaded prior to placing the concrete, or they can be positioned after the casting and curing of the concrete has been completed. In both cases the remaining space within the duct should be filled with grout to stop any moisture present setting up a corrosive action, and to assist in stress distribution. A typical arrangement is shown in Fig. 10.1.5. For comparison of anchorages and their performance under test, reference can be made to BS EN 13391: *Mechanical tests for post-tensioning systems*. System types and comparisons are shown in the next chapter.

Post-tensioning is the method usually employed where stressing is to be carried out on site. Curved tendons can be used where the complete member is to be formed by joining together a series of precast concrete units, or where negative bending moments are encountered. Figure 10.1.6 shows diagrammatically various methods of overcoming negative bending moments at fixed ends and for continuous spans. Figure 10.1.7 shows a typical example of the use of curved tendons in the cross-members of a girder bridge. Another application of post-tensioning is in the installation of ground anchors.

▨▨■ GROUND ANCHORS

A ground anchor is basically a prestressing tendon embedded and anchored into the soil to provide resistance to structural movement of a member by acting on a 'tying back' principle. Common applications include anchoring or tying back retaining walls and anchoring diaphragm walls, particularly in the context of deep excavations. This latter application also has the advantage of providing a working area entirely free of timbering members such as struts and braces. Ground anchors can also be used in basement and similar constructions for anchoring the foundation slab to resist uplift pressures and to prevent flotation especially during the early stages of construction.

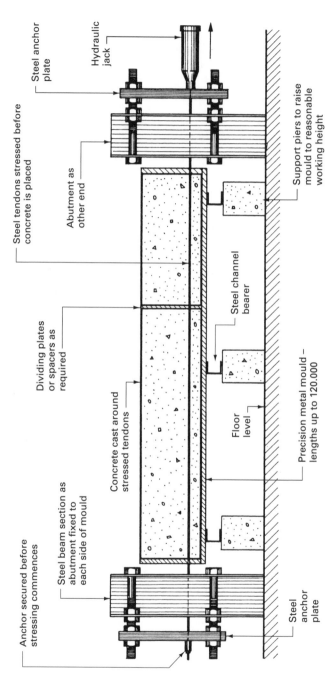

Figure 10.1.4   Typical pre-tensioning arrangement

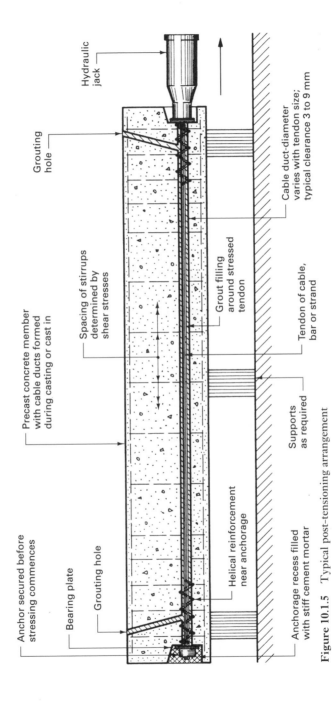

Figure 10.1.5 Typical post-tensioning arrangement

Hydraulic jack

Grouting hole

Cable duct-diameter varies with tendon size; typical clearance 3 to 9 mm

Precast concrete member with cable ducts formed during casting or cast in

Spacing of stirrups determined by shear stresses

Grout filling around stressed tendon

Tendon of cable, bar or strand

Supports as required

Anchor secured before stressing commences

Bearing plate

Grouting hole

Helical reinforcement near anchorage

Anchorage recess filled with stiff cement mortar

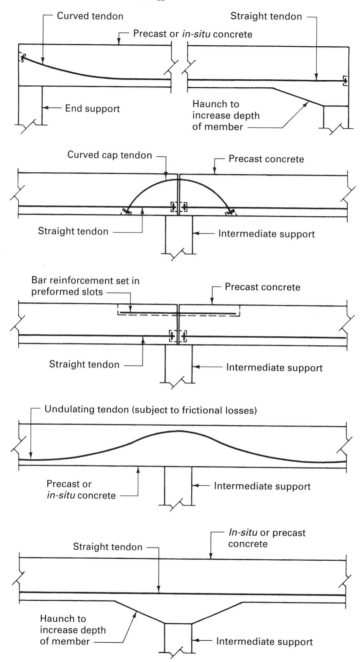

Figure 10.1.6   Prestressing: overcoming negative bending moments

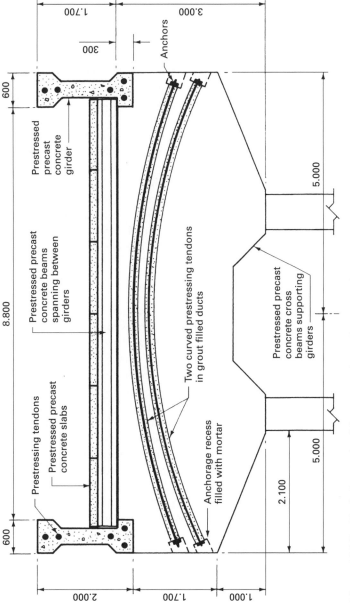

Figure 10.1.7   Typical example of the structural use of prestressed concrete

Ground anchors are known by their method of installation, such as grouted anchors, or by the nature of the subsoil into which they are embedded, such as rock anchors.

ROCK ANCHORS

These have been used successfully for many years, and can be formed by inserting a prestressing bar into a pre-drilled hole. The leading end of the bar has expanding sleeves that grip the inside of the bored hole when the bar is rotated to a recommended torque to obtain the desired grip. The anchor bar is usually grouted over the fixed or anchorage length before being stressed and anchored at the external face.

Alternatively the anchorage of the leading end can be provided by grout injection, relying on the bond developed between a ribbed sleeve and the wall of the bored hole. A dense high-strength grout is required over the fixed length to develop sufficient resistance to pull out when the tendon is stressed. The unbonded or elastic length will need protection against corrosion, and this can be provided by protective coatings such as bitumen or rubberised paint, casings of PVC, or wrappings of greased tape. Alternatively a full-length protection can be given by filling the void with grout after completion of the stressing operation (see Fig. 10.1.8).

INJECTION ANCHORS

The knowledge and experience gained in the use of rock anchors has led to the development of suitable ground anchorage techniques for most subsoil conditions, except for highly compressible soils such as alluvial clays and silts. Injection-type ground anchorages have proved to be suitable for most cohesive and non-cohesive soils. Basically a hole is bored into the soil using a flight auger with or without water-flushing assistance; casings or linings can be used where the borehole would not remain open if unlined. The prestressing tendon or bar is placed into the borehole and pressure-grouted over the anchorage length. For protection purposes the unbonded or elastic length can be grouted under gravity for permanent ground anchors, or covered with an expanded polypropylene sheath for temporary anchors. Anchor boreholes in clay soils are usually multi-under-reamed to increase the bond, using special expanding cutter or brush tools. Gravel placement ground anchors can also be used in clay and similar soils for lighter loadings. In this method an irregular gravel is injected into the borehole over the anchorage length. A small casing with a non-recoverable point is driven into the gravel plug to force the aggregate to penetrate the soil around the borehole. The stressing tendon is inserted into the casing and pressure-grouted over the anchorage length as the casing is removed. Typical ground anchor examples are shown in Fig. 10.1.9.

▓■■ ADVANTAGES AND DISADVANTAGES

The preceding text clarifies that in pre-tensioning it is the bond between the tendon and the concrete that prevents the prestressing wire from returning to its original

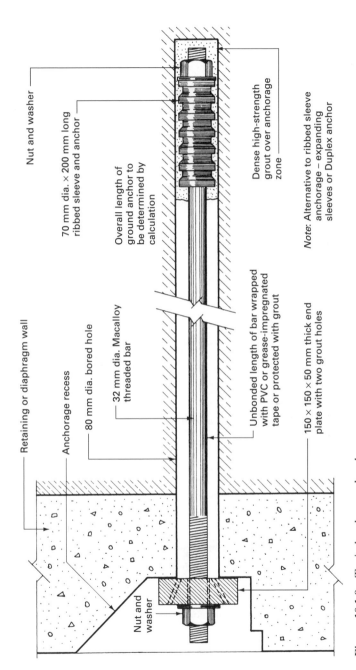

Nut and washer

70 mm dia. × 200 mm long ribbed sleeve and anchor

Overall length of ground anchor to be determined by calculation

Dense high-strength grout over anchorage zone

Note: Alternative to ribbed sleeve anchorage – expanding sleeves or Duplex anchor

Retaining or diaphragm wall

Anchorage recess

80 mm dia. bored hole

32 mm dia. Macalloy threaded bar

Unbonded length of bar wrapped with PVC or grease-impregnated tape or protected with grout

150 × 150 × 50 mm thick end plate with two grout holes

Nut and washer

Figure 10.1.8 Typical rock ground anchor

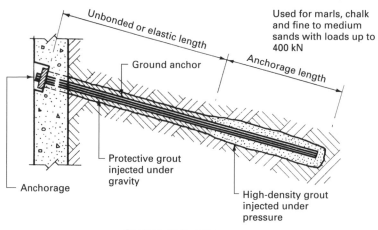

Used for marls, chalk
and fine to medium
sands with loads up to
400 kN

Unbonded or elastic length

Anchorage length

Ground anchor

Protective grout
injected under
gravity

Anchorage

High-density grout
injected under
pressure

Straight shaft anchor

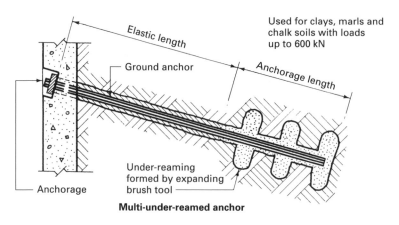

Used for clays, marls and
chalk soils with loads
up to 600 kN

Elastic length

Anchorage length

Ground anchor

Under-reaming
formed by expanding
brush tool

Anchorage

Multi-under-reamed anchor

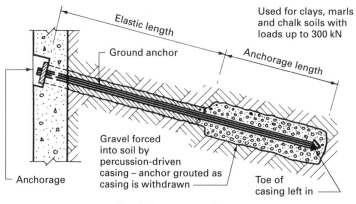

Used for clays, marls
and chalk soils with
loads up to 300 kN

Elastic length

Anchorage length

Ground anchor

Gravel forced
into soil by
percussion-driven
casing – anchor grouted as
casing is withdrawn

Anchorage

Toe of
casing left in

Gravel placement anchor

Figure 10.1.9   Typical injection ground anchors

length, and in post-tensioning it is the anchorages that prevent the stressing tendon returning to its original length.

The advantages and disadvantages of prestressed concrete when compared with conventional reinforced concrete can be summarised as follows:

Advantages

- Makes full use of the inherent compressive strength of concrete.
- Makes full use of the special alloy steels used to form prestressing tendons.
- Eliminates tension cracks, thus reducing the risk of corrosion of steel components.
- Reduces shear stresses.
- For any given span and loading condition a member with a smaller cross-section can be used, giving a reduction in weight and height.
- Individual units can be joined together to act as a single member.

Disadvantages

- High degree of control of materials, design and workmanship is required.
- Special alloy steels are dearer than mild steels.
- Extra cost of special equipment required to carry out stressing activities.

As a general comparison between the two structural media under consideration it is usually found that:

- For spans of up to 6.000 m traditional reinforced concrete is the most economic method.
- For spans between 6.000 and 9.000 m the two mediums are compatible.
- For spans of over 9.000 m prestressed concrete is generally more economical than reinforced concrete.

10.2

PRESTRESSED CONCRETE SYSTEMS

The prestressing of concrete is usually carried out by a specialist contractor or alternatively by the main contractor using a particular system and equipment. The basic conception and principles of prestressing are common to all systems; it is only the type of tendon, type of anchorage and stressing equipment that vary. The following established systems are typical and representative of the methods available.

■■■ BBRV (SIMON-CARVES LTD)

This is a unique system of prestressing developed by four Swiss engineers, namely Birkenmaier, Brandestini, Ros and Vogt, whose initials are used to name the system. It differs from other systems in that multi-wire cables capable of providing prestressing forces from 300 to 7,800 kN are used, and each wire is anchored at each end by means of enlarged heads formed on the wire. The cables or tendons are purpose-made to suit individual requirements, and may comprise up to 121 wires. The high-tensile steel wire conforming to the recommendations of BS 5896: *Specification for high tensile steel wire and strand for the prestressing of concrete* is cut to the correct length, and any sheathing required is threaded on together with the correct anchorage before the button heads are formed using special equipment. The completed tendons can be coiled or left straight for delivery to site. Four types of stressing anchor are available, the choice being dependent on the prestressing force being induced, whereas the fixed anchors come in three forms. The finished tendon is fixed in the correct position to the formwork before concreting; when the concrete has hardened the tendons are stressed, and grout is injected into the sheathing. If unsheathed tendons are to be drawn into preformed ducts the anchor is omitted from one and is fixed after drawing through the tendon, the button heads being formed with a portable machine.

Tendons in this system are tensioned by using a special hydraulic jack of the centre hole type with capacities ranging from 30 to 800 tonnes. A pull rod or pull sleeve, depending on anchor type, is coupled to the basic element carrying the wires. A locknut, stressing stool, hydraulic jack and dynamometer are then threaded on. The applied stressing force can be read off the dynamometer, whereas the actual extension achieved can be seen by the scale engraved on the jack.

After stressing, the locknut is tightened up and the jack released before grouting takes place. Losses due to friction, shrinkage and creep can be overcome by restressing at any time after the initial stressing operation so long as the tendons have not been grouted.

In common with most other prestressing systems, tendons in long continuous members can be stressed in stages without breaking the continuity of the design. The first tendon length is stressed and grouted before the second tendon length is connected to it using a coupling anchor, after which it is stressed and grouted, before repeating the procedure for any subsequent lengths. Alternatively, lengths of unstressed tendons can be coupled together and the complete tendon stressed from one or both ends. Figure 10.2.1 shows a typical BBRV stressing arrangement.

▨▧■■ CCL (CCL SYSTEMS LTD – NOW ANCON CCL LTD)

To form the tendon, this post-tensioning system uses a number of strands, ranging from 4 to 31, according to the system being employed, giving a range of prestressing forces from 450 to 5,000 kN. CCL established two basic systems, namely Cabco and Multiforce. These have since been further developed by the company, but are retained here to illustrate a principle of post-tensioning. In the Cabco system each strand or wire is stressed individually, the choice of hydraulic jack being governed by the size of strand being used. The system is fast and, being manually operated, eliminates the need for lifting equipment. By applying the total tendon force in stages, problems such as differential elastic shortening and out-of-balance forces are reduced. Curved tendons are possible using this system without the need for spacers, but spacers are recommended for tendons over 30.000 m long.

The alternative Multiforce system uses the technique of simultaneously stressing all the strands forming the tendon. In both systems the basic anchorages are similar in design. The fixed or dead-end anchorage has a tube unit to distribute the load, a bearing plate and a compressing grip fitted to each strand. If the fixed anchorage is to be totally embedded in concrete a compressible gasket and a bolted-on retaining plate are also used, a grout vent pipe being inserted into the grout hole of the tube unit. The compression grips consist of an outer sleeve with a machined and hardened insert, which can be fitted while making up the tendon or installed after positioning the tendon. The stressing anchor is a similar device, consisting of a tube unit, bearing plate and wedges working on a collet principle to secure the strands. The wedges are driven home to a 'set' by the hydraulic jack used for stressing the tendon before the jack is released and the stress is transferred. Figure 10.2.2 shows a typical CCL Cabco stressing arrangement.

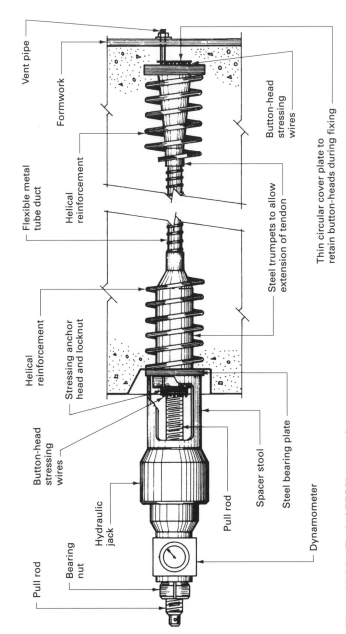

Figure 10.2.1 Typical BBRV prestressing arrangement

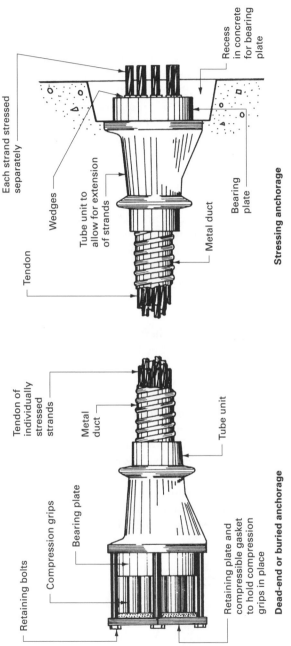

Each strand stressed separately

Tendon

Wedges

Tube unit to allow for extension of strands

Metal duct

Recess in concrete for bearing plate

Bearing plate

Stressing anchorage

Tendon of individually stressed strands

Retaining bolts

Compression grips

Bearing plate

Metal duct

Tube unit

Retaining plate and compressible gasket to hold compression grips in place

Dead-end or buried anchorage

Figure 10.2.2   Typical details of the CCL Cabco system of prestressing

■■■ DYWIDAG (DIVIDAG STRESSED CONCRETE LTD)

This post-tensioning system uses single- or multiple-bar tendons with diameters ranging from 12 to 36 mm for single-bar applications and 16 mm diameter threaded bars for multiple-bar tendons, giving prestressing forces up to 950 kN for single-bar tendons and up to 2,000 kN for multiple-bar tendons. Both forms of tendon are placed inside a thin-wall corrugated sheathing, which is filled with grout after completion of the stressing operation. The single-bar tendon can be of a smooth bar with cold-rolled threads at each end to provide the connection for the anchorages, or it can be a threadbar that has a coarse thread along its entire length, providing full mechanical bond. A threadbar is used for all multiple-bar tendons.

Two forms of anchorage are available: the bell anchor and the square or rectangular plate anchor. During stressing, using a hydraulic jack acting against the bell or plate anchor, the tendon is stretched and at the same time the anchor nut is being continuously screwed down to provide the transfer of stress when the specified stress has been reached and the jack is released. A counter shows the number of revolutions of the anchor nut and the amount of elongation of the tendon. As in other similar systems the tendons can be restressed at any time before grouting. Figure 10.2.3 shows typical details of a single-bar tendon arrangement.

■■■ MACALLOY (BRITISH STEEL CORPORATION – NOW CORUS UK LTD)

Like the method previously described, the Macalloy system uses single- or multiple-bar tendons. The bars used are of a cold-worked high-strength alloy steel threaded at each end to provide an anchorage connection, and are available in lengths up to 18.000 m with nominal diameters ranging from 20 to 40 mm, giving prestressing forces up to 875 kN for single-bar tendons and up to 3,500 kN using four 40 mm diameter bars to form the tendon.

The fixed anchorage consists of an end plate drilled and tapped to receive the tendon, whereas the stressing anchor consists of a similar plate complete with a grouting flange and anchor nut. Stressing is carried out using a hydraulic jack operating on a drawbar attached to the tendon, the tightened anchor nut transferring the stress to the member upon completion of the stressing operation. The prestressing is completed by grouting in the tendon after all the stressing and any necessary restressing has been completed. Figure 10.2.4 shows typical details of a single-bar arrangement.

■■■ PSC FREYSSINET LTD

This post-tension system uses strand to form the tendon and is available in three forms: the Freyssi-monogroup, Freyssinet multistrand and PSC monostrand. Monogroup tendons are composed of 7, 13, 15 or 19 wire strands stressed in a single pull by the correct model of hydraulic jack, giving prestressing forces of up to 5,000 kN. The anchorages consist of a cast iron guide shaped to permit the deviation of the strands to their positions in the steel anchor block, where they are

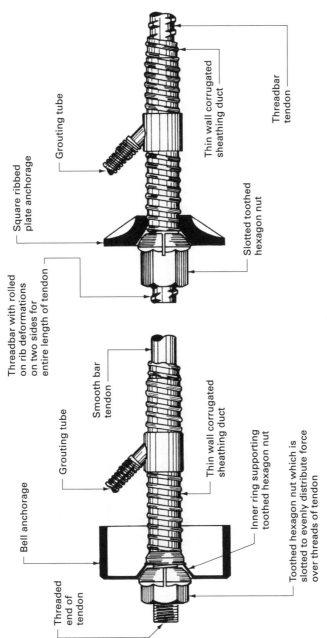

Figure 10.2.3 Typical stressing anchors for Dywidag post-tensioning system

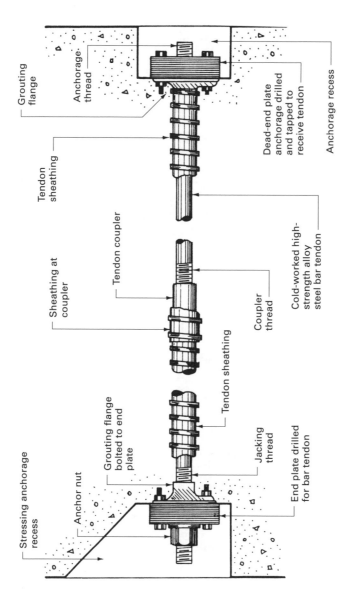

Figure 10.2.4   Typical Macalloy prestressing system details

secured by collet-type jaws. During stressing, 12 wires of the tendon are anchored to the body of the jack, the remaining wires passing through the jack body to an anchorage at the rear.

The multistrand system is based upon the first system of post-tension ever devised, and consists of a cable tendon made from 12 high-tensile steel wires laid parallel to one another and taped together, resulting in a tendon that is flexible and compact. Two standard cable diameters are produced, giving tendon sizes of 29 and 33 mm capable of taking prestressing forces up to 750 kN. The anchorages consist of two parts: the outer reinforced concrete cylindrical body has a tapered hole to receive a conical wedge, which is grooved or fluted to receive the wires of the tendon. The 12 wires of the cable are wedged into tapered slots on the outside of the hydraulic jack body during stressing, and when this operation has been completed the jack drives home the conical wedge to complete the anchorage.

Monostrand uses a four- or seven-strand tendon for general prestressing or a three-strand tendon developed especially for prestressing floor and roof slabs. This system is intended for the small to medium range of prestressing work requiring a prestressing force not exceeding 2,000 kN, and as its title indicates each strand in the group is stressed separately, requiring only a light compact hydraulic jack. All PSC system tendons are encased in a steel sheath and grouted after completion of the stressing operation. Typical details are shown in Fig. 10.2.5.

■ ■ ■ SCD (STRESSED CONCRETE DESIGN LTD)

This post-tensioning system offers three variations of tendon and/or stressing: multigrip circular, monogrip circular and monogrip rectangular. The multigrip circular system uses a tendon of 7, 12, 13 or 19 strands, forming a cable tendon capable of accepting prestressing forces up to 5,000 kN. The anchorages consist of a cast iron guide plate, enabling the individual strands of the cable to fan out and pass through a bearing plate, where they are secured with steel collet-type wedges. The strands pass through the body of the hydraulic jack to a rear anchorage and are therefore stressed simultaneously.

The monogrip circular tendons are available as a single-, 4-, 7- or 12-strand cable in which each strand is stressed individually, and the whole tendon is capable of taking prestressing forces up to 3,200 kN. Each strand in the tendon is separated from adjacent strands by means of circular spacers at 2.000 m centres, or less if the tendon is curved. The anchorages are similar in principle to those already described for multigrip tendons.

The monogrip rectangular system, by virtue of using a rectangular tendon composed of 3 to 27 strands capable of taking a prestressing force of up to 3,900 kN, affords maximum eccentricity with a wide range of tendon sizes. The anchorage guide plate and bearing plates work on the same principle as described above for the multigrip method. In all methods the tendon is encased in a sheath unless preformed ducts have been cast; upon completion of the stressing operation all tendons are grouted in. See Fig. 10.2.6 for typical details.

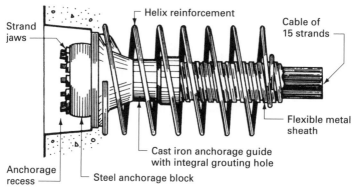

Typical Freyssi-monogroup anchorage

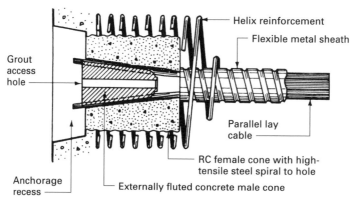

Typical Freyssinet multistrand anchorage

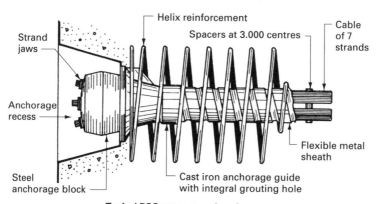

Typical PSC monostrand anchorage

Figure 10.2.5   Typical PSC prestressing system

Note: In all cases anchorage zone helix reinforcement would
be used according to design

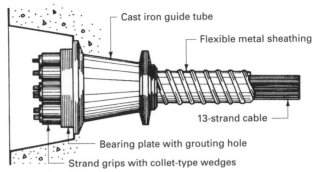

- Cast iron guide tube
- Flexible metal sheathing
- 13-strand cable
- Bearing plate with grouting hole
- Strand grips with collet-type wedges

Typical multigrip circular cable anchorage

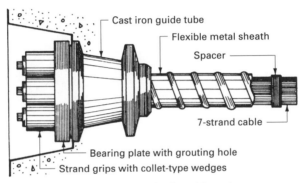

- Cast iron guide tube
- Flexible metal sheath
- Spacer
- 7-strand cable
- Bearing plate with grouting hole
- Strand grips with collet-type wedges

Typical monogrip circular cable anchorage

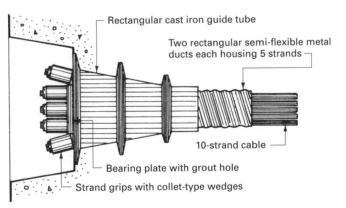

- Rectangular cast iron guide tube
- Two rectangular semi-flexible metal ducts each housing 5 strands
- 10-strand cable
- Bearing plate with grout hole
- Strand grips with collet-type wedges

Typical rectangular cable anchorage

Figure 10.2.6 Typical SCD prestressing system details

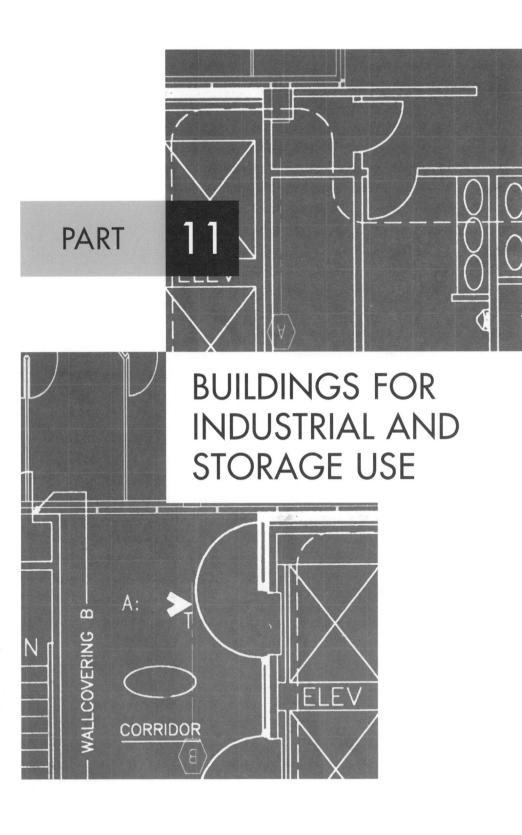

PART **11**

BUILDINGS FOR INDUSTRIAL AND STORAGE USE

FACTORY BUILDINGS: ROOFS

For a review of the general concepts of a roof's function, construction techniques and methods of covering, the reader is recommended to review Parts 6 and 10 in the associated volume, *Construction Technology*, before proceeding with this chapter. This advanced level of study is concerned mainly with particular building types such as those suitable as small to large industrial or factory buildings.

Buildings of this nature set the designer two main problems:

- production layout at floor level and the consequent need for large unobstructed areas by omitting as far as practicable internal roof supports such as loadbearing walls and columns;
- provision of natural daylight from the roof over the floor area below.

The amount of useful daylight that can penetrate into a building from openings in side and end walls is very limited and depends upon such factors as height of windows above the working plane or surface, sizes of windows and the arrangement of windows. Buildings with spans in excess of 18.000 m will generally need some form of overhead supplementary lighting. In single-storey buildings this can take the form of glazed rooflights, but in multi-storey buildings the floors beneath the top floor will have to have the natural daylighting from the side and end walls supplemented by permanent artificial lighting.

The factors to be considered when designing or choosing a roof type or profile for a factory building in terms of rooflighting are:

- amount of daylight required;
- spread of daylight required over the working plane;
- elimination of solar heat gain and solar glare.

The amount of daylight required within a building is usually based upon the **daylight factor**, which can be defined as 'the illumination at a specific point indoors expressed as a percentage of the simultaneous horizontal illumination

Table 11.1.1   Daylight factors

Building type	Daylight factor (%)	Recommendations
Dwellings:		
kitchen	2	Over at least 50% of floor area (min. approx. 4.5 m^2)
living room	1	Over at least 50% of floor area (min. approx. 7.0 m^2)
bedroom	0.5	Over at least 75% of floor area (min. approx. 5.5 m^2)
Schools	2	Over all teaching areas and kitchens
Hospitals	1	Over all wards
Offices:		
general	1	With side lighting at approx. 3.75 m^2 penetration
general	2	With top light over whole area
drawing	6	On drawing boards
drawing	2	Over remainder of work areas
typing and computing	4	Over whole working area
Laboratories	3–6	Depending on dominance of side or top lighting
Factories	5	General recommendation
Art galleries	6	Max. on walls or screens where no special problems of fading
Churches	1	General, over whole area
Churches	1.5–2	In sanctuary areas
Public buildings	1	Depending on function, figure given can be taken as minimum

outdoors under an unobstructed overcast sky'. The minimum daylight factor for factories depends on usage and work function within the building. Table 11.1.1 provides some comparison. A figure of 3 to 6% may be acceptable in most factory assembly situations, depending on the intensity of work.

More accurate determination of daylight factor can be achieved by plotting the particular roof profile and glass area with the use of a **daylight protractor**. These are made available in literature pack AP 68 from the Building Research Establishment, with further information in literature pack AP 220. The BRE Report BR 288: *Designing buildings for daylight* is also helpful, with worked examples and practical exercises relating to daylight design.

An estimation of glazed area can be made using a rule of thumb that gives a ratio of one-fifth glass to floor area. For example, assuming a daylight factor of 3% is required over a floor area of 500 m^2, then the area of glass can be calculated as:

$$\text{Daylight factor} \times \text{floor area} \times 5 = \frac{3}{100} \times 500 \times 5$$

$$= 75 \text{ m}^2$$

It must be emphasised that this one-fifth rule of thumb is a preliminary design aid, and the result obtained should be checked by a more precise method. It should also be compared with current thermal insulation requirements for maximum areas of glazing to industrial and storage buildings. This is considered later in the chapter and defined in Building Regulations, Approved Document L2.

For an even spread of light over the working plane the ratio between the spacing and height of the rooflights is particularly important unless a monitor roof is used, which gives a reasonable even spread of light by virtue of its shape. For typical ratios see Fig. 11.1.1. If the ratios shown in Fig. 11.1.1 are not adopted the result could give a marked difference in illumination values at the working plane, resulting in light and darker areas.

Another factor to be considered when planning rooflighting is the amount of obstruction to natural daylighting that could be caused by services and equipment housed in the roof void. Also, the effectiveness of the glazed areas will gradually deteriorate as dirt collects on the surfaces.

The elimination of solar heat gain and solar glare can be achieved in various ways, such as fitting reflective louvres to the glazed areas and treating the surface with a special thin paint wash to act as a diffusion agent. Probably the best solution is to choose a roof profile such as the northlight roof, which is orientated away from the southern aspect. However, other factors such as site size, site position and/or planning requirements may mean that the last solution is not always possible.

▨▨■ NORTHLIGHT ROOFS

This form of roof profile is asymmetric: the north-orientated face is pitched at an angle of between 50° and 90° and is covered with glass, usually in the form of patent glazing. The south-orientated face of the roof is pitched at an angle of between 20° and 30° and is covered with profiled fibre cement sheeting or similar lightweight covering attached to purlins. The structural roof members can be of timber, steel or precast concrete formed as a plane frame in single or multi-span format and spaced at 4.500 to 6.000 m centres according to the spanning properties of the purlins (see Fig. 11.1.2).

Single-span northlight roofs in excess of 12.000 m span are generally unacceptable because the void formed by the roof triangulation is very large, and as it is not divorced from the main building it has to be included in the total volume to be heated. The solution is to use a series of smaller plane frames to form a multiple northlight roof, which will reduce the total volume of the roof considerably. Note that there is almost no difference in the total roof area to be covered, whichever method is adopted; however, multiple roofs will require more fittings in the form of ridge pieces, eaves closures and gutters.

Using a system of multiple northlight frames a valley is formed between each plane frame, and this must be designed to collect and discharge the surface water run-off from both sides of the valley (see Fig. 11.1.3). Internal support for the ends of adjacent frames can be obtained by using a valley beam spanning over internal columns, the spacing of these supporting columns being determined by the spanning properties of the chosen beam. If the number of internal columns required becomes unacceptable in terms of floor layout and circulation, an alternative arrangement can be used consisting of a lattice girder housed between the apex and bottom tie of the roof frame (see Fig. 11.1.2). This method will enable

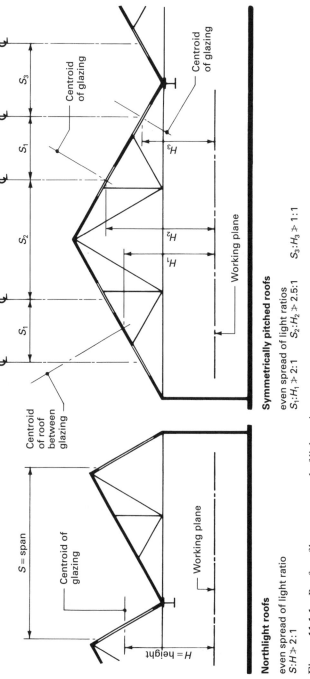

Northlight roofs

even spread of light ratio
$S : H \not> 2 : 1$

Symmetrically pitched roofs

even spread of light ratios
$S_1 : H_1 \not> 2 : 1$ $S_2 : H_2 \not> 2.5 : 1$ $S_3 : H_3 \not> 1 : 1$

Figure 11.1.1   Roof profiles: even spread of light ratios

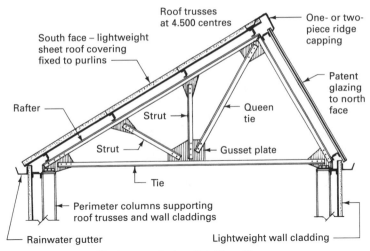

Roof trusses at 4.500 centres

One- or two-piece ridge capping

South face – lightweight sheet roof covering fixed to purlins

Patent glazing to north face

Rafter

Queen tie

Strut

Strut

Gusset plate

Tie

Perimeter columns supporting roof trusses and wall claddings

Rainwater gutter

Lightweight wall cladding

Single-span steel northlight roof truss

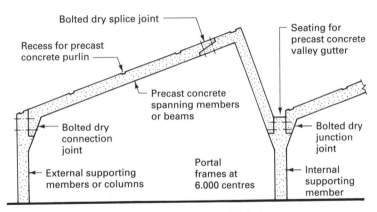

Bolted dry splice joint

Seating for precast concrete valley gutter

Recess for precast concrete purlin

Precast concrete spanning members or beams

Bolted dry connection joint

Bolted dry junction joint

External supporting members or columns

Portal frames at 6.000 centres

Internal supporting member

Multi-span precast concrete northlight frames

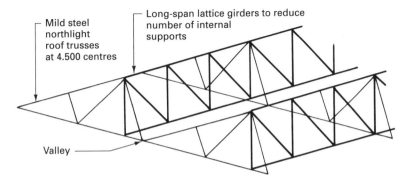

Mild steel northlight roof trusses at 4.500 centres

Long-span lattice girders to reduce number of internal supports

Valley

Figure 11.1.2 Typical northlight roof profiles

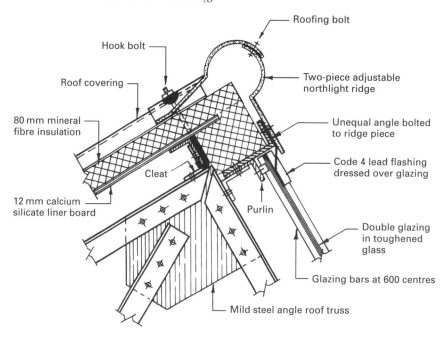

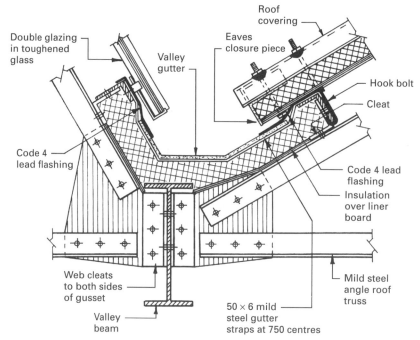

Figure 11.1.3   Multiple northlight roof: typical details

economic spans of up to 30.000 m to be achieved. The inclusion of a lattice girder in this position will create a cantilever northlight roof truss, and if the same principle is adopted for a multi-span symmetrically pitched roof truss it is usually called an **umbrella roof**. The only real disadvantages of this alternative method are a slight increase in shadow casting caused by the lattice beam member, and a small increase in long-term maintenance such as painting.

▦ ▦ ▦ MONITOR ROOFS

This form of roof is basically a flat roof with raised portions glazed on two faces, which are called **monitor lights**. The roof covering is usually some form of lightweight metal decking covered with asphalt or built-up roofing felt. The glazed areas, like those in a northlight roof, are usually of patent glazing and are generally pitched at an angle of between 70° and 90°.

Monitor lights give a uniform distribution of natural daylight with a daylight factor of between 5% and 8%. Their near-vertical pitch does not give rise to solar glare problems, and therefore orientation is not of major importance. The void is considerably less than either the symmetrically pitched or northlight roof, which gives a more economic solution to heating design problems. The flat ceiling areas below the monitor lights will give a better distribution of artificial lighting than pitched roofs; also the flat roof areas surrounding the projecting monitors will give better and easier access to the glazed areas for maintenance and cleaning purposes.

The formation of the projecting monitor lights can be of light steelwork supported by lattice girders or standard universal beam sections; alternatively they can be constructed of cranked and welded universal beam sections supported on internal columns. To give long clear internal spans, deep lattice beams of lightweight construction can be incorporated within the depth of the monitor framing in a similar manner to that used in northlight roofs (see Fig. 11.1.4). Precast concrete monitor portal frames can also be constructed for both single- and multi-span applications (see Fig. 11.1.5).

▦ ▦ ▦ ENERGY CONSERVATION

The control of fuel energy consumption and measures to reduce the associated atmospheric contamination by CO_2 emissions from industrial buildings are included in the Approved Document to Building Regulation L2: *Conservation of fuel and power in buildings other than dwellings*. The Approved Document contains reference to other technical publications produced by industry, research organisations and the professions. Amongst other provisions, the Approved Document and the additional references contain guidance for:

- limiting heat losses from hot water pipes, hot water storage vessels and warm air ducting;
- limiting heat gains by chilled water and refrigeration vessels;
- energy-efficient heating and hot water systems;
- limiting exposure to solar over-heating;

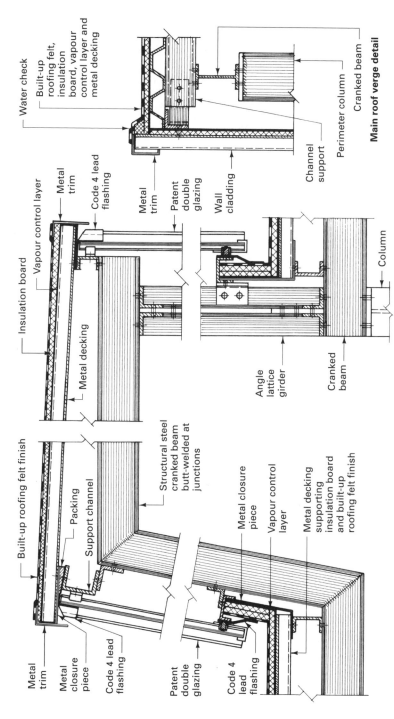

Water check

Built-up roofing felt, insulation board, vapour control layer and metal decking

Main roof verge detail

Perimeter column

Cranked beam

Channel support

Wall cladding

Patent double glazing

Metal trim

Code 4 lead flashing

Metal trim

Vapour control layer

Insulation board

Metal decking

Column

Angle lattice girder

Cranked beam

Structural steel cranked beam butt-welded at junctions

Built-up roofing felt finish

Packing

Support channel

Metal closure piece

Vapour control layer

Metal decking supporting insulation board and built-up roofing felt finish

Metal trim

Metal closure piece

Code 4 lead flashing

Patent double glazing

Code 4 lead flashing

Figure 11.1.4   Typical cranked beam monitor roof details

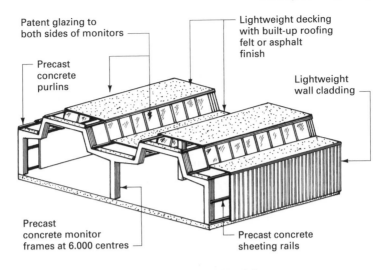

Patent glazing to both sides of monitors

Precast concrete purlins

Lightweight decking with built-up roofing felt or asphalt finish

Lightweight wall cladding

Precast concrete monitor frames at 6.000 centres

Precast concrete sheeting rails

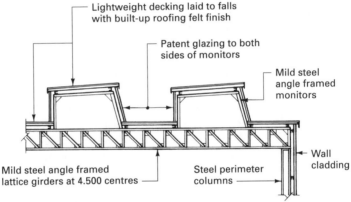

Lightweight decking laid to falls with built-up roofing felt finish

Patent glazing to both sides of monitors

Mild steel angle framed monitors

Mild steel angle framed lattice girders at 4.500 centres

Steel perimeter columns

Wall cladding

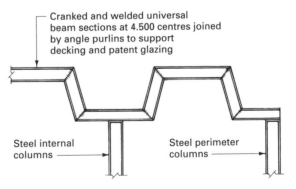

Cranked and welded universal beam sections at 4.500 centres joined by angle purlins to support decking and patent glazing

Steel internal columns

Steel perimeter columns

Figure 11.1.5   Typical monitor roof profiles

- controlling energy use in air conditioning and ventilation systems;
- energy-efficient artificial lighting and controls;
- documenting building and plant operating details to enable the enclosure to be maintained in an energy-efficient manner;
- avoidance of thermal bridges in construction and gaps in insulation layers;
- incorporating reasonably airtight construction with satisfactory sealing methods at construction interfaces;
- maximum acceptable elemental thermal transmittance coefficients (U-values);
- maximum acceptable window, door and other opening areas;
- calculation methodology for establishing carbon emissions.

Under the European Union's Energy Performance of Buildings Directive (EPBD), Member States are required to have in place a methodology for calculating the energy performance of buildings. This calculation must incorporate every aspect of energy efficiency, not least heat losses through the fabric, daylight calculations, airtightness of the external envelope, heating and ventilating system efficiency and internal lighting efficiency. In the UK the methodology can be satisfied by a standard carbon emission calculation. For dwellings (AD.L1), this is effected through the government's recommended system for energy rating, known as **Standard Assessment Procedure** or SAP. For non-domestic buildings a National Calculation Method (NCM) has been devised by the Building Research Establishment to show compliance with Approved Document L2 in accordance with the EPBD. This is known as the **Simplified Building Energy Method** (SBEM). The SBEM incorporates all information relating to a building's design, principally its geometry, method of construction, function and energy-consuming services systems. The final documentation is in the form of an asset-rating certificate. This may be used as part of the application for local authority building control approval and as an energy guide for building users and prospective purchasers.

The Approved Document L2 is not a package of prescriptive measures. It sets overall targets or benchmarks with objectives to be satisfied under the NCM. Nevertheless, there are some parameters for construction practice, which can be gauged. These include:

- poorest acceptable fabric U-values;
- air permeability, particularly at construction interfaces;
- area allowances for windows, doors, roof windows and rooflights;
- solar controls – maximum solar energy transmittance or window energy rating.

Poorest acceptable fabric U-values

In addition to limiting heat losses through the building fabric, establishment of maximum U-values will reduce the possibility of condensation. Table 11.1.2 shows the worst acceptable area-weighted elemental U-values and the worst acceptable individual elemental values. Some objective U-values are also included for guidance. Actual values can vary quite significantly, depending on the effectiveness of other energy conservation measures.

Table 11.1.2 U-values

Element of construction	Limiting area-weighted average (W/m²K)	Limiting individual element (W/m²K)	Objective value (W/m²K)
Floor	0.25	0.70	0.22
Wall	0.35	0.70	0.27
Roof	0.25	0.35	0.20[a]
Windows/rooflights[b]	2.20	3.30	1.80
Curtain wall		See note[c]	
Personnel doors	2.20	3.00	–
Vehicle access and other large doors	1.50	4.00	–
High-use entrance doors	6.00	6.00	–
Roof and smoke vents	6.00	6.00	–

[a] Pitched roof with insulation between rafters; between joists 0.16.
[b] Not including display windows.
[c] The limiting U-value for windows should be applied to glazed areas; the limiting U-value for walls should be applied to opaque areas.

Notes:
Area-weighted allows for the thermal properties of various materials in a component of construction and the area that they occupy, e.g. glass and framing materials in a window. Individual element refers to facilities that interrupt the main fabric of construction, e.g. a meter cupboard enclosure.

Air permeability

No infiltration of air should occur at intersections of elements of construction, e.g. wall/roof. Guidance for conformity is provided in Approved Document L2, with reference to the Chartered Institution of Building Services Engineers (CIBSE) publication TM23: *Testing buildings for air leakage*.

Currently, conformity can be shown as follows:

■ Buildings of less than 1,000 m² gross floor area: a report from a competent inspector to show that the design details, material specifications and quality of construction are to an acceptable standard.
■ Buildings of gross floor area greater than 1,000 m²: also the subject of a report from a competent inspector, plus documented evidence that the air permeability does not exceed 10 m³/h/m² of external surface, at an applied pressure of 50 Pa (1 Pa = 1 N/m²).

Air permeability can be determined by pressurising a building interior with portable fans. Smoke capsules are discharged to provide visual indication of air leakage.

Area allowances for windows, doors, roof windows and rooflights

The window and door opening area for industrial and storage buildings should not exceed 15% of the internal area of exposed wall. Residential buildings (hotels, institutional, etc.) may have up to 30% window and door opening area; places of assembly, offices and shops up to 40%. Rooflight and dormer window area is a maximum of 20% of the roof area. Purpose groups of buildings other than industrial (dwellings excluded) have the same roof openings limitations.

Solar controls

Windows, roof windows and rooflights should be fitted with means for controlling solar gain to prevent overheating of the building interior. Solar control or solar heat block can be measured by solar energy transmittance or solar factor expressed in terms of a g-value. g-values are stated as a number between zero and one. For purposes of an approximate comparison, 0.48 equates to no curtains, 0.43 curtains open, and 0.17 curtains closed. Table 11.1.3 shows the required standards for various building orientations.

Table 11.1.3   Solar energy transmittance factors

Windows/rooflight orientation	Maximum g-value
N	0.81
NE, NW and S	0.65
E, SE, SW and W	0.52
Horizontal	0.39

The heat loss from within a building is affected by the temperature difference between the internal and external environments, and to comply with various Acts and to create good working conditions it is desirable to maintain a minimum temperature for various types of activity.

The ideal working temperature for any particular task is a subjective measure, but as a guide the following internal temperatures recommended by the Chartered Institution of Building Services Engineers are worth considering.

- Sedentary work: 18.4 °C minimum.
- Light work: 15.6 °C minimum.
- Heavy work: 12.8 °C minimum.

The Factories Act 1961 requires a minimum temperature, after the first hour, of 15.6 °C to be maintained except where sedentary work takes place, where a higher temperature would be required.

The advantages that can be gained by having a well-insulated roof (see Fig. 11.1.6) are:

- lower fuel bills;
- reduced capital outlay on heating equipment;
- better working conditions for employees and hence better working relationships.

The initial cost of a building will be higher if a good standard of thermal insulation is specified and installed but, in the long term, an overall saving is usually experienced, the increased capital outlay being recovered within the first five or so years by the savings made in running costs.

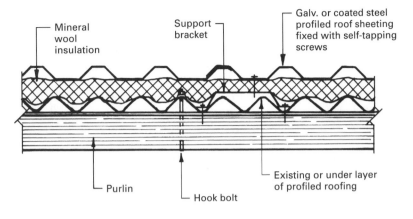

Mineral wool insulation

Support bracket

Galv. or coated steel profiled roof sheeting fixed with self-tapping screws

Purlin

Hook bolt

Existing or under layer of profiled roofing

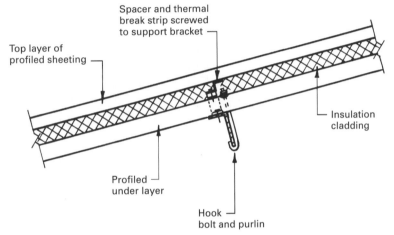

Top layer of profiled sheeting

Spacer and thermal break strip screwed to support bracket

Insulation cladding

Profiled under layer

Hook bolt and purlin

Upgraded or double-skin roof

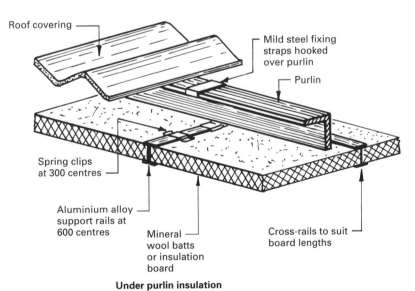

Roof covering

Mild steel fixing straps hooked over purlin

Purlin

Spring clips at 300 centres

Aluminium alloy support rails at 600 centres

Mineral wool batts or insulation board

Cross-rails to suit board lengths

Under purlin insulation

Figure 11.1.6   Industrial premises: typical insulation details

The inclusion of certain materials within a factory roof to comply with the thermal insulation requirements of this Act may introduce into the structure a fire hazard. To this end the Act stipulates that the exposed surfaces of insulation materials used, even if within a cavity, must be at least class 1 spread of flame, as defined in BS 476-7. Furthermore, where structural steel sections are used to make up the roof frame and provide lateral stability to external walls, they will also need resistance to fire by encasement in plaster profiles, sprayed vermiculite cement or other acceptable material (see Part B to the Building Regulations and the Loss Prevention Certification Board's Standards). By using suitable materials or combinations of materials the risk of fire and fire spread in factories can be reduced considerably. The main objective, in common with all fires in buildings, is to confine the fire to the vicinity of the outbreak.

Factories can have compartment-type walls with automatic closing fire-resistant doors as previously described (see Fig. 7.3.11), but if open planning is required other precautions will be necessary. The roof volume can be divided into cells by fitting permanent fire barriers within the triangulated profile of the roof structure with non-combustible materials such as fibre cement sheet or galvanised wire mesh reinforced mineral wool suitably fire-stopped within the profile of the roof covering. Beneath these fire barriers can be rolled curtains of fire fibre cloth controlled by a fusible link or similar fire-detection device, which will allow the curtain to fall forming a fire screen from roof to floor in the advent of a fire. A similar curtain could also be positioned in the longitudinal direction under a valley beam.

Using the above method a fire can be contained for a reasonable period within the confined area, but it can also create another problem or hazard, that of smoke logging. The smoke generated by a fire will rise to the roof level and then start to circulate within the screened compartment, completely filling the volume of the confined section within a short space of time, which, apart from the hazard to people trying to escape, can present the firefighters with the following problems:

- difficulty in breathing;
- prevention from seeing the source of the fire;
- detecting nature of outbreak;
- assessing extent of outbreak.

A method of overcoming this problem is to have automatic high-level ventilators that will allow the smoke to escape rapidly and thus give the firefighters a chance to see clearly and enable them to deal with the outbreak. The use of ventilators, to overcome the problems of smoke logging, will of course introduce more air, which aids combustion, but this does not have the same negative effect as smoke logging, because the volume of air in this type of building is usually so vast that the introduction of more air will have very little effect on the intensity of the outbreak.

The design and position of automatic roof ventilators are normally the prerogative of a specialist designer, but as a guide the total area of opening ventilators should be between 0.5% and 5% of the floor area, depending on the likely area of fire. The essential requirements for an automatic ventilator are:

- It must open in the event of a fire, a common specification being when the heat around the ventilator reaches a temperature of 68 °C.
- It must be weatherproof in normal circumstances.
- It must be easy to fix, and blend with the chosen roof covering material and profile.

Many automatic fire ventilators are designed to act as manually controlled ventilators under normal conditions – typical examples are shown in Fig. 11.1.7.

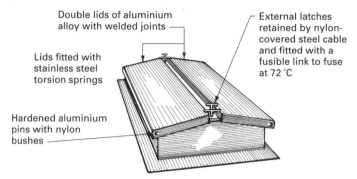

Double lids of aluminium
alloy with welded joints

Lids fitted with
stainless steel
torsion springs

Hardened aluminium
pins with nylon
bushes

External latches
retained by nylon-
covered steel cable
and fitted with a
fusible link to fuse
at 72 °C

Double flap automatic fire ventilator

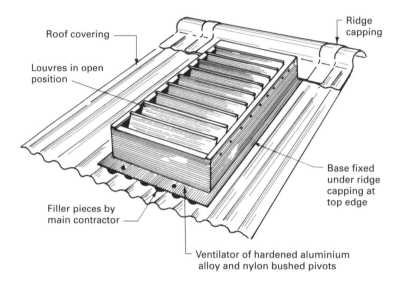

Roof covering

Louvres in open
position

Filler pieces by
main contractor

Ridge
capping

Base fixed
under ridge
capping at
top edge

Ventilator of hardened aluminium
alloy and nylon bushed pivots

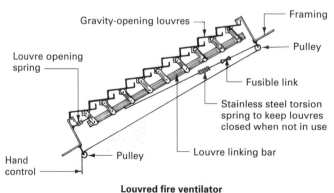

Gravity-opening louvres

Louvre opening
spring

Hand
control

Pulley

Framing

Pulley

Fusible link

Stainless steel torsion
spring to keep louvres
closed when not in use

Louvre linking bar

Louvred fire ventilator

Figure 11.1.7   Automatic roof fire ventilators

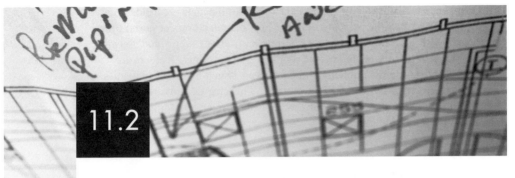

FACTORY BUILDINGS: WALLS

The walls of factory buildings have to fulfil the same functions as any enclosing wall to a building:

- protection from the elements;
- provision of the required sound and thermal insulation;
- provision of the required degree of fire resistance;
- provision of access to and exit from the interior;
- provision of natural daylighting to the interior;
- provision of reasonable security protection to the premises;
- resistance to anticipated wind pressures;
- reasonable durability to keep long-term maintenance costs down to an acceptable level;
- reasonable appearance, in keeping with the occupant's business image.

Most contemporary factory buildings are constructed as framed structures using a three-dimensional frame or a system of portal frames, which means that the enclosing walls can be considered as non loading-bearing claddings supporting their own dead weight plus any wind loading. The wall can be designed as a complete envelope masking the structural framework entirely using brick or block walling, precast concrete panels, curtain-walling techniques or lightweight profiled steel wall claddings. Alternatively, an infill panel technique could be used, making a feature of the structural members.

The choice will depend on such factors as appearance, local planning requirements, short- and long-term costs and personal preference. If the factory is small and contains both production and offices within the same building, presentation of the company's image to business associates and clients is an important design consideration.

The topic of brick and block panel walls in the context of framed buildings can be found in Chapter 10.6 of the associated volume, *Construction Technology*. The use of precast concrete cladding panels and lightweight infill panels is considered in Chapters 8.1 and 8.2 of the present volume. Many manufacturers of portal frame buildings provide a service that includes design, fabrication, supply and assembly of the complete structure, including the roof and wall coverings. Readers are recommended to study the publicity and data sheets issued by these companies for a comprehensive analysis of current practice. This chapter provides details of lightweight external wall claddings only, as applied to framed buildings in popular application to factory buildings.

■■■ LIGHTWEIGHT EXTERNAL WALL CLADDINGS

In common with other cladding methods for framed buildings, lightweight external wall claddings do not require high compressive strength because they only have to support their own dead load and any imposed wind loading, which will become more critical as the height and/or exposure increases. The subject of wind pressures is dealt with in greater detail in the next chapter. Lightweight claddings are usually manufactured from impervious materials, which means that the run-off of rainwater can be high, particularly under storm conditions, when the discharge per minute could reach 2 litres per square metre of wall area exposed to the rain.

A wide variety of materials can be used as a cladding medium, most being profiled to a corrugated or trough form because the shaping will increase the strength of the material over its flat sheet form. Flat sheet materials are available but are rarely applied to large buildings because of the higher strength obtained from a profiled sheet of similar thickness. Special contoured sheets have been devised by many manufacturers to give the designer a wide range of choice in the context of aesthetic appeal. Claddings in two layers or sandwich construction are also available to provide reasonable degrees of thermal insulation and sound insulation, and to combat the condensation hazard that can occur with lightweight claddings of any nature.

The sheets are fixed in a similar manner to that for sheet roof coverings, except that support purlins are replaced in walls by similar steel angles called **sheeting rails** fixed by angle cleats to the vertical structural frame. Alternatively, sheeting rails can be of zed or sigma profile for direct fixing to the structural frame. A significant difference occurs with the position of the sheet fixings, which in wall claddings are usually specified as being positioned in the trough of the profile as opposed to the crest when fixing roof coverings. This change in fixing detail is to ensure that the wall cladding is pulled tightly up to the sheeting rail or lining tray.

Plastic protective caps for the heads of fixings are available, generally of a colour and texture that will blend with the wall cladding. A full range of fittings and trims are usually obtainable for most materials and profiles to accommodate openings, returns, top edge and bottom edge closing. Typical cladding details are shown in Figs 11.2.1 and 11.2.2.

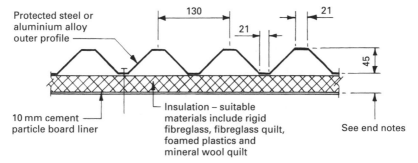

Protected steel or aluminium alloy outer profile

130

21

21

45

10 mm cement particle board liner

Insulation – suitable materials include rigid fibreglass, fibreglass quilt, foamed plastics and mineral wool quilt

See end notes

Typical cladding profile

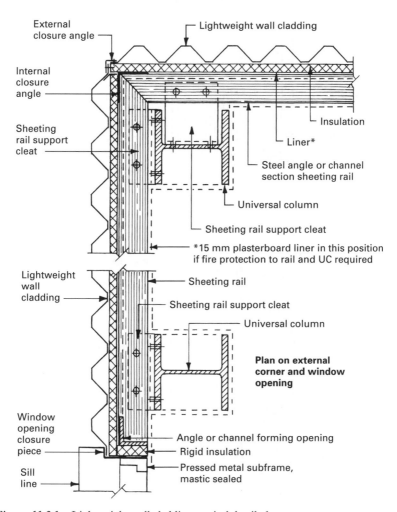

External closure angle

Lightweight wall cladding

Internal closure angle

Sheeting rail support cleat

Insulation

Liner*

Steel angle or channel section sheeting rail

Universal column

Sheeting rail support cleat

*15 mm plasterboard liner in this position if fire protection to rail and UC required

Lightweight wall cladding

Sheeting rail

Sheeting rail support cleat

Universal column

Plan on external corner and window opening

Window opening closure piece

Angle or channel forming opening

Rigid insulation

Pressed metal subframe, mastic sealed

Sill line

Figure 11.2.1   Lightweight wall cladding: typical details 1

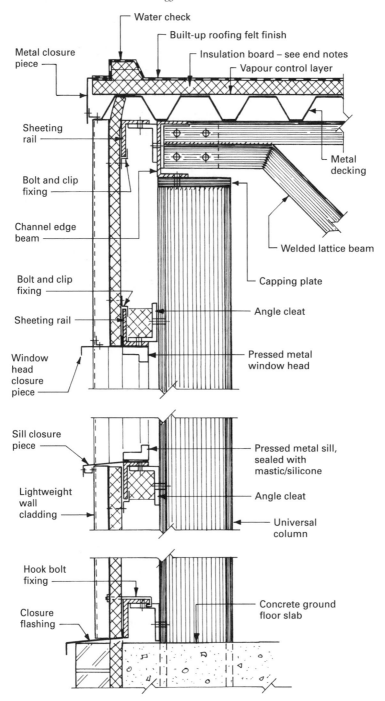

Water check

Built-up roofing felt finish

Metal closure piece

Insulation board – see end notes

Vapour control layer

Sheeting rail

Metal decking

Bolt and clip fixing

Channel edge beam

Welded lattice beam

Bolt and clip fixing

Capping plate

Sheeting rail

Angle cleat

Window head closure piece

Pressed metal window head

Sill closure piece

Pressed metal sill, sealed with mastic/silicone

Lightweight wall cladding

Angle cleat

Universal column

Hook bolt fixing

Concrete ground floor slab

Closure flashing

Note: See Fig. 11.2.1 re location of fire protection lining

Figure 11.2.2   Lightweight wall cladding: typical details 2

Common materials used for lightweight wall claddings are:

■ **Fibre cement**   This is a non-combustible material in corrugated and troughed sheets, which are generally satisfactory when exposed to the weather but are susceptible to impact damage. Average life is about 20 years, which can be increased considerably by paint protection. Unpainted sheets lose their surface finish at the exposed surface by carbonation, and become ingrained with dirt. To achieve adequate thermal insulation standards a lining material of mineral wool or similar will be required. This is normally sandwiched between the outer sheet cladding and an inner lining.

■ **Coated steel sheet**   A non-combustible material with a wide range of profiles produced by various manufacturers. The steel sheet forms the core of the cladding, providing its strength, and this is covered with various forms of coatings to give weather protection, texture and colour. A typical specification would be a galvanised 0.9 mm thick core, covered on both sides with silicone polyester for improved appearance and durability.

■ **Aluminium alloy sheet**   A non-combustible material in corrugated and troughed profiles, which are usually made to the recommendations of CP 143-1 and BS 4868 respectively. Other profiles are also available as manufacturers' standards. Durability will depend on the alloy specification, but this can be increased by specialist paint applications or plastic coatings as for steel. If unpainted or otherwise protected, regular cleaning may be necessary to maintain its natural bright appearance. Aluminium alloys are unsuited to coastal locations, where the salty atmosphere will react and corrode the metal. Fixings, fittings and the availability of linings are as indicated for other cladding materials.

■ **Polyvinylchloride sheet**   This is generally supplied in a corrugated profile with an embedded wire reinforcement to provide a cladding with a surface spread of flame classification of class 1 in accordance with BS 476-7. The extent of this type of external cladding may be limited by combustibility and fire spread regulations defined in the Building Regulations, Approved Document B4, Section 11. For most applications the critical building height is 18.000 m, with boundary proximity 1.000 m. As many industrial buildings are single storey of limited height and in excess of the boundary threshold, materials such as profiled rigid PVC and polycarbonate sheeting may be considered viable. The durability of this form of cladding is somewhat lower than of those previously considered, and the colours available are limited. The usual range of fittings and trims is available.

The importance of adequate design, detail and fixing of all forms of lightweight cladding cannot be overstressed, because the primary objective of these claddings is to provide a lightweight envelope to the building, giving basic weather protection and internal comfort at a reasonable cost. Claddings that will fulfil these objectives are very susceptible to wind damage unless properly secured to the structural frame.

Notes: Figs 11.2.1 and 11.2.2

1. Wall insulation thickness as required by national building regulations and building purpose. Approximate values using mineral fibre insulants:

Thickness (mm)	*U-value* (W/m² K)
None	2.80
60	0.55
80	0.45
100	0.35
150	0.25

2. Roof insulation thickness as required by national building regulations and building purpose. Approximate values using mineral fibre insulants:

Thickness (mm)	*U-value* (W/m² K)
None	2.60
60	0.54
80	0.40
100	0.33
130	0.25
150	0.22

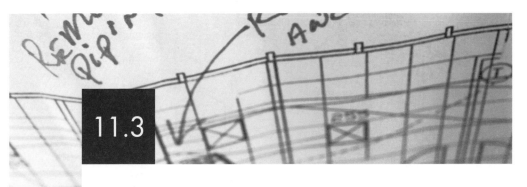

11.3

WIND PRESSURES

Wind can be defined as a movement of air, the full nature of which is not fully understood, but two major contributory factors that can be given are:

- convection currents caused by air being warmed at the earth's surface, becoming less dense, rising and being replaced by colder air;
- transference of air between high- and low-pressure areas.

The speed with which the air moves in replacement or transfer is termed its **velocity**, and can be from 0 to 1.5 metres per second, when it is hardly noticeable, to speeds in excess of 24 metres per second, when considerable damage to property and discomfort to persons could be the result.

The physical nature of the ground or topography over which the wind passes will have an effect on local wind speeds, because obstructions such as trees and buildings can set up local disturbances by forcing the wind to move around the sides of the obstruction or funnel between adjacent obstacles. Where funnelling occurs, the velocity and therefore the pressure can be increased considerably. Experience and research have shown that the major damage to buildings is caused not by a wind blowing at a constant velocity but by short-duration bursts or gusts of wind of greater intensity than the prevailing mean wind speed. The durations of these gusts are usually measured in 3, 5 and 15 second periods, and information on the likely maximum gust speeds for specific long-term time durations of 50 years or more is available from the Meteorological Office.

When the wind encounters an object in its path, such as the face of a building, it is usually rebuffed and forced to turn back on itself. This has the effect of setting up a whirling motion or eddy, which eventually finds its way around or over the obstruction. The pressure of the wind is normally in the same direction as the path of the wind, which tends to push the wall of the building inwards, and indeed will do so if sufficient resistance is not built into the structure. The effect of local eddies, however, is very often opposite in direction and force to that of the prevailing wind, producing a negative or suction force (see Fig. 11.3.1).

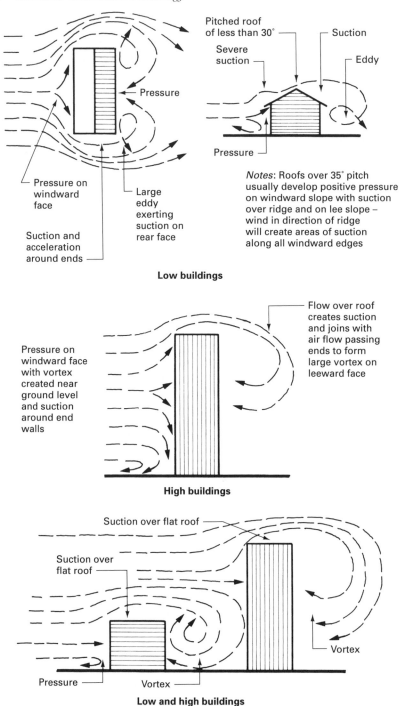

Figure 11.3.1   Typical effect of wind pressures around buildings

Many factors must be taken into account before the magnitude and direction of wind pressures can be determined: these include the height-to-width ratio of the building, the length-to-width ratio of the building, the plan shape of the building, the approach topography, exposure of the building, and the proximity of surrounding structures. Account must also be taken of any likely openings in the building, because the entry of wind will exert a positive pressure on any walls or ceilings encountered. These internal pressures must be added to or subtracted from the type of pressure anticipated acting on the external face at the same point.

All buildings are at some time subjected to wind pressures, but some are more vulnerable than others because of their shape, exposure or method of construction. One method of providing adequate resistance to wind pressures is to use materials of high density: it follows, therefore, that buildings that are clad with lightweight coverings are more susceptible to wind damage than those using the heavier, traditional materials. Factory buildings using lightweight claddings have therefore been taken to serve as an illustration of providing suitable means of resistance to wind pressures.

To overcome the problem of uplift or suction on roofs caused by the negative wind pressures, adequate fixing or anchorage of the lightweight coverings to the structural frame is recommended. Generally, sufficient resistance to uplift of the frame is inherent in the material used for the structural members; the problem is therefore to stop the covering being pulled away from the supporting member. This can be achieved by the quality or holding power of the fixings used, or by the number of fixings employed, or by a combination of both. Note that the whole roof considered as a single entity is at risk, and not merely individual sheets.

If the supporting member does not have sufficient self-dead load to overcome the suction forces, such as a timber plate bedded onto a brick wall, then it will be necessary to anchor the plate adequately to the wall. This can be carried out by means of bolts or straps fixed to the plate and wall in such a manner that part of the dead load of the wall can be added to that of the supporting member.

Positive wind pressures tend to move or bend the wall forwards in the same direction of the wind; this tendency is usually overcome when using light structural framing by adding stiffeners called **wind bracing** to the structure. Wind braces are usually of steel angle construction fitted between the structural members where unacceptable pressures are anticipated. They take the form of cross-bracing, forming what is in fact a stiffening lattice within the frame (see Fig. 11.3.2). Although each area of the country has a predominant prevailing wind, buildings requiring wind bracing are usually treated the same at all likely vulnerable positions to counteract changes in wind direction and the effect of local eddies.

The immense destructive power of the wind, in the context of building works, cannot be overemphasised, and careful consideration is required from design stage to actual construction on site. Readers should also appreciate that temporary works and site hutments are just as vulnerable to wind damage as the finished structure. Great care must be taken, therefore, when planning site layouts, plant positioning, erection of scaffolds and hoardings if safe working conditions are to be obtained on building sites.

Steel roof trusses have simple
or non-rigid connections to
columns and thus tend to
deflect under wind
pressures

Side
wall

Gable
end wall

Wind bracing fixed
to all edges of building
to provide resistance to
wind pressures from all
directions

Wind bracing of
mild steel angles
fixed between truss
ties at ceiling level
to work in conjunction
with bracing in walls

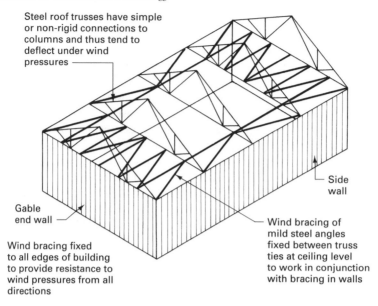

Purlins

Roof
truss

Alternative form
– single diagonal
brace

Diagonal wind bracing
up slope of roof fixed
between purlins

Side
wall

Mild steel angle diagonal wind bracing
fixed in end bays to work in conjunction
with bracing at ceiling level and bracing
within roof slope

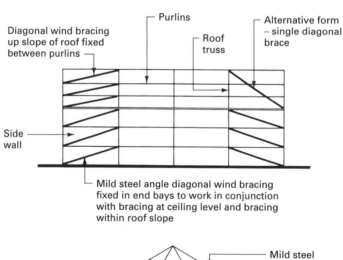

Mild steel
angle truss

Wind bracing to
end bays of gable
end wall to work
in conjunction
with other wind
bracing

Column or
similar support
member

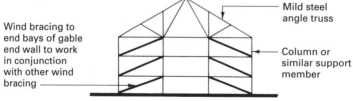

Figure 11.3.2   Typical wind-bracing arrangements

Further details of the effects and assessment of wind loads on buildings can be obtained from the Building Research Establishment publication *Wind, floods and climate pack*.

An additional reference for the determination of wind load values is BS 6399-2: *Loading for buildings. Code of practice for wind loads*. This publication provides two methods for calculation:

- **Standard method** This contains many graphs, tables and simple formulae for basic calculations, which produce a general result on the conservative side of the directional method. It is most suited to the orthogonal wind direction and load examples normal to the building face shown in Fig. 11.3.1. However, wind directions up to 45° either side of the building face can be considered.
- **Directional method** This gives a more complex but detailed assessment of wind load analysis, with charts and formulae for a variety of situations. It is particularly appropriate where wind direction and load are at an angle to the building face, or where the building has an irregular profile or plan shape of geometric or curved proportions. Data in this section of the BS are suited more to computer processing than to simple calculations.

For both applications, calculations are derived from basic wind speeds typical of those shown for the UK and Ireland in Fig. 11.3.3. However, in accordance with guidance in the BS, these figures will need adjustment for:

- the altitude of the building;
- wind direction and building orientation;
- season;
- a risk or probability factor of basic wind speed being exceeded.

Loading must be assessed for external and internal surface pressures, and a balance obtained for both. Detailed calculation involves numerous computations from figures, tables and formulae in BS 6399-2, which is beyond the brief of this text. To pursue this subject further, the reader is recommended to refer to Sections 2 and 3 of the British Standard for the standard and directional methods respectively.

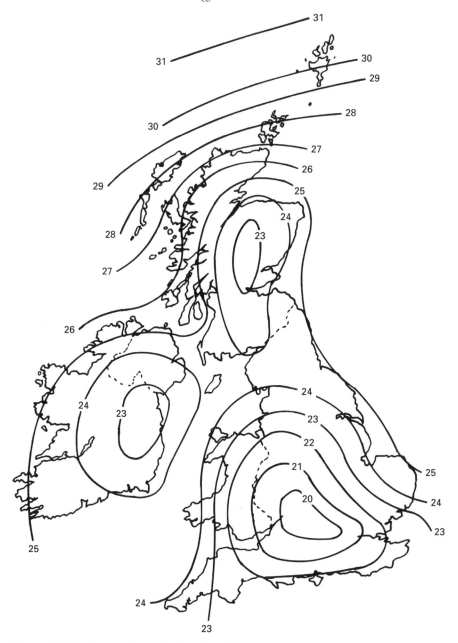

Figure 11.3.3   Zones of basic wind speed (m/s)

DRIVING RAIN

Wind-driven rain can cause material damage by dampness penetration into the structure, and loss of insulation by penetration of lightweight components. It can cause the building fabric to become sufficiently moist for the growth of fungi, mildew, mosses, moulds and other plant and insect life.

The quantity of rain falling on the vertical surfaces of domestic, industrial and commercial buildings is an important consideration for exposure design considerations. The source of rain penetration is not necessarily defective or damaged materials, as many building components such as tiles and window parts rely on sufficient overlapping to prevent moisture access. Total exclusion of moisture is optimum, but the nature of some building materials, such as tiles and porous bricks and stone, will inevitably permit some rain penetration in extreme conditions: hence the need for undertiling layers and purpose-made cavities in wall construction.

The choice of external finishing may be influenced by architectural trends, but research into exposure conditions should be undertaken before specification of wall-facing products. The Building Research Establishment has published maps providing guidance on the **driving rain index** (DRI) for particular locations in the UK and Ireland, shown in Fig. 11.4.1. Values of DRI are obtained by taking the mean annual wind speed (m/s), multiplying by the mean annual rainfall (mm) and dividing by 1,000 (m²/s). The result is a series of contours linking areas with similar DRIs throughout the country. Correction factors may be applicable for topography, exposure, shelter, roughness of terrain, altitude and building height. Figure 11.4.2 shows annual mean driving rain **roses** for different parts of the country. The length of each radiating line indicates the mean DRI received from that direction. The direction of prevailing wind is therefore the longest of the lines. See also BS 5618: *Code of practice for thermal insulation of cavity walls (with masonry or concrete inner and outer leaves) by filling with urea-formaldehyde foam systems*, and BRE Report 59: *Directional driving rain indices for the UK – computation and mapping*.

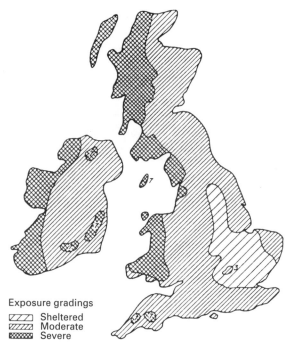

Exposure gradings

ZZZ Sheltered
ZZZZ Moderate
XXXXX Severe

Sheltered exposure zone: areas where the DRI is 3 or less.
Moderate exposure zone: areas where the DRI is between 3 and 7.
Severe exposure zone: areas where the DRI is 7 or more.

Figure 11.4.1　Driving rain index (DRI)

An alternative approach is detailed in BS 8104: *Code of practice for assessing exposure of walls to wind driven rain*. This provides formulae, tables and maps to enable calculation of the quantity of wind-driven rain (l/m^2 per spell) at a given location on a wall. It is known as the **wall spell index**, where spell is the period for which wind-driven rain occurs on a vertical surface. The quantity of wind-driven rain can also be calculated on a yearly basis (l/m^2 per year), and this is known as the **wall annual index**. From these a **local spell index** and a **local annual index** can be established – the former useful for assessing the resistance of the structure to water penetration and the latter appropriate when considering the quantity of moisture in a wall and its possible base for lichen, mosses and other growths that can have a deteriorating effect.

The procedure for assessment can be followed using worked examples provided in BS 8104. Specialist maps for the UK are endorsed with data on subregions, spell and annual rose values and geographical incremental data. A typical rose is shown in Fig. 11.4.3 with numbers corresponding to building orientation: for example, if a building faces west, the rose value in Fig. 11.4.3 is 7. Geographical increments will vary with location, and may add or reduce the value by a nominal amount.

From this information the **airfield spell** and **annual indices** can be established from tables. These are a measure of driving rain (l/m^2) occurring 10.000 m above

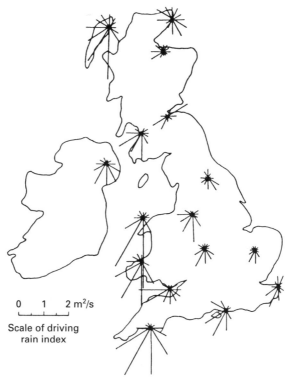

Figure 11.4.2   Driving rain rose diagram

Driving rain rose (BS 5618)
prevailing wind – south westerly

Figure 11.4.3   Rose diagrams

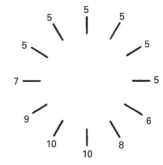

Spell and annual rose values (BS 8104)
e.g. building facing west = 7

ground in the middle of an airfield for the worst spell in 3 years and for an average annually, respectively. Factors for terrain roughness, topography, obstructions and wall characteristics compound the calculation to provide an assessment in litres/m^2 for the respective calculations. Further calculations can be used to determine the run-off in litres per metre width of wall, which is useful for assessing local flooding potential and the need for adjacent subsoil and surface drainage.

BRE Report No. 59 and BS 8104 are complementary, whereas BS 5618 applies specifically to UF foam filled cavity walls. Therefore, when considering information, source identification is essential to avoid confusion.

The importance of categorising driving rain, whether by indices or by roses, is to determine suitable specification of construction details. Details appropriate for sheltered exposure zones are unlikely to be acceptable in areas of severe exposure. Establishing standards for construction should be by consultation with the local building control authority, because different areas of the country will have specific climatic conditions and possibly microclimates within those areas. For example, simple masonry cavity wall construction in a zone of severe exposure should maintain a 50 mm air space beyond any insulation included in the overall cavity width. This is essential to prevent dampness bridging the cavity.

ROOF STRUCTURES

Roof design and construction is an extensive topic. It is therefore usual practice to develop an understanding by including specific aspects of roofing techniques at each element or level of study. In this chapter the roofs considered are those suitable for a variety of large, clear-span applications using structural timber, steel and reinforced concrete. Roof construction techniques for lesser spans are described in Chapter 11.1 of this volume and Part 6 of the accompanying volume, *Construction Technology*.

■■■ LARGE-SPAN TIMBER ROOFS

A wide variety of timber roofs are available for both medium and large spans, and these can be classified under three headings:

- pitched trusses;
- flat top girders;
- bowstring trusses.

Pitched trusses

These are two-dimensional triangulated designed frames spaced at 4.500 to 6.000 m centres with spans up to 30.000 m. The pitch should have a depth to span ratio of 1:5 or steeper and be chosen to suit roof coverings. The basic construction follows that of an ordinary domestic small-span roof truss, as shown in Fig. 6.1.7 of *Construction Technology*, except that the arrangement and number of struts and ties will vary according to the type of truss being used (see Fig. 11.5.1).

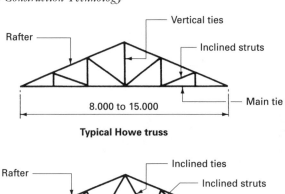

Rafter

Vertical ties

Inclined struts

Main tie

8.000 to 15.000

Typical Howe truss

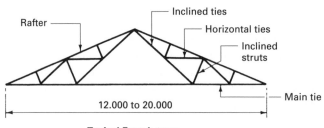

Rafter

Inclined ties

Inclined struts

Main tie

8.000 to 15.000

Typical Fink or Belgian truss

Inclined ties

Rafter

Horizontal ties

Inclined struts

Main tie

12.000 to 20.000

Typical French truss

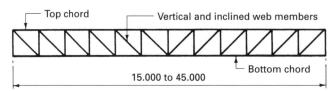

Top chord

Vertical and inclined web members

Bottom chord

15.000 to 45.000

Typical 'N' or Pratt truss

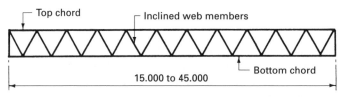

Top chord

Inclined web members

Bottom chord

15.000 to 45.000

Typical Warren girder

Figure 11.5.1   Typical large-span truss and girder types

Flat top girders

Basically these are lattice beams of low pitch spaced at 4.500 to 6.000 m centres and can be economically used for spans up to 45.000 m with a depth to span ratio of 1:8 to 1:10. Construction details are similar to those of the roof truss in that the joints and connections are usually made with timber connectors and bolts. The main advantage of this form of roof structure is the reduction in volume of the building, which should result in savings in the heating installation required and in the running costs. Typical flat top girder outlines are shown in Fig. 11.5.1 and typical construction details are shown in Fig. 11.5.2.

Bowstring trusses

These trusses are basically a lattice girder with a curved upper chord, and are spaced at 4.500 to 6.000 m centres with an economic span range of up to 75.000 m. The depth-to-span ratio is usually 1:6 to 1:8 with the top chord radius approximately equal to the span. They can be constructed from solid segmental timber pieces, but the chords are usually formed from laminated timber with solid timber struts and ties forming the lattice members (see Fig. 11.5.3 for typical details). The older form of bowstring truss, known as a **belfast truss**, which has interlacing struts and ties, is not very often specified because it is difficult to analyse fully the stresses involved, and although relatively small-section timbers can be used they are very expensive in labour costs.

CHOICE OF TRUSS AND TIMBER

To decide upon the most suitable and economic truss to be specified for any given situation the following should be considered:

- availability of suitable timber in the sizes required;
- cost of alternative timber;
- design and fabrication costs;
- transportation problems and costs;
- on-site assembly and erection problems and costs;
- roof covering material availability and costs;
- architectural design considerations.

It may be possible that the consideration given to the last item may well outweigh some of the economic solutions found for the preceding items.

The timber specified should comply with the recommendations of BS 5268, the *Structural use of timber* codes of practice, i.e. BS 5268-2: *Code of practice for permissible stress design, materials and workmanship* and BS 5268-3: *Code of practice for trussed rafter roofs*. The specified timber must also satisfy the structural stability requirements of Building Regulation A1 and the recommendations set out in the supportive Approved Document A. This means that unless a designer, manufacturer or builder makes special arrangements to obtain approval, each piece of timber used in a strength application in any form of building will have to be

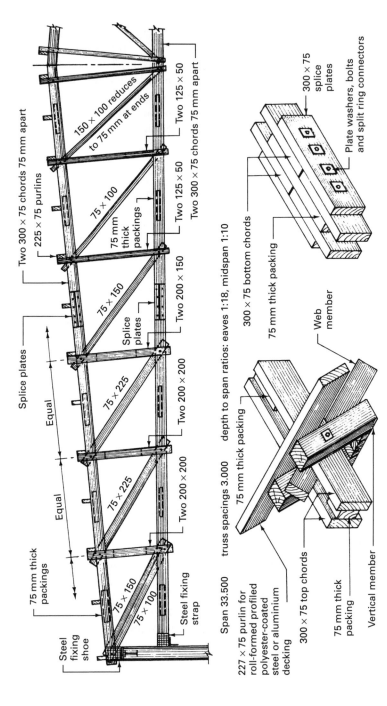

Figure 11.5.2   Typical timber flat top Pratt truss details

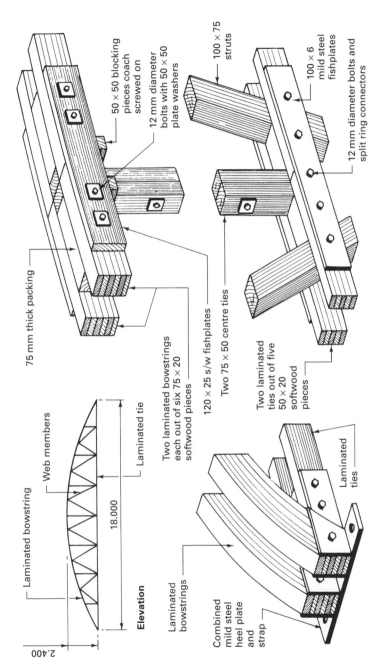

Figure 11.5.3 Typical bowstring truss details

stress graded to the grades set out in BS 5268-2. Reference should also be made to BS 4978: *Specification for visual strength grading of softwood*. This standard was published after collaboration between engineers, scientists and the timber trades of the United Kingdom and the main timber-exporting countries such as Canada, Finland and Sweden. The stress grading can be carried out either visually or by machine, and the permissible stresses for the various grades and species are set out in BS 5268-2 – see extract in Chapter 3.8 of *Construction Technology*.

Visual stress grading is carried out by the knot area ratio (KAR) method, in which the proportion of cross-section occupied by the projected area of knots is assessed. Any pieces of timber where the KAR is less than one-fifth are graded **special structural** (SS), and pieces of timber where the KAR is between one-fifth and one-half are graded as **general structural** (GS) or SS depending on whether a margin condition exists. Any piece of timber with a KAR exceeding one-half is automatically rejected for structural work.

Machine stress grading relies on the correlation between the strength of timber and its stiffness. These grading machines work on one of two principles: those that apply a fixed load and measure the deflection, and those that measure the load required to cause a fixed deflection. The gradings obtained are designated MGS (machine general structural) or MSS (machine special structural), and these grades are comparable to the visual grades given above. BS 5268 also describes two further grades, M50 and M75, which means that these pieces of timber have been graded as having 50% or 75% of the strength of a clear specimen of a similar species.

All graded timber must be marked so that it can be immediately identified by the specifier, supplier and user. Visually graded timber is marked at least once within the length of each piece with GS or SS, together with a mark to indicate the company or grader responsible for the grading. Machine stress-graded timber should be marked MGS, MSS, M50 or M75 at least once within the length of each piece together with a mark to indicate species, grading machine used, BSI Kitemark and the relevant BS number, namely BS 4978. Machine stress-graded timber can also be colour coded with a coloured dye at one end or a series of dashes throughout the length, the colour coding used being:

Green   MGS
Purple MSS
Blue   M50
Red   M75

The most readily available species of structural softwoods are imported redwood, imported whitewood and imported commercial western hemlock. Other suitable structural softwoods are generally available only in small quantities. Structural softwoods are supplied in standard lengths, commencing at 1.800 m and increasing by 300 mm increments to lengths rarely exceeding 6.300 m. When specifying stress-graded timber the following points should be considered:

■ species;
■ section size;
■ length;

- preparation requirements;
- stress grade;
- moisture content.

Additional timber grading methods in accordance with European standards are incorporated into BS EN 518: *Structural timber. Grading. Requirements for visual strength grading standards*, and BS EN 519: *Structural timber. Grading. Requirements for machine strength graded timber and grading machines*, respectively. This and other more recent categorisation of timber, including the North American gradings, are detailed in Chapter 3.8 of the accompanying volume, *Construction Technology*.

JOINTING

Connections between structural timber members may be made by:

- nails;
- screws;
- glue and nails or screws;
- truss plates;
- bolts;
- bolts and timber connectors.

Nails and screws are usually used in conjunction with plywood gussets and like the truss or gangnail plates are usually confined to the small to medium span trusses.

The usual fixings, such as nails, screws and bolts, have their own limitations. Cut nails will generally have a greater holding power than round wire nails because of the higher friction set up by their rough sharp edges and also the smaller disturbance of the timber grain. The joint efficiency of nails may be as low as 15% because of the difficulty in driving sufficient numbers of them within a given area to obtain the required shear value. Screws have a greater holding power than nails but are dearer in both labour and material costs.

Joints made with a rigid bar such as a bolt usually have low efficiency because of the low shear strength of timber parallel to the grain and the unequal distribution of bearing stress along the shank of the bolt. The weakest point in these connections is where the high stresses are set up around the bolt, and various methods have been devised to overcome this problem. The solution lies in the use of timber connectors that are designed to increase the bearing area of the bolts, as described below.

Toothed plate timber connector

Sometimes called **bulldog connectors**, these are used to form an efficient joint without the need for special equipment, and are suitable for all types of connection, especially when small sections are being used. To form the connection the timber members are held in position and drilled for the bolt to provide a bolt hole with 1.5 mm clearance. If the timber is not too dense the toothed connectors can be embedded by tightening up the permanent bolts. In dense timbers, or where more than three connectors are used, the embedding pressure is provided by a high-tensile steel rod threaded at both ends. Once the connectors have been embedded, the rod is removed and replaced by the permanent bolt.

Split ring timber connectors

These connectors are suitable for any type of structure, and timbers of all densities; they are very efficient and develop a high-strength joint. The split ring is a circular band of steel with a split tongue and groove vertical joint. A special boring and cutting tool is required to form the bolt hole and the grooves in the face of the timber into which the connector is inserted, making the ring independent of the bolt itself. The split in the ring ensures that a tight fit is achieved on the timber core but at the same time it is sufficiently flexible to give a full bearing on the timber outside the ring when under heavy load.

Shear plate timber connector

These are counterparts of the split ring connectors. They are housed flush into the timber members and are used for demountable structures. See Fig. 11.5.4 for typical timber connector details.

■■■ LARGE-SPAN STEEL ROOFS

The roof types given for large-span timber roofs can also be designed and fabricated using standard structural steel sections. Span ranges and the spacings of the frames or lattice girders are similar to those given for timber roofs. Connections can be of traditional gusset plates to which the struts and ties would be bolted or welded; alternatively an all-welded construction is possible, especially if steel tubes are used to form the struts and ties. Large-span steel roofs can also take the form of space decks and space frames.

SPACE DECKS

A space deck is a structural roofing system designed to give large clear spans with wide column spacings. It is based on a simple repetitive unit consisting of an inverted pyramid frame, which can be joined to similar frames to give spans of up to 22.000 m for single spanning designs and up to 33.000 m for two-way spanning roofs. These basic units are joined together at the upper surface by bolting adjacent angle framing together and by fixing threaded tie bars between the apex couplers (see Fig. 11.5.5).

Edge treatments include vertical fascia, mansard and cantilever. Rooflights can be fixed directly over the 1.200 m × 1.200 m modular upper framing (see Fig. 11.5.6). The roof can be laid or screeded to falls, or alternatively a camber to form the falls can be induced by tightening up the main tie bars. The space deck roof can be supported by steel or concrete columns or fixed to padstones or ring beams situated at the top of loadbearing perimeter walls.

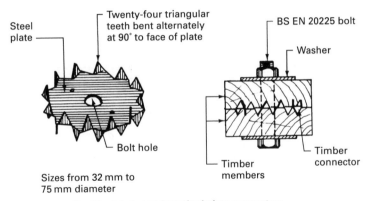

Steel plate

Twenty-four triangular teeth bent alternately at 90° to face of plate

Bolt hole

Sizes from 32 mm to 75 mm diameter

BS EN 20225 bolt

Washer

Timber connector

Timber members

Double-sided round-toothed plate connectors

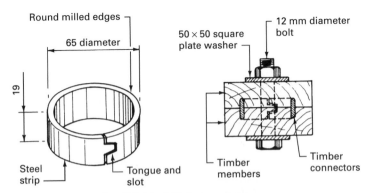

Round milled edges

65 diameter

19

Steel strip

Tongue and slot

50 × 50 square plate washer

12 mm diameter bolt

Timber members

Timber connectors

Parallel side split ring connector

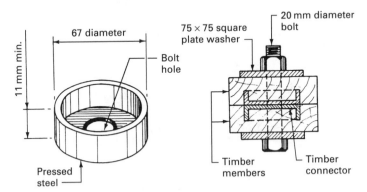

67 diameter

11 mm min.

Bolt hole

Pressed steel

75 × 75 square plate washer

20 mm diameter bolt

Timber members

Timber connector

Shear plate connectors

Figure 11.5.4 Typical timber connectors. See also BS EN 912: *Timber fasteners. Specification for connections to timber.*

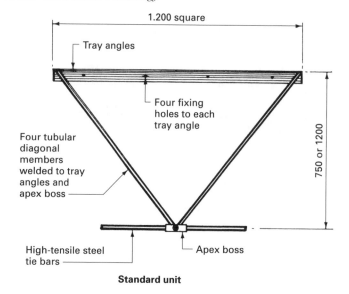

Standard unit

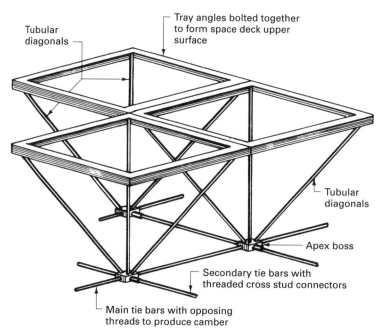

Figure 11.5.5   Typical space deck standard units

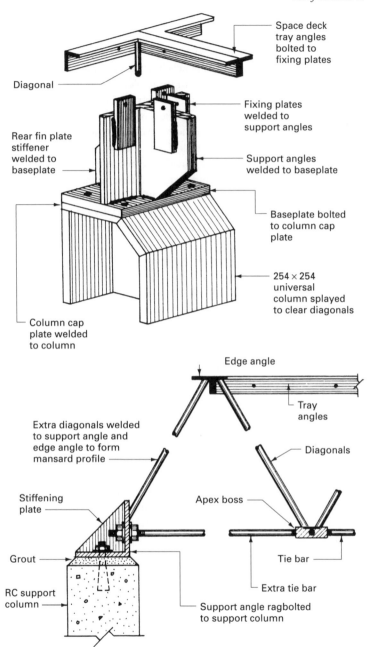

Figure 11.5.6 Typical space deck edge-fixing details

The most usual and economic roof covering for space decks is a high-density mineral wool insulated roofing board covered with three layers of built-up roofing felt with a layer of reflective chippings. Alternatively, composite double-skin decking may be used. This comprises panels of rigid urethane core with profiled coated steel or aluminium sheet facings. In principle, similar to that shown in Fig. 11.1.6. The void created by the space deck structure can be used to house all forms of services, and the underside can be left exposed or covered in with an attached or suspended ceiling. The units are usually supplied with a basic protective paint coating applied by a dipping process after the units have been degreased, shot-blasted and phosphate-treated to provide the necessary key.

The simplicity of the space deck unit format eliminates many of the handling and transportation problems encountered with other forms of large-span roof. A complete roof can usually be transported on one lorry by stacking the pyramid units one inside the other. Site assembly and erection are usually carried out by a specialist subcontractor, who must have access to the whole floor area. Assembly is rapid with a small labour force which assembles the units as beams in the inverted position, turns them over and connects the whole structure together by adding the secondary tie bars. Two general methods of assembly and erection can be used:

1. The deck is assembled on the completed floor immediately below the final position and lifted directly into its final position.
2. The deck is assembled outside the perimeter of the building and lifted in small sections to be connected together in the final position. This is generally more expensive than method 1.

Prior to assembly and erection the main contractor has to provide a clear, level and hard surface to the perimeter of the proposed building as well as to the whole floor area to be covered by the roof. These surfaces must be capable of accepting the load from a 25 tonne mobile crane. The main contractor is also responsible for unloading, checking, storing and protecting the units during and after erection, and for providing all necessary temporary works and plant such as scaffolds, ladders and hoists. The site procedures and main contractor responsibilities set out above in the context of space decks are generally common to all roofing contractors involving specialist subcontractors and materials.

SPACE FRAMES

Space frames are similar to space decks in their basic concept, but they are generally more flexible in their design and layout possibilities, because their main component is the connector joining together the chords and braces. Space frames are usually designed as a double-layer grid as opposed to the single-layer grid used mainly for geometrical shapes such as a dome. The depth of a double-layer grid is relatively shallow when compared with other structural roof systems of similar loadings and span: the span-to-depth ratio for a space frame supported on all its edges would be about 1:20, whereas a space frame supported near the corners would require a ratio

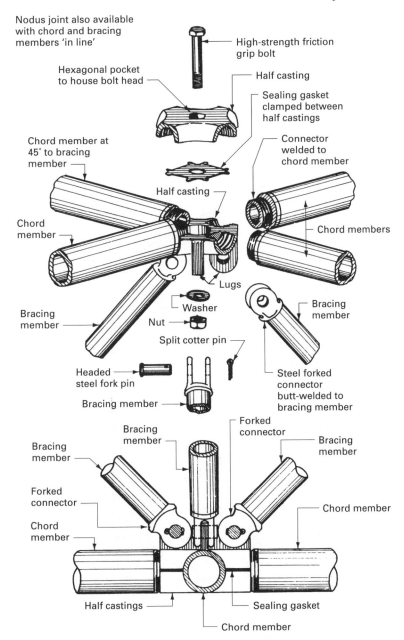

Nodus joint also available with chord and bracing members 'in line'

High-strength friction grip bolt

Hexagonal pocket to house bolt head

Half casting

Sealing gasket clamped between half castings

Connector welded to chord member

Chord member at 45° to bracing member

Half casting

Chord member

Chord members

Lugs

Bracing member

Washer

Bracing member

Nut

Split cotter pin

Steel forked connector butt-welded to bracing member

Headed steel fork pin

Bracing member

Forked connector

Bracing member

Bracing member

Bracing member

Forked connector

Chord member

Chord member

Half castings

Sealing gasket

Chord member

Figure 11.5.7 Typical BSC (now Corus UK Ltd) 'Nodus' system joint details

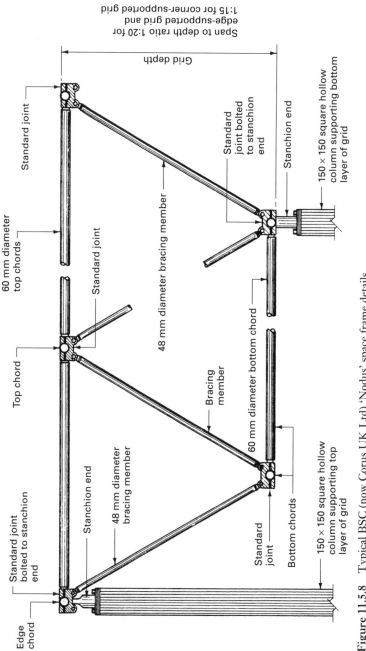

Figure 11.5.8   Typical BSC (now Corus UK Ltd) 'Nodus' space frame details

of about 1:15. A variety of systems are available to the architect and builder. The concept is illustrated in Figs 11.5.7 and 11.5.8. These show the 'Nodus' system established by the British Steel Corporation (now Corus UK Ltd).

The claddings used in conjunction with a space frame roof should not be unduly heavy, and normally any lightweight profiled decking would be suitable. As with the space decks described above, some of the main advantages of these systems are:

- They are constructed from simple standard prefabricated units.
- Units can be mass produced.
- The roof can be rapidly assembled and erected on site using semi-skilled labour.
- Small sizes of components make storage and transportation easy.

Site works consist of assembling the grid at ground level, lifting the completed space frame, and fixing it to its supports. The grid can be assembled on a series of blocks to counteract any ground irregularities, and during assembly the space frame will automatically generate the correct shape and camber. The correct procedure is to start assembling the space frame at the centre of the grid and work towards the edges, ensuring that there is sufficient ground clearance to enable the camber to be formed. Generally the space frame is assembled as a pure roof structure, but it is possible to install services and fix the cladding prior to lifting and fixing. Mobile cranes are usually employed to lift the completed roof structure, holding it in position while the columns are erected and fixed. Alternatively the grid can be constructed in an offset position around the columns that pass through the spaces in the grid; the completed roof structure is then lifted and moved sideways onto the support seatings on top of the columns.

▨▪■ SHELL ROOFS

A shell roof may be defined as a structural curved skin covering a given plan shape and area. The main points are:

- It is primarily a structural element.
- The basic strength of any particular shell is inherent in its shape.
- The quantity of material required to cover a given plan shape and area is generally less than for other forms of roofing.

The basic materials that can be used in the formation of a shell roof are concrete, timber and steel. Concrete shell roofs consist of a thin curved reinforced membrane cast *in situ* over timber formwork, whereas timber shells are usually formed from carefully designed laminated timber, and steel shells are generally formed using a single-layer grid. Concrete shell roofs, although popular, are very often costly to construct because the formwork required is usually purpose-made from timber and is, in itself, a shell roof and has little chance of being reused to enable the cost of the formwork to be apportioned over several contracts.

A wide variety of shell roof shapes and types can be designed and constructed, but they can be classified under three headings:

■ Domes.
■ Vaults.
■ Saddle shapes and conoids.

Domes

In their simplest form these consist of a half sphere, but domes based on the ellipse, parabola and hyperbola are also possible. Domes have been constructed by architects and builders over the centuries using individually shaped wedge blocks or traditional timber roof construction techniques. It is therefore the method of construction together with the materials employed rather than the geometrical setting out that has changed over the years.

Domes are double-curvature shells, which can be rotational and are formed by a curved line rotating around a vertical axis, or they can be translational domes, which are formed by a curved line moving over another curved line (see Figs 11.5.9 and 11.5.10). Pendentive domes are formed by inscribing within the base circle a polygon and cutting vertical planes through the true hemispherical dome.

Any dome shell roof will tend to flatten because of the loadings, and this tendency must be resisted by stiffening beams or similar to all the cut edges. As a general guide, domes that rise in excess of one-sixth of their diameter will require a ring beam. Timber domes, like their steel counterparts, are usually constructed on a single-layer grid system and covered with a suitable thin skin membrane.

Vaults

These are shells of single curvature and are commonly called **barrel vaults**. A barrel vault is basically a continuous arch or tunnel, and was first used by the Romans and later by the Norman builders in this country. Geometrically a barrel vault is a cut half cylinder, which presents no particular setting-out problems. When two barrel vaults intersect the lines of intersection are called **groins**. Barrel vaults, like domes, tend to flatten unless adequately restrained, and in vaults restraint will be required at the ends in the form of a diaphragm and along the edges (see Fig. 11.5.11).

From a design point of view barrel vaults act as a beam with the length being considered as the span which, if it is longer than its width or chord distance, is called a **long span barrel vault**, or conversely if the span is shorter than the chord distance is termed a **short barrel vault**. Short barrel vaults with their relatively large chord distances – and, consequently, large radii to their inner and outer curved surfaces – may require stiffening ribs to overcome the tendency to buckle. The extra stresses caused by the introduction of these stiffeners or ribs will necessitate the inclusion of extra reinforcement at the rib position; alternatively the shell could be thickened locally about the rib for a distance of about one-fifth of the rib spacing (see Fig. 11.5.11).

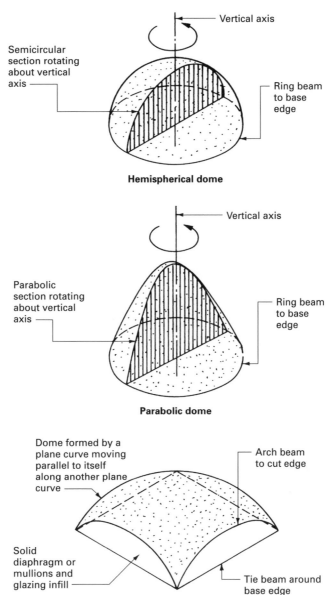

Vertical axis

Semicircular section rotating about vertical axis

Ring beam to base edge

Hemispherical dome

Vertical axis

Parabolic section rotating about vertical axis

Ring beam to base edge

Parabolic dome

Dome formed by a plane curve moving parallel to itself along another plane curve

Arch beam to cut edge

Solid diaphragm or mullions and glazing infill

Tie beam around base edge

Translational dome

Figure 11.5.9 Typical dome roof shapes: 1

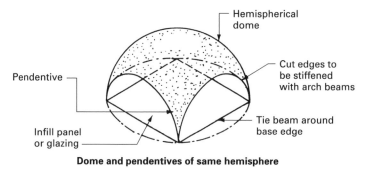

Hemispherical
dome

Pendentive

Cut edges to
be stiffened
with arch beams

Tie beam around
base edge

Infill panel
or glazing

Dome and pendentives of same hemisphere

Hemispherical
dome

Hemispherical
dome

Cut edges to
be stiffened
with arch beams

Infill panel
or glazing

Pendentive

Tie beam around
base edge

Dome and pendentives of different hemispheres

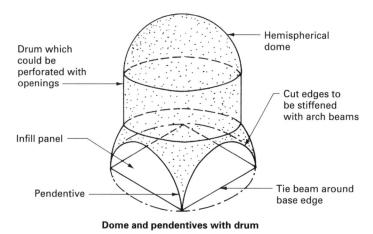

Drum which
could be
perforated with
openings

Hemispherical
dome

Cut edges to
be stiffened
with arch beams

Infill panel

Pendentive

Tie beam around
base edge

Dome and pendentives with drum

Figure 11.5.10   Typical dome roof shapes: 2

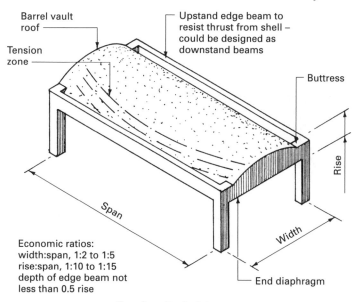

Barrel vault roof

Tension zone

Upstand edge beam to resist thrust from shell – could be designed as downstand beams

Buttress

Rise

Span

Width

End diaphragm

Economic ratios:
width:span, 1:2 to 1:5
rise:span, 1:10 to 1:15
depth of edge beam not less than 0.5 rise

Barrel vault principles

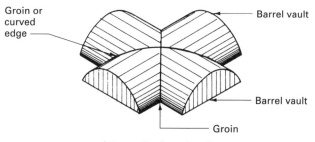

Groin or curved edge

Barrel vault

Barrel vault

Groin

Intersecting barrel vaults

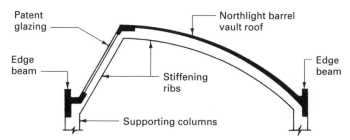

Patent glazing

Northlight barrel vault roof

Edge beam

Edge beam

Stiffening ribs

Supporting columns

Figure 11.5.11 Typical barrel vaults

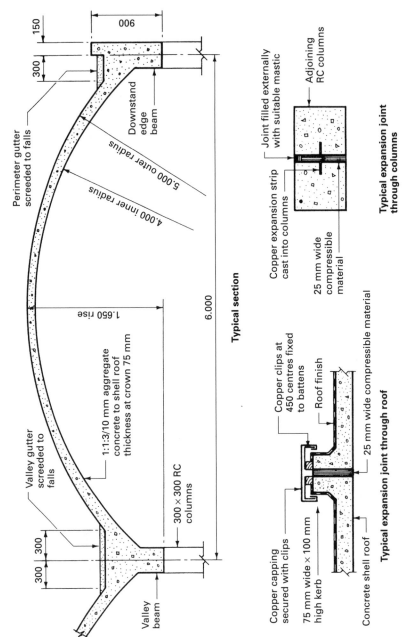

Figure 11.5.12   Typical barrel vault details

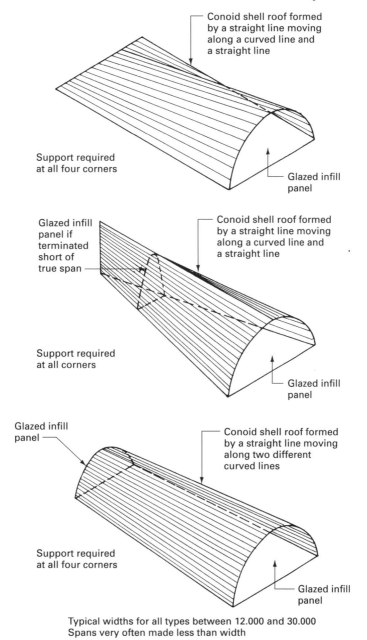

Conoid shell roof formed by a straight line moving along a curved line and a straight line

Support required at all four corners

Glazed infill panel

Glazed infill panel if terminated short of true span

Conoid shell roof formed by a straight line moving along a curved line and a straight line

Support required at all corners

Glazed infill panel

Glazed infill panel

Conoid shell roof formed by a straight line moving along two different curved lines

Support required at all four corners

Glazed infill panel

Typical widths for all types between 12.000 and 30.000
Spans very often made less than width

Figure 11.5.13 Typical conoid shell roof types

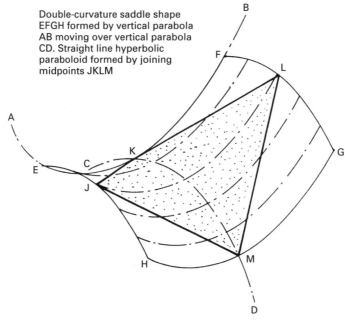

Double-curvature saddle shape
EFGH formed by vertical parabola
AB moving over vertical parabola
CD. Straight line hyperbolic
paraboloid formed by joining
midpoints JKLM

Double-curvature shell

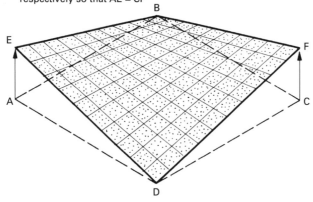

Straight line limited hyperbolic paraboloid formed by
raising corners A and B of square ABCD to E and F
respectively so that AE = CF

Hyperbolic paraboloid shell

Figure 11.5.14   Hyperbolic paraboloid roof principles

In large barrel vault shell roofs allowances must be made for thermal expansion, and this usually takes the form of continuous expansion joints, as shown in Fig. 11.5.12, spaced at 30.000 m centres along the length. This will in fact create a series of individually supported abutting roofs weather-sealed together.

Conoid shells

These are similar to barrel vaults but are double-curvature shells as opposed to the single curvature of the barrel vault. Two basic geometrical forms are encountered:

- A straight line is moved along a curved line at one end and a straight line at the other end, the resultant shape being cut to the required length.
- A straight line is moved along a curved line at one end and a different curved line at the other end.

Typical shapes are shown in Fig. 11.5.13.

Hyperbolic paraboloids

These are double-curvature saddle-shaped shells formed geometrically by moving a vertical parabola over another vertical parabola set at right angles to the moving parabola (see Fig. 11.5.14). The saddle shape created is termed a **hyperbolic paraboloid** because horizontal sections taken through the roof will give a hyperbolic outline and vertical sections will result in a parabolic outline. To obtain a more practical shape than the true saddle the usual shape is that of a warped parallelogram or straight line limited hyperbolic paraboloid, which is formed by raising or lowering one or more corners of a square, as shown in Figs 11.5.14 and 11.5.15. By virtue of its shape this form of shell roof has a greater resistance to buckling than dome shapes.

Hyperbolic paraboloid shells can be used singly or in conjunction with one another to cover any particular plan shape or size. If the rise – that is, the difference between the high and low points of the roof – is small, the result will be a hyperbolic paraboloid of low curvature acting structurally like a plate, which will have to be relatively thick to provide the necessary resistance to deflection. To obtain full advantage of the in-built strength of the shape the rise to diagonal span ratio should not be less than 1:15; indeed, the higher the rise the greater will be the strength and the shell can be thinner.

By adopting a suitable rise-to-span ratio it is possible to construct concrete shells with diagonal spans of up to 35.000 m with a shell thickness of only 50 mm. Timber hyperbolic paraboloid roofs can also be constructed using laminated edge beams with three layers of 20 mm thick tongued and grooved boards. The top and bottom layers of boards are laid parallel to the edges but at right angles to one another, and the middle layer is laid diagonally. This is to overcome the problem of having to twist the boards across their width and at the same time bend them in their length.

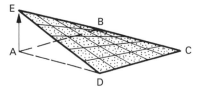

Straight line limited hyperbolic paraboloid formed by raising
corner A of square ABCD to E

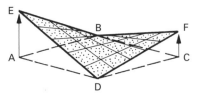

Straight line limited hyperbolic paraboloid formed by raising
corners A and C of square ABCD to E and F respectively
so that AE ≠ CF

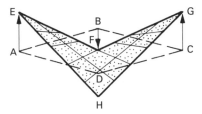

Straight line limited hyperbolic paraboloid formed by raising
corners A and C and lowering corners B and D of square
ABCD to EFG and H respectively so that AE = CG and BF = DH

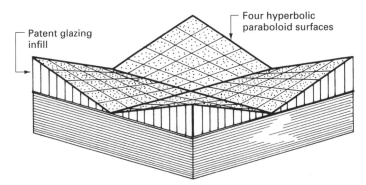

Hyperbolic paraboloids combined to form single roof

Figure 11.5.15   Typical hyperbolic paraboloid roof shapes

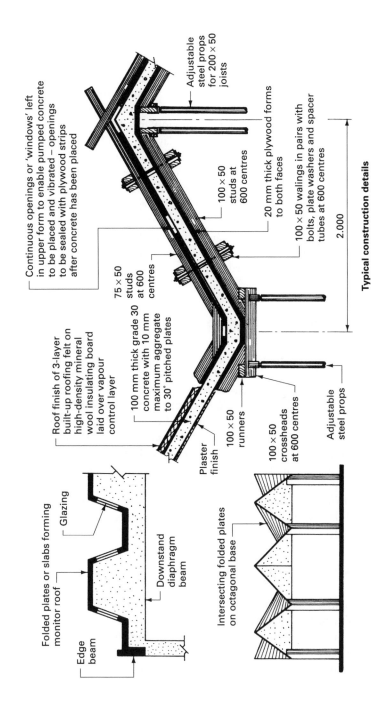

Continuous openings or 'windows' left in upper form to enable pumped concrete to be placed and vibrated – openings to be sealed with plywood strips after concrete has been placed

Adjustable steel props for 200 × 50 joists

100 × 50 studs at 600 centres

20 mm thick plywood forms to both faces

100 × 50 walings in pairs with bolts, plate washers and spacer tubes at 600 centres

2.000

Typical construction details

75 × 50 studs at 600 centres

Roof finish of 3-layer built-up roofing felt on high-density mineral wool insulating board laid over vapour control layer

100 mm thick grade 30 concrete with 10 mm maximum aggregate to 30° pitched plates

Plaster finish

100 × 50 runners

100 × 50 crossheads at 600 centres

Adjustable steel props

Folded plates or slabs forming monitor roof

Glazing

Downstand diaphragm beam

Edge beam

Intersecting folded plates on octagonal base

Figure 11.5.16 Typical folded plate roof details

CONSTRUCTION OF SHELL ROOFS

Concrete shell roofs are constructed on traditional formwork adequately supported to take the loads. When casting barrel vaults it is very often convenient to have a movable form consisting of birdcage scaffolding supporting curved steel ribs to carry the curved plywood or steel forms. Top formwork is not usually required unless the angle of pitch is greater than 45°. Reinforcement usually consists of steel fabric and bars of small diameter, the bottom layer of reinforcement being welded steel fabric followed by the small-diameter trajectory bars following the stress curves set out on the formwork and finally a top layer of steel fabric. The whole reinforcement arrangement is wired together, and spacer blocks of precast concrete or plastic are fixed to maintain the required cover of concrete.

The concrete is usually specified as a mix with a characteristic strength of 25 or 30 N/mm². Preferably the concrete should be placed in one operation in 1.000 m wide strips, commencing at one end and running from edge beam to edge beam over the crown of the roof. A wet mix should be placed around the reinforcement, followed by a floated drier mix. Thermal insulation can be provided by laying high-density mineral wool insulation boards over the completed shell prior to laying the roof covering.

■■■ FOLDED PLATE ROOFS

This is another form of stressed skin roof, and is sometimes called **folded slab construction**. The basic design concept is to bend or fold a flat slab so that the roof will behave as a beam spanning in the direction of the fold. To create an economic roof the overall depth of the roof should be related to span and width so that it is between 1/10 and 1/15 of the span or 1/10 of the width, whichever is the greater. The fold may take the form of a pitched roof, a monitor roof, or a multi-fold roof in single or multiple bays with upstand or downstand diaphragms at the supports to collect and distribute the slab loadings (see Fig. 11.5.16). Formwork may be required to both top and bottom faces of the slabs. To enable concrete to be introduced and vibrated, openings or 'windows' can be left in the upper surface formwork, and these will be filled in with slip-in pieces after the concrete has been placed and vibrated.

■■■ TENSION ROOF STRUCTURES

Suspended or tensioned roof structures can be used to form permanent or temporary roofs and are generally a system or network of cables, or in the temporary form they could be pneumatic tubes, which are used to support roof covering materials of the traditional form or continuous sheet membranes. With this form of roof the only direct stresses that are encountered are tensile stresses, and this – apart from aesthetic considerations – is their main advantage. Because of their shape and lightness, tension roof structures can sometimes present design problems in the context of negative wind pressures, and this is normally overcome by having a second system of curved cables at right angles to the main suspension cables. This will in effect prestress the main suspension cables. Figure 11.5.17 shows an outline principle in one direction only.

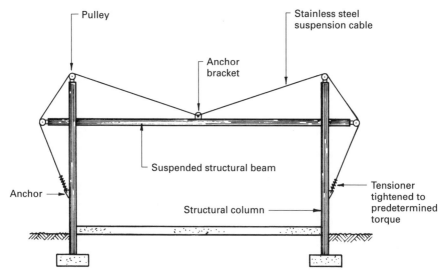

Figure 11.5.17 Tensioned roof structure

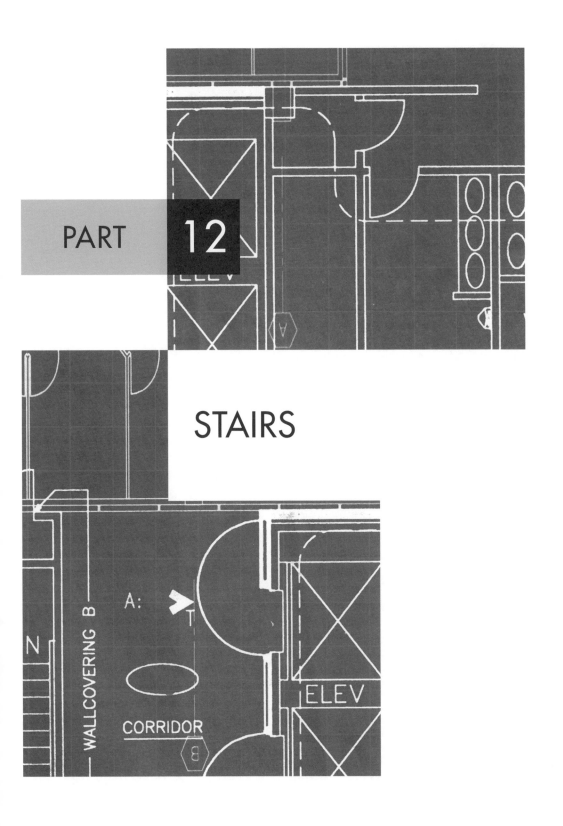

PART **12**

STAIRS

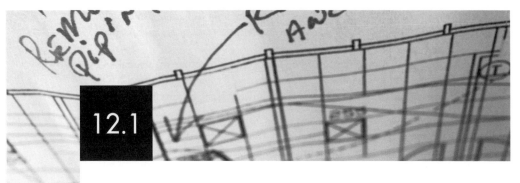

CONCRETE STAIRS

Any form of stairs is primarily a means of providing circulation and communication between the various levels within a building. Apart from this primary function, stairs may also be classified as a means of escape in case of fire; if this is the case the designer is severely limited by necessary regulations as to choice of materials, position and sizing of the complete stairway. Stairs that do not fulfil this means of escape function are usually called **accommodation stairs** and as such are not restricted by the limitations given above for escape stairs.

Escape stairs have been covered in a previous section of this volume, and it is necessary only to reiterate the main points:

1. Constructed from non–combustible materials.
2. Stairway protected by a fire-resisting enclosure.
3. Separated from the main floor area by a set or sets of self-closing fire-resisting doors.
4. Limitations as to riser heights, tread lengths and handrail requirement usually based upon use of building.
5. In common with all forms of stairs, all riser heights must be equal throughout the rise of the stairs.

It should be appreciated that points 2 and 3 listed above can prove to be inconvenient to persons using the stairway for general circulation within the building, such as having to pass through self-closing doors, but in the context of providing a safe escape route this is unavoidable.

Generally, concrete escape stairs are designed in straight flights with not more than 16 risers per flight, particularly if the stairs serve a shop or an assembly area. A turn of at least 30° is required after any flight that has more than 36 risers. In special circumstances such as low usage (not more than 50 persons), minimum overall diameter of 1.600 m, i.e. 0.800 m width, spiral stairs may be acceptable as a suitable means of escape in case of fire. To satisfy the Building Regulations,

Approved Document B1, for escape routes, spiral stairs should adhere to the requirements of BS 5395-2: *Stairs, ladders and walkways. Code of practice for the design of helical and spiral stairs.*

The construction of simple reinforced concrete stairs, together with suitable formwork is considered in Chapter 7.5 of *Construction Technology.* Figure 12.1.1 shows another example of formwork, and the following basic requirements will serve as a reminder of the construction:

- Concrete mix usually specified as 1:1.5:3/3–10 mm aggregate (25–30 N/mm^2).
- Minimum cover of concrete over reinforcement 15 mm or bar diameter, whichever is the greater to give a 1-hour fire resistance.
- Waist thickness usually between 100 and 250 mm depending on stair type.
- Mild steel or high-yield steel bars can be used as reinforcement.
- Continuous handrails of non-combustible materials at a height of between 900 mm and 1,000 mm above the pitch line are required to all stairs and to both sides if the stair width exceeds 1.000 m.

▨ ▨ ■ TYPES OF STAIR

The main area of study at advanced level is consideration of the different stair types or arrangements, which can range from a single straight flight to an open spiral stairway.

SINGLE STRAIGHT FLIGHT STAIRWAY

This form of stairway, although simple in design and construction, can be unpopular because of the plan space it occupies. In this arrangement the flight behaves as a simply supported slab spanning from landing to landing. The effective span or total horizontal going is usually taken as being from landing edge to edge by providing a downstand edge beam to each landing. If these edge beams are not provided, the effective span would be taken as overall of the landings, resulting in a considerably increased bending moment and hence more reinforcement. Typical details are shown in Fig. 12.1.2.

INCLINED SLAB STAIR WITH HALF-SPACE LANDINGS

These stairs have the usual plan format for reinforced concrete stairs, giving a more compact plan layout and better circulation than the single straight flight stairs. The half-space or 180° turn landing is usually introduced at the mid-point of the rise, giving equal flight spans, thus reducing the effective span and hence the bending moment considerably. In most designs the landings span crosswise onto a loadbearing wall or beam and the flights span from landing to landing. The point of intersection of the soffits to the flights with the landing soffits can be detailed in one of two ways:

- Soffits can be arranged so that the intersection or change of direction is in a straight line: this gives a better visual appearance from the underside but will

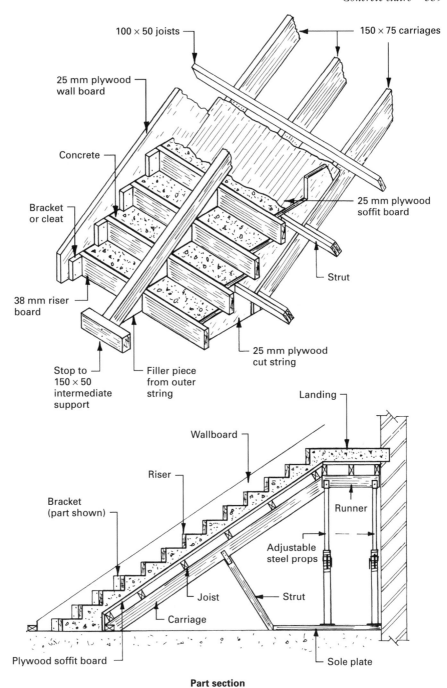

100 × 50 joists

150 × 75 carriages

25 mm plywood wall board

Concrete

25 mm plywood soffit board

Bracket or cleat

Strut

38 mm riser board

25 mm plywood cut string

Stop to 150 × 50 intermediate support

Filler piece from outer string

Landing

Wallboard

Riser

Bracket (part shown)

Runner

Adjustable steel props

Joist

Strut

Carriage

Plywood soffit board

Sole plate

Part section

Figure 12.1.1 Typical formwork to reinforced concrete stair

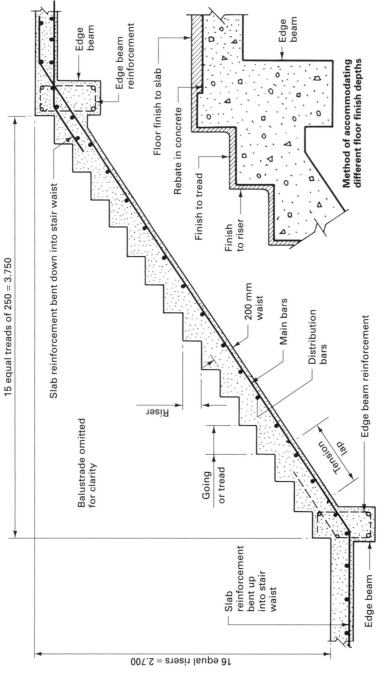

Figure 12.1.2   Straight flight concrete stairs

mean that the riser lines of the first and last steps in consecutive flights are offset in plan.

■ Flights and landing soffit intersections are out of line on the underside by keeping the first and last risers in consecutive flights in line on plan (see Fig. 12.1.3).

Note from the reinforcement pattern shown in the detail in Fig. 12.1.3 that a tension lap is required at the top and bottom of each flight to overcome the tension induced by the tendency of the external angles of the junctions between stair flights and landings to open out.

STRING BEAM STAIRS

These stairs are an alternative design for the stairs described above. A string or edge beam is used to span from landing to landing to resist the bending moment, with the steps spanning crosswise between them: this usually results in a thinner waist dimension and an overall saving in the concrete volume required, but this saving in material is usually offset by the extra formwork costs. The string beams can be either upstand or downstand in format and to both sides if the stairs are free standing (see Fig. 12.1.4).

CRANKED SLAB STAIRS

These stairs are very often used as a special feature because the half-space landing has no visible support, being designed as a cantilever slab. Bending, buckling and torsion stresses are induced with this form of design, creating the need for reinforcement to both faces of the landing and slab or waist of the flights; indeed, the amount of reinforcement required can sometimes create site problems with regard to placing and compacting the concrete. The problem of deciding upon the detail for the intersection line between flight and landing soffits, as described above for inclined slab stairs with half-space landings, also occurs with this arrangement. Typical details of a cranked slab stairs, which is also known as continuous stairs, scissor stairs or jack-knife stairs, are shown in Fig. 12.1.5.

CANTILEVER STAIRS

Sometimes called spine wall stairs, these consist of a central vertical wall from which the flights and half-space landings are cantilevered. The wall provides a degree of fire resistance between the flights, and they are therefore used mainly for escape stairs. Because both flights and landings are cantilevers, the reinforcement is placed in the top of the flight slab and in the upper surface of the landing to counteract the induced negative bending moments. The plan arrangement can be a single straight flight or, as is usual, two equal flights with an intermediate half-space landing between consecutive stair flights (see Fig. 12.1.6 for typical details).

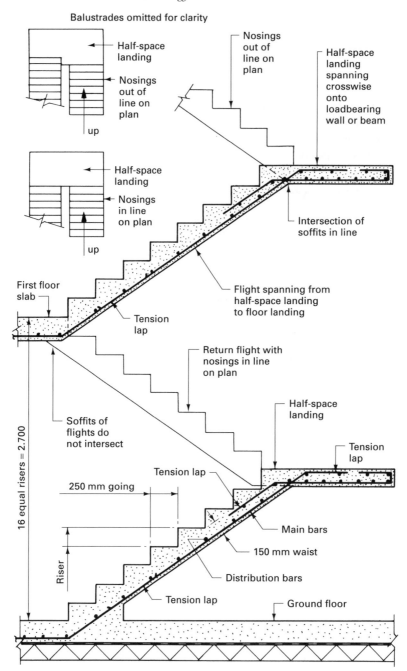

Balustrades omitted for clarity

Half-space landing

Nosings out of line on plan

up

Half-space landing

Nosings in line on plan

up

Nosings out of line on plan

Half-space landing spanning crosswise onto loadbearing wall or beam

Intersection of soffits in line

First floor slab

Tension lap

Flight spanning from half-space landing to floor landing

Return flight with nosings in line on plan

Half-space landing

Tension lap

Soffits of flights do not intersect

Tension lap

Main bars

150 mm waist

Distribution bars

Ground floor

16 equal risers = 2.700

250 mm going

Riser

Tension lap

Figure 12.1.3   Inclined slab concrete stairs with half-space landings

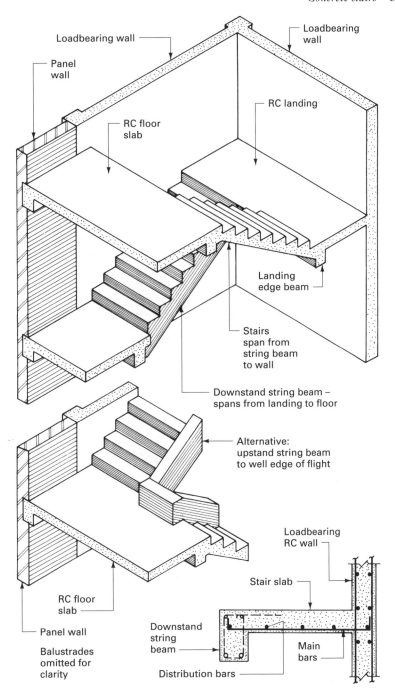

Figure 12.1.4 String beam concrete stairs

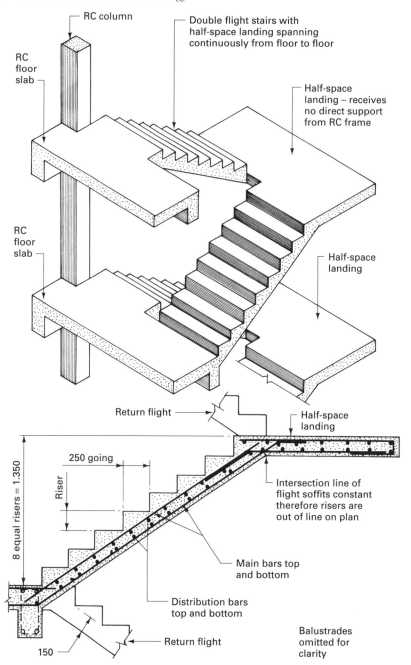

Figure 12.1.5 Cranked or continuous concrete stairs

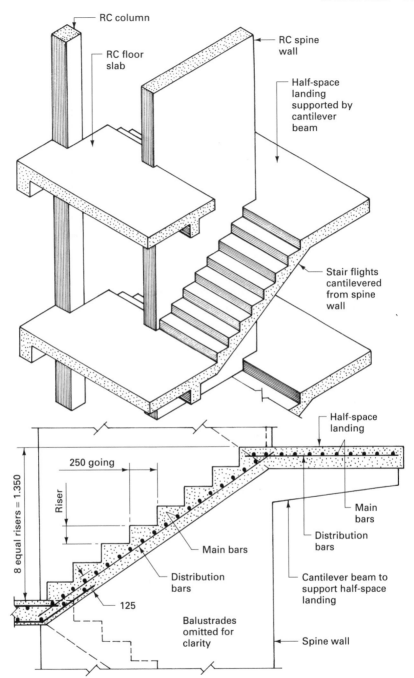

Figure 12.1.6 Cantilever concrete stairs

SPIRAL STAIRS

These are used mainly as accommodation stairs in the foyers of prestige buildings such as theatres and banks. They can be expensive to construct, being normally at least seven times the cost of conventional stairs. The plan shape is generally based on a circle, although it is possible to design an open spiral stairway with an elliptical core. The spiral stairway can be formed around a central large–diameter circular column in a similar manner to that described for cantilevered stairs or, as is usual, designed with a circular open stair well. Torsion, tension and compressive stresses are induced in this form of stairway, which will require reinforcement to both faces of the slab in the form of radial main bars bent to the curve of the slab with distribution bars across the width of the flight (see Fig. 12.1.7). Formwork for spiral stairs consists of a central vertical core or barrel to form the open stair well, to which the soffit and riser boards are set out and fixed, the whole arrangement being propped and strutted as required from the floor level in a conventional manner.

Notes

1. For stairs in buildings for institutional and assembly purposes, the minimum going is 280 mm unless the building floor area is less than 100 m^2, when 250 mm is acceptable. For other stairs the minimum going is 250 mm except for private/domestic use, where it is 220 mm minimum.
2. Maximum rises are 180 mm in institutional and assembly buildings, 190 mm elsewhere except in private/domestic use, where it is 220 mm maximum.

■ ■ ■ PRECAST CONCRETE STAIRS

Most of the concrete stair arrangements previously described can be produced as precast concrete components, which can have the following advantages:

- Better quality control of the finished product.
- Saving in site space, because formwork storage and fabrication space are no longer necessary.
- Stairway enclosing shaft can be utilised as a space for hoisting or lifting materials during the major construction period.
- Can usually be positioned and fixed by semi-skilled operatives.

In common with the use of all precast concrete components the stairs must be repetitive and in sufficient quantity to justify their use and be an economic proposition.

The straight flight stairs spanning between landings can have a simple bearing or, by leaving projecting reinforcement to be grouted into preformed slots in the landings, can be given a degree of structural continuity: this latter form is illustrated in Chapter 7.5 of *Construction Technology*. Straight flight precast concrete stairs with a simple bearing require only bottom reinforcement to the slab and extra reinforcement to strengthen the bearing rebate or nib. The bearing location is a rebate cast in the *in-situ* floor slab or landing, leaving a tolerance gap of 8 to 12 mm, which is filled with a compressible material to form a flexible joint. The decision as

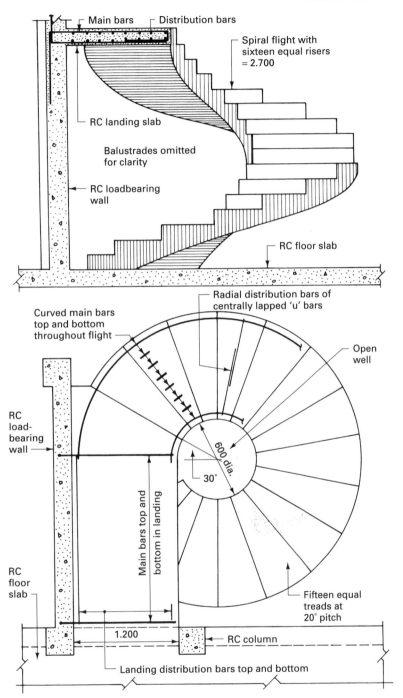

Figure 12.1.7 Open spiral concrete stairs

to whether the stair and landing soffits will be in line or, alternatively, the first and last risers kept in line on plan remains, and as the bearing rebates are invariably cast in a straight line to receive both upper and lower stair flights the intersection design is detailed with the precast units. A typical example is shown in Fig. 12.1.8.

Cranked slab precast concrete stairs are usually formed as an open well stairway. The bearing for the precast landings to the *in-situ* floor or to the structural frame is usually in the form of a simple bearing, as described above for straight flight precast concrete stairs. The infill between the two adjacent flights, in an open well plan arrangement at floor and intermediate landing levels, can be of *in-situ* concrete with structural continuity provided by leaving reinforcement projecting from the inside edge of the landings (see Fig. 12.1.9). Remember that, when precast concrete stair flights are hoisted into position, different stresses may be induced from those that will be encountered in the fixed position. To overcome this problem the designer can either reinforce the units for both conditions or, as is more usual, provide definite lifting points in the form of projecting lugs or by utilising any holes cast in to receive the balustrading.

Precast open riser stairs are a form of stairs that can be both economic and attractive, consisting of a central spine beam in the form of a cut string supporting double cantilever treads of timber or precast concrete. The foot of the lowest spine beam is located and grouted into a preformed pocket cast in the floor, whereas the support at landing and floor levels is a simple bearing located in a housing cast into the slab edge (see Fig. 12.1.10). Anchor bolts or cement in sockets are cast into the spine beam to provide the fixing for the cantilever treads. The bolt heads are recessed below the upper tread surface and grouted over with a matching cement mortar for precast concrete treads or concealed with matching timber pellets when hardwood treads are used. The supports for the balustrading and handrail are located in the holes formed at the ends of the treads and secured with a nut and washer on the underside of the tread. The balustrade and handrail details have been omitted from all the stair details so far considered: this has been done for reasons of clarity and as this aspect of stairwork is considered in Chapter 7.5 of the associated volume, *Construction Technology*.

Spiral stairs in precast concrete work are based upon the stone stairs found in many historic buildings such as Norman castles and cathedrals, consisting essentially of steps that have a 'keyhole' plan shape rotating round a central core. Precast concrete spiral stairs are usually open riser stairs with a reinforced concrete core or alternatively a concrete-filled steel tube core. Holes are formed at the extreme ends of the treads, to receive the handrail supports in such a manner that the standard passes through a tread and is fixed to the underside of the tread immediately below. A hollow spacer or distance piece is usually incorporated between the two consecutive treads (see typical details in Fig. 12.1.11). In common with all forms of this type of stairs, precast concrete spiral stairs are limited as to minimum diameter and total rise when being considered as escape stairs, and therefore they are usually installed as accommodation stairs.

The finishes that can be applied to a concrete floor can also be applied in the same manner to an *in-situ* or precast concrete stairway. Care must be taken, however, with the design and detail, because the thickness of finish given to stairs

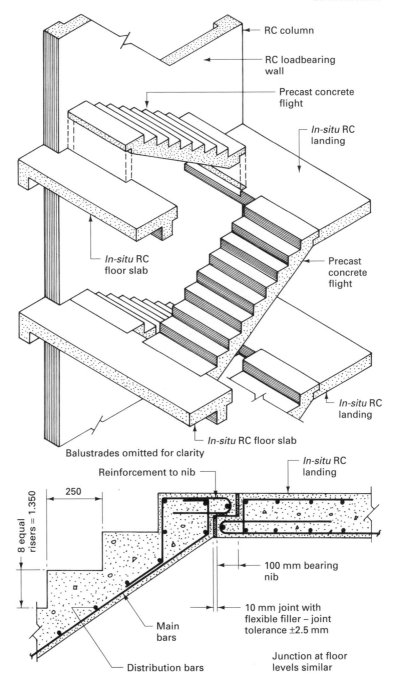

RC column

RC loadbearing wall

Precast concrete flight

In-situ RC landing

In-situ RC floor slab

Precast concrete flight

In-situ RC landing

In-situ RC floor slab

Balustrades omitted for clarity

Reinforcement to nib

In-situ RC landing

250

8 equal risers = 1.350

100 mm bearing nib

10 mm joint with flexible filler – joint tolerance ±2.5 mm

Main bars

Distribution bars

Junction at floor levels similar

Figure 12.1.8   Precast concrete straight flight stairs

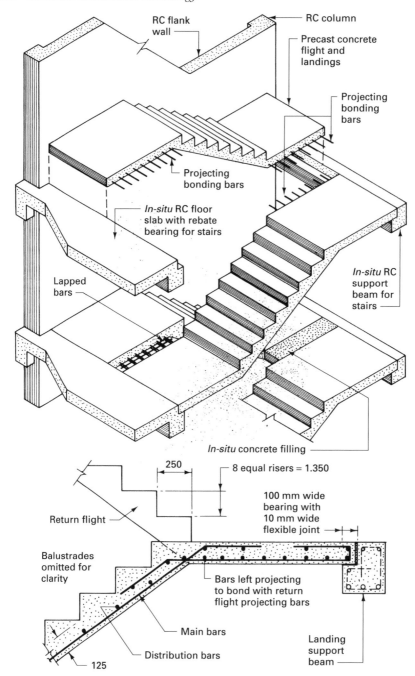

Figure 12.1.9   Precast concrete cranked slab stairs

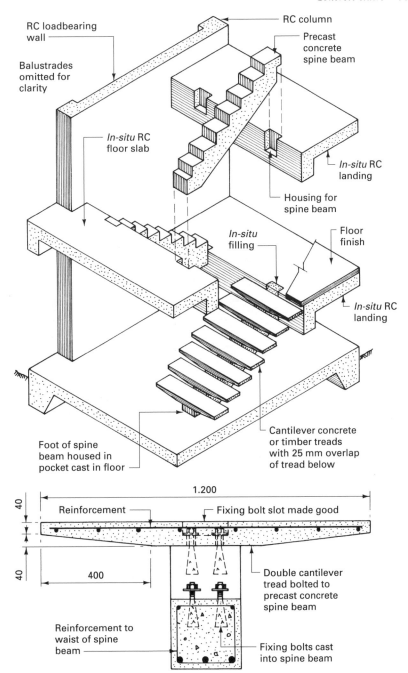

RC loadbearing wall

RC column

Precast concrete spine beam

Balustrades omitted for clarity

In-situ RC floor slab

In-situ RC landing

Housing for spine beam

In-situ filling

Floor finish

In-situ RC landing

Foot of spine beam housed in pocket cast in floor

Cantilever concrete or timber treads with 25 mm overlap of tread below

1.200

40

Reinforcement

Fixing bolt slot made good

40

400

Double cantilever tread bolted to precast concrete spine beam

Reinforcement to waist of spine beam

Fixing bolts cast into spine beam

Figure 12.1.10 Precast concrete open riser stairs

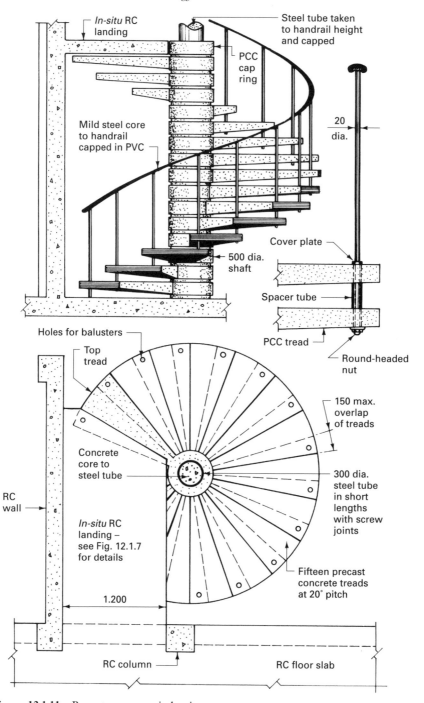

Figure 12.1.11   Precast concrete spiral stairs

is generally less than the thickness of a similar finish to floors. It is necessary, for reasons of safety, to have equal height risers throughout the stairway rise: therefore it may be necessary when casting the stairs to have the top and bottom risers of different heights from the remainder of the stairs. An alternative method is to form a rebate at the last nosing position to compensate for the variance in floor and stair finishes (see Fig. 12.1.2).

If the stairs are to be left as plain concrete, an anti-slip surface should be provided by trowelling into the upper surfaces of the treads some carborundum dust or casting in rubber or similar material grip inserts to the leading edge, or by fixing a special nosing covering of aluminium alloy or other suitable metal with a grip patterned surface or containing non-slip inserts. Metal nosing coverings with an upper grip surface can also be used in conjunction with all types of applied finish to stairs.

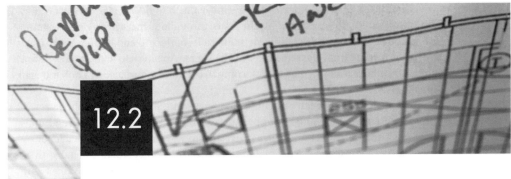

METAL STAIRS

Metal stairs can be constructed to be used as escape stairs or accommodation stairs, both internally and externally. Most metal stairs are manufactured from mild steel with treads of cast iron or mild steel and in straight flights with intermediate half-space landings. Spiral stairs in steel are also produced, but their use as escape stairs is limited by size and the number of persons likely to use the stairway in the event of a fire. Aluminium alloy stairs are also made, and are used almost exclusively as internal accommodation stairs.

As the layout of most buildings is different, stairs are very often purpose made to suit the particular situation. Concrete, being a flexible material at the casting stage, generally presents little or no problems in this respect, whereas purpose-made metal stairs can be more expensive and take longer to fabricate in the workshop. Metal spiral stairs have the distinct advantage that the need for temporary support and hoisting equipment is eliminated. All steel stairs have the common disadvantage of requiring regular maintenance in the form of painting as a protection against corrosion.

Most metal stairs are supplied in a form that requires some site fabrication, and this is usually carried out by the supplier's site erection staff, the main contractor having been supplied with the necessary data as to foundation pads, holding-down bolts, any special cast-in fixings and any pockets to be left in the structural members or floor slabs to enable this preparatory work to be completed before the stairs are ready to be fixed.

■■■ STEEL ESCAPE STAIRS

These have already been considered in the context of means of escape in case of fire, and a typical general arrangement is illustrated in Fig. 7.3.12, which shows a structural steel support frame and a stairway composed of steel plate strings with preformed treads giving an open riser format. The treads for this type of stair are bolted to the strings and can be of a variety of types, ranging from perforated cast iron to patterned steel treads with renewable non-slip nosings. Handrail balustrades or standards can be of steel square or tubular sections bolted to the upper surface of a channel string or to the side of a channel or steel plate string. Figure 12.2.1 shows typical steel escape stair components, and should be read in conjunction with the stair arrangement shown in Fig. 7.3.12.

■ ■ ■ STEEL SPIRAL STAIRS

These may be allowed as an internal or external means of escape stairs if they are to accommodate no more than 50 persons. The minimum overall diameter is 1.600 m and the design must be in accordance with BS 5395-2: *Stairs, ladders and walkways. Code of practice for the design of helical and spiral stairs*, to satisfy Building Regulation classification as an escape stair. Spiral stairs give a very compact arrangement, and can be the solution in situations where plan area is limited. In common with all steel external escape stairs the tread and landing plates should have a non-slip surface and be self-draining, with the stairway circulation width completely clear of any opening doors. Two basic forms are encountered: those with treads that project from the central pole or tube, and those that have riser legs. The usual plan format is to have 12 or 16 treads to complete one turn around the central core and terminating at floor level with a quarter-circle landing or square landing. The standards, like those used for precast concrete spiral stairs, pass through one tread and are secured on the underside of the tread immediately below, giving strength and stability to both handrail and steps. Handrails are continuous and usually convex in cross-section of polished metal, painted metal or plastic covered. Typical details are shown in Fig. 12.2.2.

■ ■ ■ STRING BEAM STEEL STAIRS

These are used mainly to form accommodation stairs that need to be light and elegant in appearance: this is achieved by using small sections and an open riser format. The strings can be of mild steel tube, steel channel, steel box or small universal beam sections fixed by brackets to the upper floor surfaces or landing edges to act as inclined beams. The treads, which can be of hardwood timber, precast concrete or steel, are supported by plate, angle or tube brackets welded to the top of the string beam. Balustrading can be fixed through the ends of the treads or alternatively supported by brackets attached to the outer face of the string beam. Typical details are shown in Fig. 12.2.3.

■ ■ ■ PRESSED STEEL STAIRS

These are accommodation stairs, made from light pressed metal such as mild steel. Each step is usually pressed as one unit with the tread component recessed to receive a filling of concrete, granolithic, terrazzo, timber or any other suitable material. The strings are very often in two pieces, consisting of a back plate to which the steps are fixed and a cover plate to form a box section string, the cover

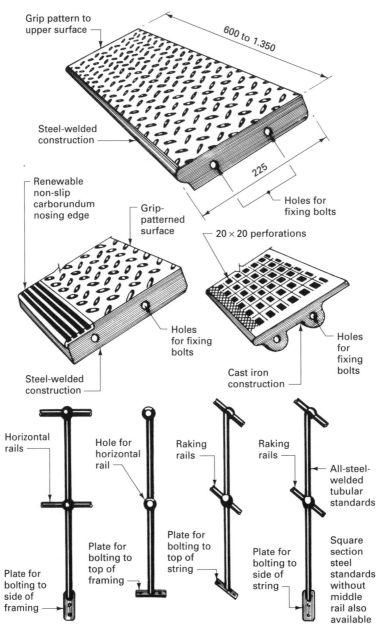

Grip pattern to
upper surface

600 to 1.350

Steel-welded
construction

225

Holes for
fixing bolts

Renewable
non-slip
carborundum
nosing edge

Grip-
patterned
surface

20 × 20 perforations

Holes
for fixing
bolts

Holes
for
fixing
bolts

Steel-welded
construction

Cast iron
construction

Horizontal
rails

Hole for
horizontal
rail

Raking
rails

Raking
rails

All-steel-
welded
tubular
standards

Plate for
bolting to
top of
framing

Plate for
bolting to
top of
string

Plate for
bolting to
side of
string

Square
section
steel
standards
without
middle
rail also
available

Plate for
bolting to
side of
framing

For typical steel stairs detail see Fig. 7.3.12

Figure 12.2.1   Typical steel stairway components

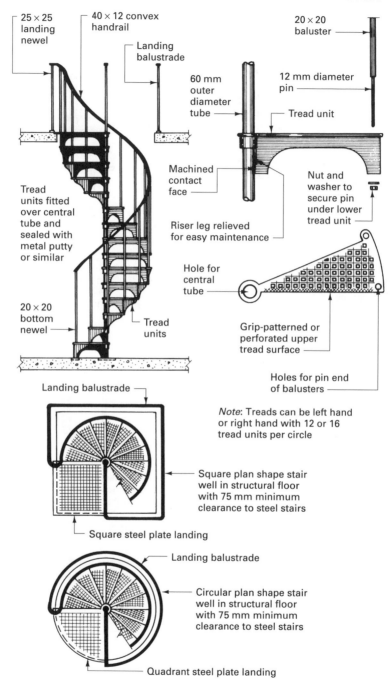

Figure 12.2.2 Typical steel spiral stairs

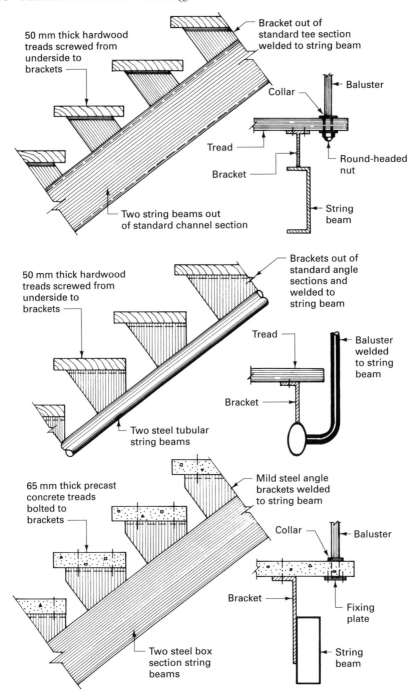

Figure 12.2.3 Typical steel string beam stairs details

plate being site-welded using a continuous MIG (metal inert gas) process. The completed strings are secured by brackets or built into the floors or landings, and provide the support for the balustrade. Stairs of this nature are generally purpose made to the required layout, and site assembled and fixed by a specialist subcontractor, leaving only the tread finishes and decoration as builder's work. Typical details are shown in Fig. 12.2.4.

▨▧■ ALUMINIUM ALLOY STAIRS

These are usually purpose made to suit individual layout requirements, with half- or quarter-space landings from aluminium alloy extrusions. They are suitable for accommodation stairs in public buildings, shops, offices and flats. The treads have a non-slip nosing with a general tread covering of any suitable floor finish material. Format can be open or closed riser, the latter having greater strength. The two-part box strings support the balustrading and are connected to one another by small-diameter tie rods that, in turn, support the tread units. The flights are secured by screwing to purpose-made baseplates or brackets fixed to floors and landings, or alternatively located in preformed pockets and grouted in. When the stairs are assembled they are very light and can usually be lifted and positioned by two operatives without the need for lifting gear. No decoration or maintenance is required except for routine cleaning. Typical details are shown in Fig. 12.2.5.

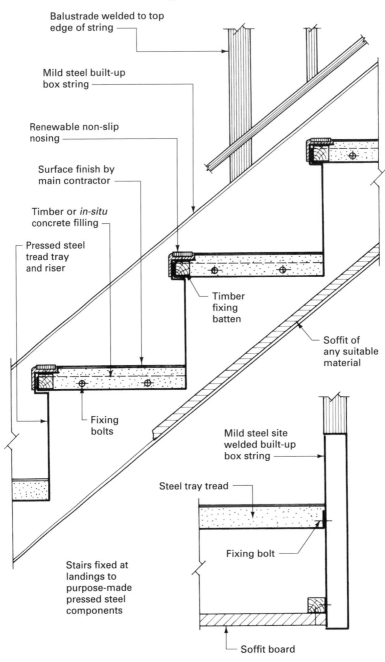

Balustrade welded to top
edge of string

Mild steel built-up
box string

Renewable non-slip
nosing

Surface finish by
main contractor

Timber or *in-situ*
concrete filling

Pressed steel
tread tray
and riser

Timber
fixing
batten

Soffit of
any suitable
material

Fixing
bolts

Mild steel site
welded built-up
box string

Steel tray tread

Fixing bolt

Stairs fixed at
landings to
purpose-made
pressed steel
components

Soffit board

Figure 12.2.4 'Prestair' internal pressed steel stairs

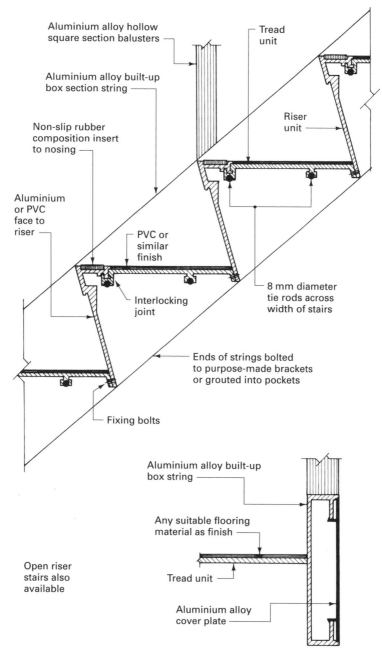

Aluminium alloy hollow
square section balusters

Tread
unit

Aluminium alloy built-up
box section string

Riser
unit

Non-slip rubber
composition insert
to nosing

Aluminium
or PVC
face to
riser

PVC or
similar
finish

8 mm diameter
tie rods across
width of stairs

Interlocking
joint

Ends of strings bolted
to purpose-made brackets
or grouted into pockets

Fixing bolts

Aluminium alloy built-up
box string

Any suitable flooring
material as finish

Open riser
stairs also
available

Tread unit

Aluminium alloy
cover plate

Figure 12.2.5 'Gradus' aluminium alloy stairs

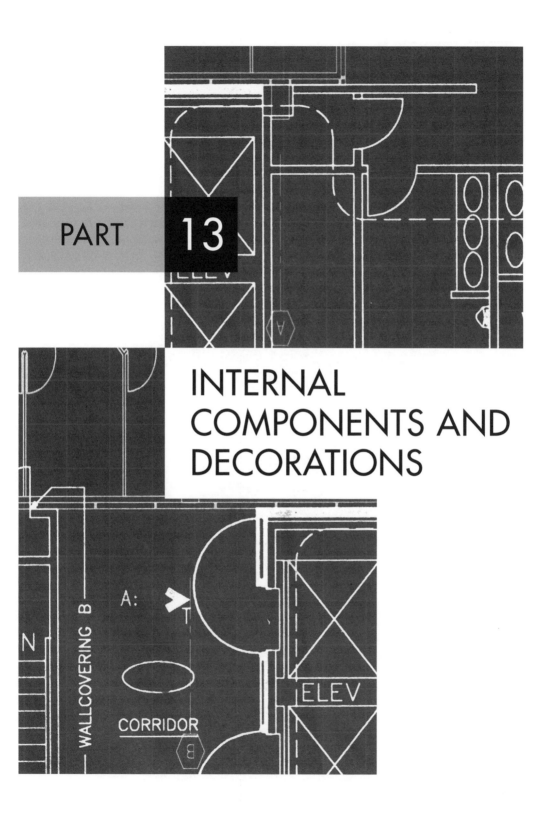

PART 13

INTERNAL COMPONENTS AND DECORATIONS

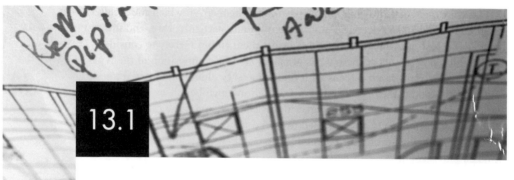

PARTITIONS, DOORS AND CEILINGS

Internal components and finishes that are normally associated with domestic dwellings such as brick walls, lightweight concrete block walls, plastering, dry linings, floor finishes and coverings, fixed partitions and the various trims are detailed in the associated volume, *Construction Technology*. Advanced studies of construction technology more appropriately concentrate on the components and finishes applied to buildings such as offices, commercial and institutional establishments. These include the partitions, sliding doors and suspended ceilings developed in this chapter.

▉▉▉ PARTITIONS

A partition is similar to a wall in that it is a vertical construction dividing the internal space of a building, but partitions are generally considered to be internal walls that are lightweight, non-loadbearing, demountable or movable. They are used in buildings such as offices where it is desirable to have a system of internal division that can be altered to suit changes in usage without excessive costs and the minimum of interference to services and ceilings. Note that any non-loadbearing partition could if necessary be taken down and moved, but in most cases moving partitions such as timber studwork with plasterboard linings will necessitate the replacement or repair of some of the materials and finishes involved. A true demountable partition can usually be taken down, moved and re-erected without any notable damage to the materials, components, finishes and surrounding parts of the building. Movable partitions usually take the form of a series of doors or door leaves hung to slide or slide and fold, and these are described later under a separate heading. Many demountable partition systems are used in conjunction with a suspended ceiling, and it is important that both systems are considered together in the context of their respective main functions, because one system may cancel out the benefits achieved by the other.

There are many patent demountable partition systems from which the architect, designer or builder can select a suitable system for any particular case, and it is not possible in a text of this nature to analyse or list them all but merely to consider the general requirements of these types of partition. The composition of the types available range from glazed screens with timber, steel or aluminium alloy frames to panel construction with concealed or exposed jointing members. Low-level screens and toilet cubicles can also be considered in this context, and these can be attached to posts fixed at floor level or suspended from the structural floor above.

When selecting or specifying a demountable partition the following points should be taken into account:

■ **Fixing and stability**   Points to be considered are bottom fixing, top fixing and joints between sections. The two common methods of bottom fixing are the horizontal base unit, which is very often hollow to receive services, and the screw jack fixing: both methods are intended to be fixed above the finished floor level, which should have a tolerance not exceeding 5 mm. The top fixing can be of a similar nature with a ceiling tolerance of the same magnitude, but problems can arise when used in conjunction with a suspended ceiling because little or no pressure can be applied to the underside of the ceiling. A brace or panel could be fixed above the partition in the void over the ceiling to give the required stability; alternatively, the partition could pass through the ceiling, but this would reduce the flexibility of the two systems should a rearrangement be required at a later date. The joints between consecutive panels do not normally present any problems as they are an integral part of the design, but unless they are adequately sealed the sound insulation and/or fire-resistance properties of the partition could be seriously impaired.

■ **Sound insulation**   The degree of insulation required will depend largely upon the usage of rooms created by the partitions. The level of sound insulation that can be obtained will depend upon three factors: the density of the partition construction, the degree of demountability, and the continuity of the partition with the floor and ceiling. Generally, the less demountable a partition, the easier it will be to achieve an acceptable sound insulation level. Partitions fixed between the structural floor and the structural ceiling usually result in good sound insulation, whereas partitions erected between the floor and a suspended ceiling give poor sound insulation properties, mainly because of the flanking sound path over the head of the partition. In the latter case two remedies are possible: either the partition could pass through the suspended ceiling, or a sound-insulating panel could be inserted in the ceiling void directly over the partition below.

■ **Fire resistance**   The main restrictions on the choice of partition lie in the requirements of the Building Regulations Part B with regard to fire-resistant properties, spread of flame, fire stopping and providing a suitable means of escape in case of fire. The weakest points in any system will be the openings and the seals at the foot and head. Provisions at openings are a matter of detail and adequate construction together with a suitable fire-resistant door or fire-resistant glazing. Fire stopping at the head of a partition used in conjunction with a suspended ceiling usually takes the form of a fire break panel or barrier fixed in the ceiling void directly above the partition below. These may be produced from 25 mm galvanised wire mesh stitched to mineral wool insulation in two layers of at least 50 mm thickness, or from two sheets of 12.5 mm plasterboard, as indicated in Fig. 13.1.1. Most proprietary demountable systems will give a half-hour or one-hour fire resistance with a class 1 or class 0 spread of flame classification.

Typical demountable partition details are shown in Figs 13.1.1 and 13.1.2.

▨▪▪ SLIDING DOORS

Sliding doors may be used in all forms of building, from the small garage to large industrial structures. They may be incorporated into a design for any of the following reasons:

- as an alternative to a swing door to conserve space or where it is not possible to install a swing door because of space restrictions;
- where heavy doors or doors in more than two leaves are to be used, and where conventional hinges would be inadequate;
- as a movable partition used in preference to a demountable partition because of the frequency with which it would be moved.

In most cases there is a choice of bottom or top sliding door gear, which in the case of very heavy doors may have to be mechanically operated. Top gear can be used only if there is a suitable soffit, beam or lintel over the proposed opening that is strong enough to take the load of the gear, track and doors. In the case of bottom gear all the loads are transmitted to the floor. Top gear has the disadvantage of generally being noisier than its bottom gear counterpart, but bottom gear either has an upstanding track, which can be a hazard in terms of tripping, or it can have a sunk channel, which can become blocked with dirt and dust unless regularly maintained. In all cases it is essential that the track and guides are perfectly aligned vertically to one another and in the horizontal direction parallel to each other to prevent stiff operation or the binding of the doors while being moved. When specifying sliding door gear it is essential that the correct type is chosen to suit the particular door combination and weight range if the partition is to be operated efficiently.

Many types of patent sliding door gear and track are available to suit all needs, but these can usually be classified by the way in which the doors operate:

- **Straight sliding**   These can be of a single leaf, double leaf or designed to run on adjacent and parallel tracks to give one parking. Generally this form of sliding door uses top gear (see Fig. 13.1.3 for a typical example).
- **End-folding doors**   These are strictly speaking sliding and folding doors that are used for wider openings than the straight doors given above. They usually consist of a series of leaves operating off top gear with a bottom guide track so that the folding leaves can be parked to one or both sides of the opening (see Fig. 13.1.4 for typical details).
- **Centre-folding doors**   Like the previous type, these doors slide and fold to be parked at one or both ends of the opening, and have a top track with a bottom floor guide channel arranged so that the doors will pivot centrally over the channel. With this arrangement the hinged leaf attached to the frame is only approximately a half-leaf (see typical details in Fig. 13.1.5).

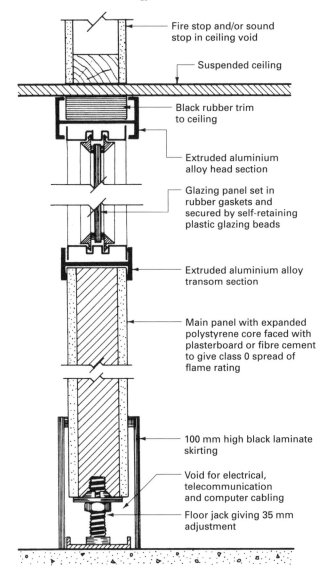

Fire stop and/or sound
stop in ceiling void

Suspended ceiling

Black rubber trim
to ceiling

Extruded aluminium
alloy head section

Glazing panel set in
rubber gaskets and
secured by self-retaining
plastic glazing beads

Extruded aluminium alloy
transom section

Main panel with expanded
polystyrene core faced with
plasterboard or fibre cement
to give class 0 spread of
flame rating

100 mm high black laminate
skirting

Void for electrical,
telecommunication
and computer cabling

Floor jack giving 35 mm
adjustment

Vertical section   Maximum panel height 3.657;
panel thickness 54 mm; overall
horizontal module 1.190

Figure 13.1.1 Typical demountable partition details: 1 (*courtesy*: Tenon Contracts)

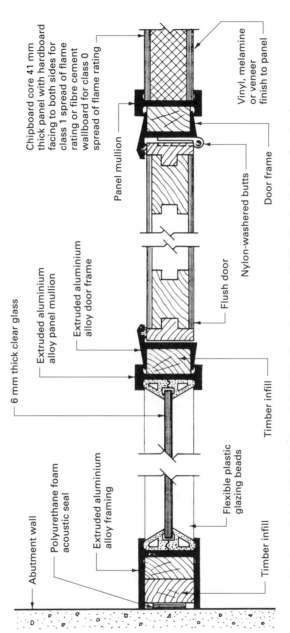

Chipboard core 41 mm thick panel with hardboard facing to both sides for class 1 spread of flame rating or fibre cement wallboard for class 0 spread of flame rating

Panel mullion

Vinyl, melamine or veneer finish to panel

Door frame

Nylon-washered butts

Flush door

Extruded aluminium alloy door frame

Extruded aluminium alloy panel mullion

Timber infill

6 mm thick clear glass

Flexible plastic glazing beads

Extruded aluminium alloy framing

Polyurethane foam acoustic seal

Abutment wall

Timber infill

Maximum height 4.577　Width module 1.200　Sound reduction 27dB　Plastic skirting at floor level
Services can be accommodated within chipboard core and along continuous base ducts.

Figure 13.1.2　Typical demountable partition details: 2 (*courtesy:* Venesta International Components)

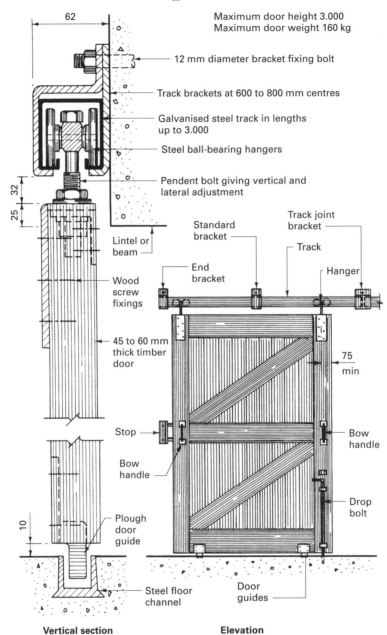

62

Maximum door height 3.000
Maximum door weight 160 kg

12 mm diameter bracket fixing bolt

Track brackets at 600 to 800 mm centres

Galvanised steel track in lengths
up to 3.000

Steel ball-bearing hangers

Pendent bolt giving vertical and
lateral adjustment

32

25

Lintel or
beam

Wood
screw
fixings

45 to 60 mm
thick timber
door

Stop

Bow
handle

Plough
door
guide

10

Steel floor
channel

Track joint
bracket

Standard
bracket

Track

End
bracket

Hanger

75
min

Bow
handle

Drop
bolt

Door
guides

Vertical section   **Elevation**

Figure 13.1.3   Typical straight sliding door details (*courtesy*: Hillaldam Coburn)

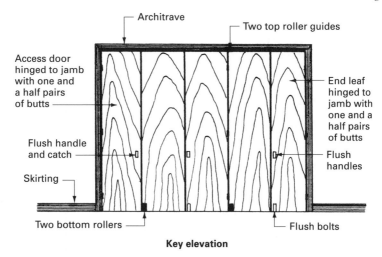

Architrave

Two top roller guides

Access door hinged to jamb with one and a half pairs of butts

End leaf hinged to jamb with one and a half pairs of butts

Flush handle and catch

Flush handles

Skirting

Two bottom rollers

Flush bolts

Key elevation

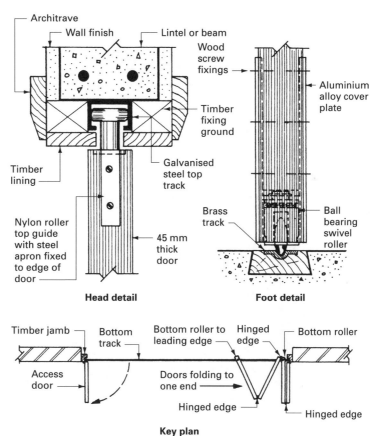

Architrave

Wall finish

Lintel or beam

Wood screw fixings

Aluminium alloy cover plate

Timber fixing ground

Timber lining

Galvanised steel top track

Nylon roller top guide with steel apron fixed to edge of door

45 mm thick door

Brass track

Ball bearing swivel roller

Head detail

Foot detail

Timber jamb

Bottom track

Bottom roller to leading edge

Hinged edge

Bottom roller

Access door

Doors folding to one end

Hinged edge

Hinged edge

Key plan

Figure 13.1.4   Typical end-folding door details (*courtesy*: Hillaldam Coburn)

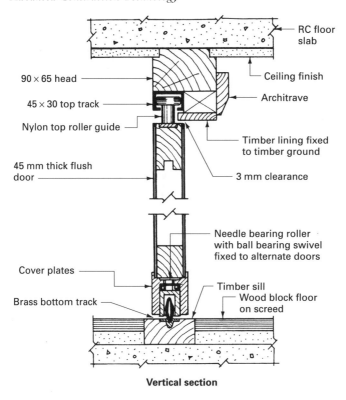

RC floor slab

90 × 65 head

Ceiling finish

45 × 30 top track

Architrave

Nylon top roller guide

Timber lining fixed to timber ground

45 mm thick flush door

3 mm clearance

Needle bearing roller with ball bearing swivel fixed to alternate doors

Cover plates

Brass bottom track

Timber sill
Wood block floor on screed

Vertical section

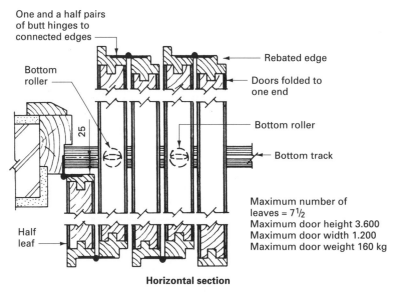

One and a half pairs of butt hinges to connected edges

Rebated edge

Bottom roller

Doors folded to one end

25

Bottom roller

Half leaf

Bottom track

Maximum number of leaves = 7½
Maximum door height 3.600
Maximum door width 1.200
Maximum door weight 160 kg

Horizontal section

Figure 13.1.5   Typical centre-folding sliding door details (*courtesy*: Hillaldam Coburn)

Other forms of sliding doors include round-the-corner sliding doors, which use a curved track and folding gear arranged so that the leaves hinged together will slide around the corner and park alongside a side wall, making them suitable for situations where a clear opening is required. The number of leaves in any one hinged sliding set should not be less than three nor more than five. Up-and-over doors operated by a counterbalance weight or spring are common for domestic garage applications and are available where they form, when raised, a partial canopy to the opening or are parked entirely within the building. Another application of the folding door technique is the folding screen, which is suitable for a limited number of lightweight leaves that can be folded back onto a side or return wall without the use of sliding gear, track or guide channels (see Fig. 13.1.6 for typical details).

■■■ SUSPENDED CEILINGS

A suspended ceiling can be defined as a ceiling that is attached to a framework suspended from the main structure, thus forming a void between the ceiling and the underside of the main structure. Ceilings that are fixed, for example, to lattice girders and trusses while forming a void are strictly speaking attached ceilings, although they may be formed by using the same methods and materials as the suspended ceilings. The same argument can be applied to ceilings that are fixed to a framework of battens attached to the underside of the main structure. Some suspended ceilings contribute to the fire resistance of the structure above, and these are defined as fire-protecting suspended ceilings. Reference should be made to Building Regulations, Approved Document B, Appendix A, Table A3 for material limitations of these fire-protecting suspended ceilings in various applications.

Further reasons for including a suspended ceiling system in a building design are that it can:

- provide a finish to the underside of a structural floor or roof, generally for purposes of concealment;
- create a void space suitable for housing, concealing and protecting services and light fittings;
- add to the sound and/or thermal insulation properties of the floor or roof above and adjacent rooms;
- provide a means of structural fire protection to steel beams supporting a concrete floor;
- provide a means of acoustic control in terms of absorption and reverberation;
- create a lower ceiling height to a particular room or space.

A suspended ceiling, for whatever reason it has been specified and installed, should fulfil the following requirements:

- Be easy to construct, repair, maintain and clean.
- Comply with the requirements of the Building Regulations and, in particular, Approved Document B2 and associated Appendix A. These are concerned with flame spread over the surface, classification of performance and the limitations imposed on the use of certain materials with regard to fire, e.g. thermoplastic light diffusers.

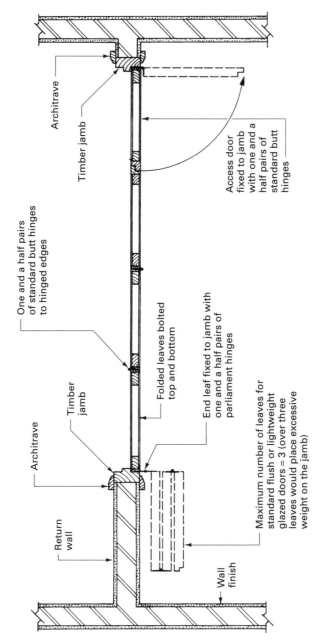

Figure 13.1.6   Typical folding screen details

Architrave

Timber jamb

One and a half pairs of standard butt hinges to hinged edges

Folded leaves bolted top and bottom

End leaf fixed to jamb with one and a half pairs of parliament hinges

Access door fixed to jamb with one and a half pairs of standard butt hinges

Maximum number of leaves for standard flush or lightweight glazed doors = 3 (over three leaves would place excessive weight on the jamb)

Architrave

Timber jamb

Return wall

Wall finish

■ Provide an adequate means of access for the maintenance of the suspension system and/or the maintenance of concealed services and light fittings.
■ Conform to a planning module that preferably should be based on the modular coordination recommendations set out in BS 6750: *Specification for modular coordination in building*, which recommends a first preference module of 300 mm.

There are many ways in which to classify suspended ceilings. They can be classified by function, e.g. sound absorbing or fire resisting, by specification of the materials involved. Another simple practical method of classification is to group the ceiling systems by their general method of construction:

■ jointless ceilings;
■ jointed ceilings;
■ open ceilings.

Jointless ceilings

These are ceilings that, although suspended from the main structure, give the internal appearance of being a conventional ceiling. The final finish is usually of plaster applied in one or two coats to plasterboard or expanded metal lathing; alternatively, a jointless suspended ceiling could be formed by applying sprayed plaster or sprayed vermiculite-cement to an expanded metal background. Typical jointless ceiling details are shown in Fig. 13.1.7.

Jointed ceilings

These suspended ceilings are the most popular and common form because of their ease of assembly, ease of installation and ease of maintenance. They consist basically of a suspended metal framework to which the finish in a board or tile form is attached. The boards or tiles can be located on a series of inverted tee bar supports with the supporting members exposed forming part of the general appearance; alternatively, by using various spring clip devices or direct fixing the supporting members can be concealed. The common ceiling materials encountered are fibreboards, metal trays, fibre cement materials and plastic tiles or trays (see Figs 13.1.8 and 13.1.9 for typical details).

Open ceilings

These suspended ceilings are largely decorative in function but by installing light fittings in the ceiling void they can act as a luminous ceiling. The format of these ceilings can be an open work grid of timber, with or without louvres or a series of closely spaced plates of polished steel or any other suitable material. The colour and texture of the sides and soffit of the ceiling void must be carefully designed if an effective system is to be achieved. It is possible to line these surfaces with an acoustic absorption material to provide another function to the arrangement – see typical details in Fig. 13.1.10.

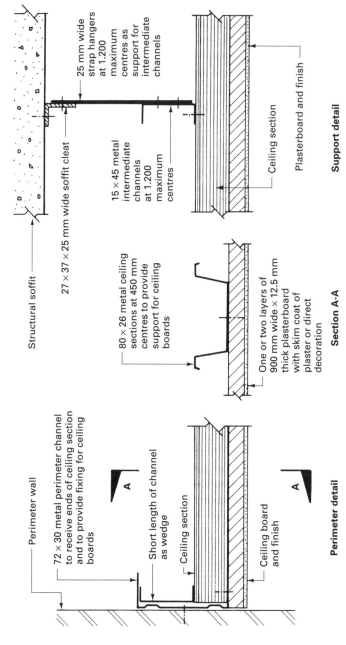

Structural soffit

27 × 37 × 25 mm wide soffit cleat

25 mm wide strap hangers at 1.200 maximum centres as support for intermediate channels

15 × 45 metal intermediate channels at 1.200 maximum centres

Ceiling section

Plasterboard and finish

Support detail

80 × 26 metal ceiling sections at 450 mm centres to provide support for ceiling boards

One or two layers of 900 mm wide × 12.5 mm thick plasterboard with skim coat of plaster or direct decoration

Section A-A

Perimeter wall

72 × 30 metal perimeter channel to receive ends of ceiling section and to provide fixing for ceiling boards

Short length of channel as wedge

Ceiling section

Ceiling board and finish

A

A

Perimeter detail

Figure 13.1.7   Typical jointless suspended ceiling details (*courtesy*: Gyproc M/F)

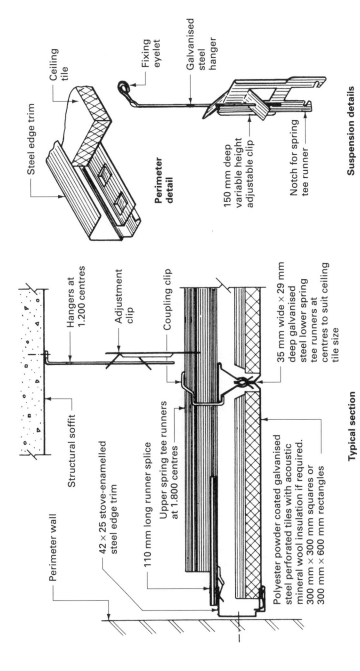

Figure 13.1.8 Typical jointed suspended ceiling details (*courtesy:* Dampa (UK) Ltd)

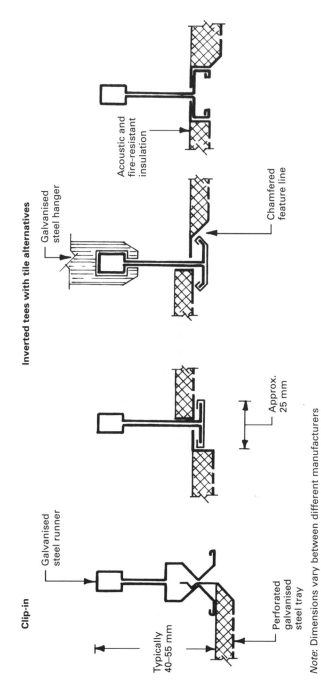

Note: Dimensions vary between different manufacturers

Figure 13.1.9   Jointed suspended ceiling support profiles

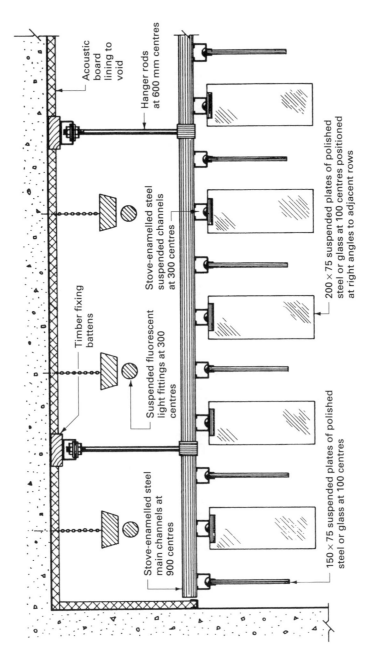

Acoustic board lining to void

Hanger rods at 600 mm centres

Stove-enamelled steel suspended channels at 300 centres

Timber fixing battens

Suspended fluorescent light fittings at 300 centres

Stove-enamelled steel main channels at 900 centres

150 × 75 suspended plates of polished steel or glass at 100 centres

200 × 75 suspended plates of polished steel or glass at 100 centres positioned at right angles to adjacent rows

Figure 13.1.10   Typical open suspended ceiling details

PAINTING AND DECORATIONS

The fundamentals of applying decorative finishes in the form of paint and wallpaper have been considered in Chapter 7.11 of *Construction Technology*. Therefore, study at this advanced level progresses beyond basic principles, particularly in the context of painting timber and metals.

▪▪▪ PAINTING TIMBER

Timber can be painted to prevent decay in the material by forming a barrier to the penetration of moisture, thus giving rise to the conditions necessary for fungal attack to begin; alternatively, timber may be painted mainly to impart colour. Timber that is exposed to the elements is more vulnerable to eventual decay than internal joinery items, but internal condensation can also give rise to damp conditions. Timber can also be protected with clear water-repellent preservatives and varnishes to preserve the natural colour and texture of the timber.

The moisture content of the timber to which paint is to be applied should be as near as possible to that at which it will stabilise in its final condition. For internal joinery this would be within an 8–12% region depending upon the internal design temperature, and for external timber within a 15–18% region according to exposure conditions. If the timber is too dry when the paint is applied, any subsequent swelling of the base material due to moisture absorption will place unacceptable stresses on the paint film, leading to cracking of the paint barrier. Conversely, wet timber drying out after the paint application can result in blistering, opening of joints and the consequential breakdown of the paint film.

New work should receive a four-coat application consisting of primer, undercoat and finishing coats. The traditional shellac knotting applied to reduce the risk of resin leaking and staining should be carried out prior to the primer application, but in external conditions it may prove to be inadequate. Therefore timber selected for exposed or external conditions should be of a high quality without or with only a small amount of knots. Knotting and priming should preferably be carried out under the ideal conditions prevailing at the place of manufacture. Filling and stopping should take place after priming but before the application of the undercoat and finishing coats. It is a general misconception that a good key is necessary to achieve a satisfactory paint surface. It is not penetration that is required but a good molecular attraction of the paint binder to the timber that is needed to obtain a satisfactory result.

If the protective paint film is breached by moisture, decay can take place under the paint coverings; indeed, the coats of paint prevent the drying out of any moisture that has managed to penetrate the timber prior to painting. Any joinery items that in their final situation could be susceptible to moisture penetration should therefore be treated with a paintable preservative. Preservation treatments include diffusion treatment using water-soluble borates, water-borne preservatives and organic solvents. The methods of application and chemical composition of these timber preservatives can be found in the following references:

- BS 1282: *Wood preservatives. Guidance on choice, use and application.*
- BS 5268-5: *Structural use of timber. Code of practice for the preservative treatment of structural timber.*
- BS 5589: *Code of practice for preservation of timber.*
- BS 8471: *Preservation of timber. Recommendations.*
- BRE: *Timber durability and treatment pack.*

▨▮▮ PAINTING METALS

The application of paint to provide a protective coat and for decorative purposes is normally confined to iron and steel, because most non-ferrous metals are left to oxidise and form their own natural protective coating. Corrosion of ferrous metals is a natural process that can cause disfiguration and possible failure of a metal component. Two aspects of painting these metals must be considered:

- initial preparation and paint application;
- maintenance of protective paint.

PREPARATION

The basic requirement is to prepare the metal surface adequately to receive the primer and subsequent coats of paint, because the mill scale, rust, oil, grease and dirt that are frequently found on metals are not suitable as a base for the applications of paint. Suitable preparation treatments are:

- **Shot and grit blasting**   This is an effective method, which can be carried out on site, although it is usually considered to be a factory process.
- **Phosphating and pickling**   This is a factory process involving immersion of the metal component in hot acid solutions to remove rust and scale.
- **Degreasing**   This is a factory process involving washing or immersion using organic solvents, emulsions or hot alkali solutions followed by washing. It is very often used as a preliminary treatment to phosphating.

- **Mechanical**   This is a site or factory process using hand-held or powered tools such as hammers, chisels, brushes and scrapers. To be effective these forms of surface preparation must be thorough.
- **Flame cleaning**   An oxyacetylene flame is used in conjunction with hand-held tools for removing existing coats of paint or loose scale and rust.

It is essential that in conjunction with the above preparation processes the correct primer is specified and used to obtain a satisfactory result. For further reference, see BS EN ISO 12944-7: *Paints and varnishes. Corrosion protection of steel structures by protective paint systems. Execution and supervision of paintwork.*

MAINTENANCE

To protect steel in the long term a rigid schedule of painting maintenance must be worked out and invoked. This is particularly true of the many areas where access is difficult, such as pipes fixed close to a wall or adjacent members in a lattice truss. Access for future maintenance is an aspect of building that should be considered during the design and construction stages, not only in the context of metal components but also for any part of a building that will require future maintenance and/or inspection. To take full advantage of the structural properties of steelwork, and at the same time avoid the maintenance problems, weathering steels could be considered for exposed conditions.

■■■ WEATHERING STEELS

These do not rust or corrode as normal steels, but interact with the atmosphere to produce a layer of sealing oxides. The colour of this protective layer will vary from a lightish brown to a dark purple grey depending on the degree of exposure, amount and type of pollution in the atmosphere and orientation. The best-known weathering steel is called 'Cor-Ten', which is derived from the fact that it is CORrosion resistant with a high TENsile strength. Cor-Ten is not a stainless steel but a low-alloy steel with a lower proportion of non-ferrous metals than stainless steel, which makes it a dearer material than mild steel but cheaper than traditional stainless steels.

Weathering steels can be used as a substitute for other steels except in wet situations, marine works or in areas of high pollution unless protected with paint or similar protective applications, which defeats the main objective of using this material. Jointing can be by welding or friction grip bolts. Note that the protective coating will not form on weathering steels in internal situations because the formation of this coat is a natural process of the wet and dry cycles encountered with ordinary weather conditions.

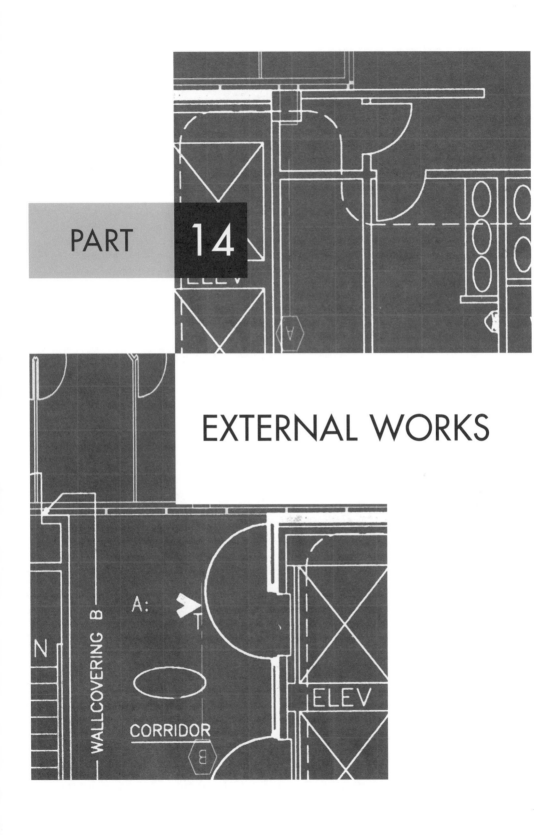

PART 14

EXTERNAL WORKS

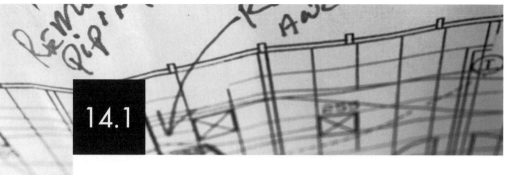

ROADS, PAVINGS AND SLABS

Building contractors are not normally engaged to construct major roads or motorways, as this is the province of the civil engineer. However, they can be involved in the laying of small estate roads, service roads and driveways, and it is within this context that the student of construction technology would consider roadworks.

Before considering road construction techniques and types it is worthwhile considering some of the problems encountered by the designer when planning road layouts. The width of a road can be determined by the anticipated traffic flow, volume and speed. Lane width dimension is calculated from an allowance of maximum vehicle width of 2.500 m plus 0.500 m minimum clearance between vehicles, to provide a minimum lane width of 3.000 m. The major layout problems occur at road junctions and at the termination of cul-de-sac roads. At right-angle junctions it is important that vehicles approaching the junction from any direction have a clear view of approaching vehicles intending to join the main traffic flow. Most planning authorities have layout restrictions at such junctions in the form of triangulated sight lines, which give a distance and area in which an observer can see an object when both are at specific heights above the carriageway. Within this triangulated area street furniture or any other obstruction is not allowed (see Fig. 14.1.1). Angled junctions, by virtue of their distinctive layout, do not present the same problems. At any junction a suitable radius should be planned so that vehicles filtering into the main road should not have to apply a full turning lock. The actual radius required will be governed by the anticipated vehicle types that will use the road.

Terminations at the end of cul-de-sacs must be planned to allow vehicles to turn round. For service roads this is usually based on the length and turning circle specifications of refuse collection vehicles, which are probably the largest vehicles to use the road. Typical examples are shown in Fig. 14.1.2.

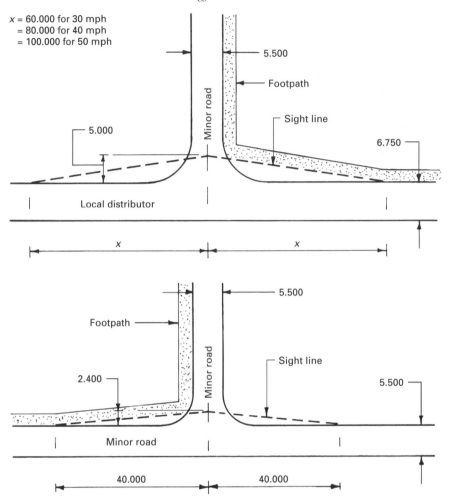

x = 60.000 for 30 mph
= 80.000 for 40 mph
= 100.000 for 50 mph

5.500

Footpath

Sight line

5.000

6.750

Minor road

Local distributor

x

x

5.500

Footpath

Sight line

2.400

5.500

Minor road

Minor road

40.000

40.000

Figure 14.1.1   Typical road junctions

▨▨■ ROADS

The construction of roads can be considered under two headings:

- preparation or earthworks;
- pavement construction.

Before any roadwork is undertaken a thorough soil investigation should be carried out to determine the nature of the subgrade, which is the soil immediately below the topsoil that will ultimately carry the traffic loads from the pavement above. Soil investigations should preferably be carried out during the winter period when subgrade conditions will be at their worst. Trial holes should be taken down to at least 1.000 m below the proposed formation level. The information required

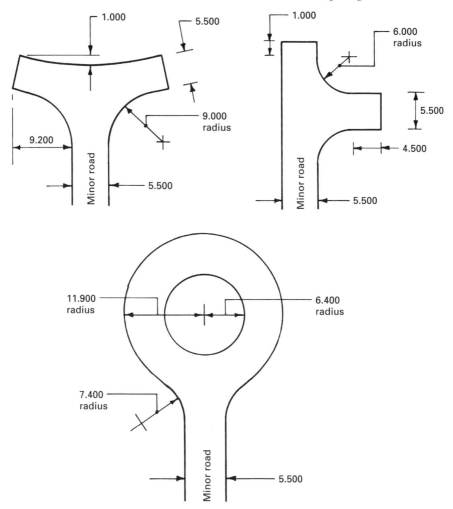

Figure 14.1.2   Typical road turn-around and terminations

from these investigations to ensure that a good pavement design can be formulated is the elasticity, plasticity, cohesion and internal friction properties of the subgrade.

EARTHWORKS

This will include removing topsoil, scraping and grading the exposed surface to the required formation level, preparing the subgrade to receive the pavement, and forming any embankments and/or cuttings. The strength of the subgrade will generally decrease as the moisture content increases. An excess of water in the subgrade can also cause damage by freezing, causing frost heave in fine sands, chalk and silty soils. Conversely, thawing may cause a reduction in subgrade strength, giving rise to failure of the pavement above. Tree roots, particularly those of fast-growing deciduous trees such as the poplar and willow, can also cause damage in heavy clay soils by extracting vast quantities of water from the subgrade down to a depth of 3.000 m.

The pavement covering will give final protection to the subgrade from excess moisture, but during the construction period the subgrade should be protected by a waterproof surfacing such as a sprayed bituminous binder with a sand cover applied at a rate of 1 litre per square metre. If the subgrade is not to be immediately covered with the sub-base of the pavement it should be protected by an impermeable membrane such as 500 gauge plastic sheeting with 300 mm side and end laps.

PAVEMENT CONSTRUCTION

'Pavement' is a general term for any paved surface, and is also the term applied specifically to the whole construction of a road. Road pavements can be classified as flexible pavements that, for the purpose of design, are assumed to have no tensile strength and consist of a series of layers of materials to distribute the wheel loads to the subgrade. The alternative form is the rigid pavement of which, for the purpose of design, the tensile strength is taken into account and which consists of a concrete slab resting on a granular base.

Flexible pavements

The sub-base for a flexible pavement is laid directly onto the formation level and should consist of a well-compacted granular material such as a quarry overburden or crushed rocks. The actual thickness of sub-base required is determined by the cumulative number of standard axles (msa) to be carried (msa = millions of standard axles, where a standard axle is taken as 8,200 kg) and the CBR (California bearing ratio) of the subgrade. The CBR is an empirical method in which the thickness of the sub-base is related to the strength of the subgrade and to the amount of traffic the road is expected to carry. A fully flexible pavement design life should be at least 20 years. As an alternative to empirical calculations, reference can be made to Volume 7 of the *Design Manual for Roads and Bridges: Part 3*, ref. HD 26/94: *Pavement Design and Maintenance*, published by the Stationery Office. This manual provides graphical presentations on the principle shown in Fig. 14.1.3 for determining road construction specification relative to design traffic load (msa) and overall bituminous layer thickness.

The subgrade is covered with a sub-base, a base course and a wearing course; the last two components are collectively known as the **surfacing**. The sub-base can consist of any material that remains stable in water, such as crushed stone, dry lean concrete or blast furnace slag. Compacted dry-bound macadam in a 75 to 125 mm thick layer with a 25 mm thick overlay of firmer material or a compacted wet mix macadam in 75 to 150 mm thick layers could also be used. The material chosen should also be unaffected by frost and be well compacted in layers, giving a compacted thickness of between 100 and 150 mm for each layer.

The base course of the surfacing can consist of rolled asphalt, dense tarmacadam, dense bitumen macadam or open-textured macadam, and should be applied to a

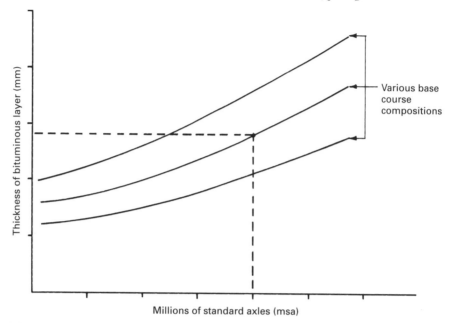

Figure 14.1.3 Design thickness for pavements (*see* publication HD 26/94, Stationery Office)

minimum thickness of 60 mm. Base courses are laid to the required finished road section providing any necessary gradients or crossfalls, ready to receive the thinner wearing course, which should be laid within three days of completing the base course. The wearing course is usually laid by machine and provides the water protection for the base layers. It should also have non-skid properties, reasonable resistance to glare, good riding properties and a good life expectancy. Materials that give these properties include hot-rolled asphalt, bitumen macadam, dense tar surfacing and cold asphalt (see Fig. 14.1.4 for typical details). Existing flexible road surfaces can be renovated quickly and cheaply by the application of a hot tar or cut-back bitumen binder with a rolled layer of gravel, crushed stone or slag chippings applied immediately after the binder and before it sets.

Pavements that contain a nominal amount of cement to bind materials in the sub-base and have bituminous materials for surfacing vary sufficiently from a truly flexible specification. These are often known appropriately as **semi-flexible** or **flexible composite**.

The above treatments are termed bound surfaces, but flexible roads or pavements with unbound surfaces can also be constructed. These are suitable for light vehicular traffic where violent braking and/or acceleration is not anticipated, such as driveways to domestic properties. Unbound pavements consist of a 100 to 200 mm thick base of clinker or hardcore laid directly onto the formation level of the subgrade, and this is covered with a well-rolled layer of screeded gravel to pass a 40 mm ring with sufficient sand to fill the small voids to form an overall consolidated thickness of 25 mm. This will give a relatively cheap flexible

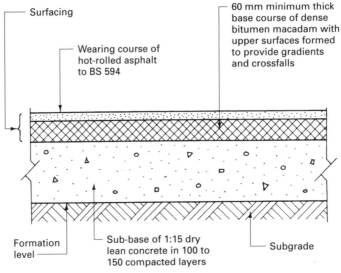

Surfacing

Wearing course of
hot-rolled asphalt
to BS 594

60 mm minimum thick
base course of dense
bitumen macadam with
upper surfaces formed
to provide gradients
and crossfalls

Formation
level

Sub-base of 1:15 dry
lean concrete in 100 to
150 compacted layers

Subgrade

Typical semi-flexible pavement

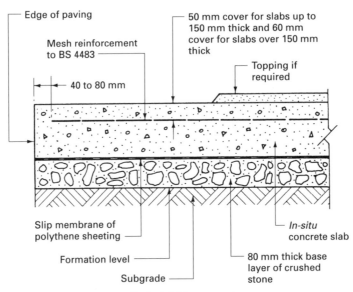

Edge of paving

Mesh reinforcement
to BS 4483

40 to 80 mm

50 mm cover for slabs up to
150 mm thick and 60 mm
cover for slabs over 150 mm
thick

Topping if
required

Slip membrane of
polythene sheeting

Formation level

Subgrade

In-situ
concrete slab

80 mm thick base
layer of crushed
stone

Typical rigid pavement

References:
BS 594-1 and 2: *Hot rolled asphalt for roads and other paved areas. Specifications*
BS 4483: *Steel fabric for the reinforcement of concrete*

Figure 14.1.4　Typical pavement details

pavement, but to be really efficient it should have adequate falls to prevent the ponding of water, and should be treated each spring with an effective weedkiller.

Rigid pavements

This is a form of road using a concrete slab laid over a base layer. The subgrade is prepared as described above for flexible pavements, and it should be adequately protected against water. The base layer is laid over the subgrade, and is required to form a working surface from which to case the concrete slab and enable work to proceed during wet and frosty weather without damage to the subgrade. Generally, granular materials such as crushed concrete, crushed stone, crushed slag and suitably graded gravels are used to form the base layer. The thickness will depend upon the nature and type of subgrade: weak subgrades normally require a minimum thickness of 150 mm whereas normal subgrades require a minimum thickness of only 80 mm.

The thickness of concrete slabs used in rigid pavement construction will depend upon the condition of the subgrade and intensity of traffic, and on whether the slab is to be reinforced. With a normal subgrade using a base layer 80 mm thick the slab thickness would vary from 125 mm for a reinforced slab carrying light traffic to 200 mm for an unreinforced slab carrying a medium to heavy traffic intensity. The usual strength specification is 28 MN/m^2 at 28 days with not more than 1% test cube failure rate: therefore the mix design should be based on a mean strength of between 40 and 50 MN/m^2, depending upon the degree of quality control possible on or off site. To minimise the damage that can be caused by frost and deicing salts, the water/cement ratio should not exceed 0.5 by weight, and air entrainment to at least the top 50 mm of the concrete should be specified. The air-entraining agent used should produce 3–6% of minute air bubbles in the hardened concrete, thus preventing saturation of the slab by capillary action.

Before the concrete is laid, the base layer should be covered with a slip membrane of polythene sheet, which will also prevent grout loss from the concrete slab. Concrete slabs are usually laid between pressed steel road forms, which are positioned and fixed to the ground with steel stakes. These side forms are designed to provide the guide for hand tamping or to provide for a concrete train consisting of spreaders and compacting units. Curved or flexible road forms have no top or bottom flange and are secured to the ground with an increased number of steel stakes (see Fig. 14.1.5 for typical road form examples).

Reinforcement generally in the form of a welded steel fabric complying with the recommendations of BS 4483: *Steel fabric for the reinforcement of concrete*, can be included in rigid pavement constructions to prevent the formation of cracks and to enable the number of expansion and contraction joints required to be reduced. If bar reinforcement is used instead of welded steel fabric it should consist of deformed bars at spacings not exceeding 150 mm. The cover of concrete over the reinforcement will depend on the thickness of concrete: for slabs under 150 mm thick the minimum cover should be 50 mm, and for slabs over 150 mm thick the minimum cover should be 60 mm.

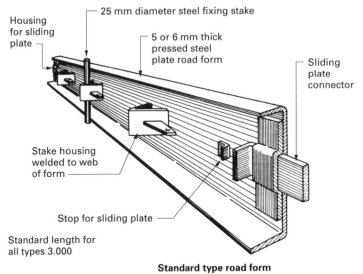

25 mm diameter steel fixing stake

Housing for sliding plate

5 or 6 mm thick pressed steel plate road form

Sliding plate connector

Stake housing welded to web of form

Stop for sliding plate

Standard length for all types 3.000

Standard type road form

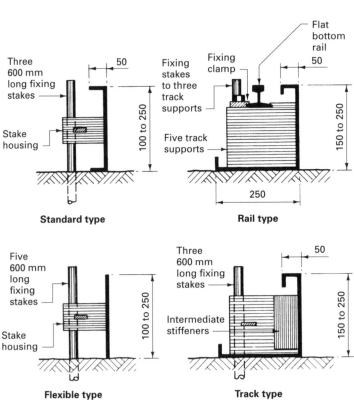

Three 600 mm long fixing stakes

50

Stake housing

100 to 250

Standard type

Fixing stakes to three track supports

Fixing clamp

Flat bottom rail

50

Five track supports

250

150 to 250

Rail type

Five 600 mm long fixing stakes

Stake housing

100 to 250

Flexible type

Three 600 mm long fixing stakes

50

Intermediate stiffeners

150 to 250

Track type

Figure 14.1.5 Typical steel road form details

Joints used in rigid pavements may be either transverse or longitudinal, and are included in the design to:

- limit the size of slab;
- limit the stresses due to subgrade restraint;
- make provision for slab movements such as expansion, contraction and warping.

The spacing of joints will be governed by several factors, namely slab thickness, presence of reinforcement, traffic intensity and the temperature at which the concrete is placed. Five types of joint are used in rigid road and pavement construction and are classed as follows:

- **Expansion joints** These are transverse joints at 36.000 to 72.000 m centres in reinforced slabs and at 27.000 to 54.000 m centres in unreinforced slabs.
- **Contraction joints** These are transverse joints that are placed between expansion joints at 12.000 to 24.000 m centres in reinforced slabs and at 4.500 to 7.500 m centres in unreinforced slabs to limit the size of slab bay or panel. Note that every third joint should be an expansion joint.
- **Longitudinal joints** These are similar to contraction joints and are required where slab width exceeds 4.500 m.
- **Construction joints** The day's work should normally be terminated at an expansion or contraction joint, but if this is impossible a construction joint can be included. These joints are similar to contraction joints but with the two portions tied together with reinforcement. Construction joints should not be placed within 3.000 m of another joint, and should be avoided wherever possible.
- **Warping joints** These are transverse joints that are sometimes required in unreinforced slabs to relieve the stresses caused by vertical temperature gradients within the slab if they are higher than the contractional stresses. The detail is similar to contraction joints but has a special arrangement of reinforcement.

Typical joint details are shown in Fig. 14.1.6.

Road joints can require fillers and/or sealers: the former need to be compressible whereas the latter should protect the joint against the entry of water and grit. Suitable materials for fillers are soft knot-free timber, impregnated fibreboard, chipboard, cork and cellular rubber. The common sealing compounds used are resinous compounds, rubber-bituminous compounds and straight-run bitumen compounds containing fillers. The sealed surface groove used in contraction joints to predetermine the position of a crack can be formed while casting the slab, or can be sawn into the hardened concrete using water-cooled circular saws. Although slightly dearer than the formed joint, sawn joints require less labour and generally give a better finish.

The curing of newly laid rigid roads and pavings is important if the concrete strength is to be maintained and the formation of surface cracks is to be avoided. Curing precautions should commence as soon as practicable after laying, preferably within 15 minutes of completion by covering the newly laid surface with a suitable material to give protection from the rapid drying effects of the sun and wind. This form of covering will also prevent unsightly pitting of the surface due to rain. Light covering materials such as waterproof paper and plastic film can be laid directly

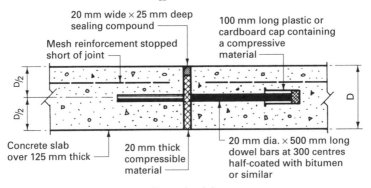

20 mm wide × 25 mm deep
sealing compound

100 mm long plastic or
cardboard cap containing
a compressive
material

Mesh reinforcement stopped
short of joint

Concrete slab
over 125 mm thick

20 mm thick
compressible
material

20 mm dia. × 500 mm long
dowel bars at 300 centres
half-coated with bitumen
or similar

Expansion joint

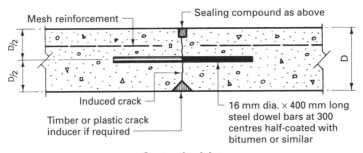

Mesh reinforcement

Sealing compound as above

Induced crack

Timber or plastic crack
inducer if required

16 mm dia. × 400 mm long
steel dowel bars at 300
centres half-coated with
bitumen or similar

Contraction joint

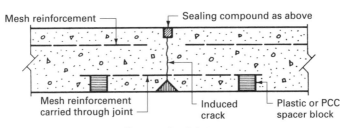

Mesh reinforcement

Sealing compound as above

Mesh reinforcement
carried through joint

Induced
crack

Plastic or PCC
spacer block

Construction joint

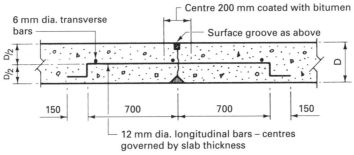

Centre 200 mm coated with bitumen

6 mm dia. transverse
bars

Surface groove as above

150   700   700   150

12 mm dia. longitudinal bars – centres
governed by slab thickness

Warping joint

Figure 14.1.6   Typical rigid road joint details

onto the concrete surface, ensuring that they are adequately secured at the edges. Plastic film can give rise to a smooth surface if the concrete is wet; this can be avoided by placing raised bearers over the surface to support the covering. Heavier coverings such as tarpaulin sheets will need to be supported on frames of timber or light metalwork so that the covering is completely clear of the concrete surface. Coverings should remain in place for about 7 days in warm weather and for longer periods in cold weather.

The design of rigid pavements follows a similar procedure to that described for flexible pavements. Graphical extracts from Volume 7 of the *Design Manual for Roads and Bridges* resemble the illustration in Fig. 14.1.3, but with concrete thickness in the vertical axis.

Rigid composite or **semi-rigid** road construction can be specified to combine the benefits of a smooth surface with a solid concrete sub-base. The concrete is reinforced throughout and surfaced with a bituminous wearing course of hot-rolled or porous asphalt.

DRAINAGE

Road drainage consists of directing the surface water to suitable collection points and conveying the collected water to a suitable outfall. The surface water is encouraged to flow off the paved area by crossfalls, which must be designed with sufficient gradient to cope with the volume of water likely to be encountered during a heavy storm to prevent vehicles skidding or aquaplaning. A minimum crossfall of 1:40 is generally specified for urban roads and motorways, whereas crossfalls of between 1:40 and 1:60 are common specifications for service roads. The run-off water is directed towards the edges of the road, where it is in turn conveyed by gutters or drainage channels at a fall of about 1:200 in the longitudinal direction to discharge into road gullies and thence into the surface water drains.

Road gullies are available in clayware and precast concrete with or without a trapped outlet; if final discharge is to a combined sewer the trapped outlet is required, and in some areas the local authority will insist upon a trapped outlet gully for all situations. Spacing of road gullies is dependent upon the anticipated storm conditions and crossfalls, but common spacings are 25.000 to 30.000 m. The gratings are usually made from cast iron, slotted and hinged to allow easy flow into the collection chamber of the gully and to allow for access for suction cleaning (see Fig. 14.1.7). Roads that are not bounded by kerbs can be drained by having subsoil drains beneath the verge or drained directly into a ditch or stream running alongside the road. The sizing and layout design of a road drainage system is based on anticipated rainfall. For example, given a rainfall (R) intensity of 40 mm/h, the quantity (Q) of water in litres/second running off a road, car-park or similar surface can be shown as:

$$Q = \frac{APR}{3,600}$$

where A is the surface area to be drained, and P is the surface permeability (asphalt and concrete $= 0.85$ to 0.95).

BS 5911 Unreinforced concrete street gully

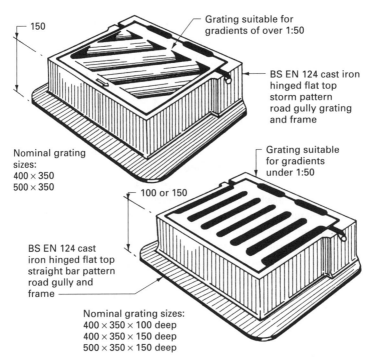

References:
BS 5911-6: *Concrete pipes and ancillary concrete products.
Specification for road gullies and gully cover slabs*
BS EN 124: *Gully tops and manhole tops for vehicular and pedestrian
areas. Design requirements, type testing, marking, quality control*

Figure 14.1.7   Typical road gully and grating details

If a road has an area of 1,000 m² and a permeability of 0.9, then:

$$Q = \frac{1,000 \times 0.9 \times 40}{3,600} = 10 \text{ litres/s} \quad \text{or} \quad 0.010 \text{ m}^3/\text{s}$$

$$Q = V \times A$$

where V is the velocity of water flowing in drain (min = 0.75 m/s for self-cleansing) and A is the area of water flowing in pipe (m²) – allow half-full bore. That is:

$$0.010 = 0.75 \times A$$

therefore

$$A = \frac{0.010}{0.75} = 0.0133 \text{ m}^2$$

Area (A) for half-bore = $\frac{1}{2}\pi r^2$; therefore r (radius) = 0.092 m and diameter of drain = 0.184 m. The nearest commercial drain is 225 mm or possibly a non-standard pipe of 200 mm.

For a more detailed perspective of drainage design, the reader is recommended to resource Chapter 8 of *Building Services, Technology and Design*, published jointly by CIOB and Pearson Education.

FOOTPATHS AND PEDESTRIAN AREAS

These can be constructed from a wide variety of materials or, for a large area, they can consist of a mixture of materials to form attractive layouts. Widths of footpaths will be determined by local authority planning requirements, but a width of 1.200 m is usually considered to be the minimum in all cases. Roads are usually separated from the adjacent footpath by a kerb of precast concrete or natural stone, which by being set at a higher level than the road not only marks the boundary of both road and footpath but also acts as a means of controlling the movement of surface water by directing it along the gutter into the road gullies (see Fig. 14.1.8).

Flexible footpaths, like roads, are those in which for design purposes no tensile strength is taken into account. They are usually constructed in at least two layers, consisting of an upper wearing course laid over a base course of tarmacadam. Wearing courses of tarmacadam and cold asphalt are used, the latter being more expensive but having better durability properties (see Fig. 14.1.8 for typical details). Gravel paths similar in construction to unbound road surfaces are an alternative method to the layered tarmacadam footpaths. Loose cobble areas can make an attractive edging to a footpath as an alternative to the traditional grass verge. The 30 to 125 mm diameter cobbles are laid directly onto a hardcore or similar bed and are handpacked to the required depth.

Unit pavings are a common form of footpath construction, consisting of a 50 to 75 mm thick base of well-compacted hardcore laid to a minimum crossfall of 1:60 if the subgrade has not already been formed to falls. The unit pavings can be of precast concrete flags or slabs laid on a 25 mm thick bed of sand, a mortar bed of 1:5 cement:sand or a weak dry cement:sharp sand mortar, or each unit can be laid

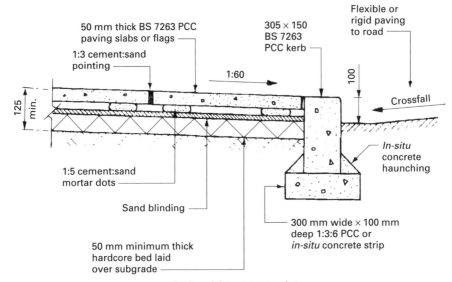

50 mm thick BS 7263 PCC
paving slabs or flags

1:3 cement:sand
pointing

305 × 150
BS 7263
PCC kerb

Flexible or
rigid paving
to road

1:60

100

Crossfall

125
min.

In-situ
concrete
haunching

1:5 cement:sand
mortar dots

Sand blinding

300 mm wide × 100 mm
deep 1:3:6 PCC or
in-situ concrete strip

50 mm minimum thick
hardcore bed laid
over subgrade

Paving slabs on mortar dots

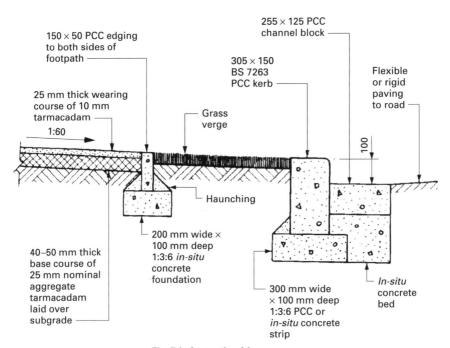

150 × 50 PCC edging
to both sides of
footpath

255 × 125 PCC
channel block

305 × 150
BS 7263
PCC kerb

Flexible
or rigid
paving
to road

25 mm thick wearing
course of 10 mm
tarmacadam

Grass
verge

1:60

100

Haunching

200 mm wide ×
100 mm deep
1:3:6 *in-situ*
concrete
foundation

40–50 mm thick
base course of
25 mm nominal
aggregate
tarmacadam
laid over
subgrade

300 mm wide
× 100 mm deep
1:3:6 PCC or
in-situ concrete
strip

In-situ
concrete
bed

Flexible footpath with grass verge

Figure 14.1.8   Typical footpath details

on five mortar dots of 1:5 cement:sand mix, one dot being located in each corner and one dot in the centre of each flag or slab (see Fig. 14.1.8). Mortar dot fixing is favoured by many designers because it facilitates easy levelling, and the slabs are easy to lift and re-lay if the need arises. Paving flags can have a dry butt joint, a 12 to 20 mm wide soil joint to encourage plant growth, or they can be grouted together with a 1:3 cement:sand grout mix. Component specification is to BS 7263-1: *Precast concrete flags, kerbs, channels, edgings and quadrants. Precast, unreinforced concrete paving flags and complementary fittings. Requirements and test methods.* See also, BS EN 1340: *Concrete kerb units. Requirements and test methods.*

Brick pavings of hard, well-burnt clay bricks or concrete pavers laid on their bed face or laid on edge and set in a bed of sharp sand or dry weak mortar with dry or filled joints can be used to create attractive patterned and coloured areas. Care must be taken with the selection of the bricks to ensure that those chosen have adequate resistance to wear, frost and sulphate attack. Bricks with a rough texture will also give a reasonably good non-slip surface.

Small granite setts of square or rectangular plan format make a very hard-wearing unit paving. Setts should be laid in a 25 mm thick sand bed with a 10 mm wide joint to a broken bond pattern. The laid setts should be well rammed, and the joints should be filled with chippings and grouted with a cement:sand grout.

Firepath pots are suitable for forming a surface required for occasional vehicle traffic such as a firefighting appliance. Firepaths consist of 100 mm deep precast concrete hexagonal or round pots with a 175 mm diameter hole in the middle, which may be filled with topsoil for growing grass or with any suitable loose filling material. The pots are laid directly onto a 150 mm thick base of sand-blinded compacted hardcore.

Cobbled pavings and footpaths can be laid to form a loose cobble surface as previously described, or the oval cobbles can be hand set into a 50 mm thick bed of 1:2:4 concrete having a small maximum aggregate size laid over a 50 mm thick compacted sand base layer.

Rigid pavements consisting of a 75 mm thick unreinforced slab of *in-situ* concrete laid over a 75 mm thick base of compacted hardcore can also be used to form footpaths. Like their road counterparts these pavings should have expansion and contraction joints incorporated into the construction. Formation of these joints would be similar to those shown in Fig. 14.1.6; for expansion joints the maximum centres would be 27.000 m and for contraction joints the maximum centres would be 3.000 m.

ACCOMMODATION OF SERVICES

The services that may have to be accommodated under a paved area could include any or all of the following:

- public or private sewers;
- electrical supply cables;
- gas mains;
- water mains;

- telecommunications cables;
- television relay cables;
- district heating mains.

In planning the layout of these services it is essential that there is adequate coordination between the various undertakings and bodies concerned if a logical and economical plan and installation programme is to be formulated.

Sewers are not generally grouped with other services, and because of their lower flexibility are given priority of position. They can be laid under the carriageway or under the footpath or verge, the latter position creating less disturbance should repairs be necessary. The specification as to the need for ducts, covers and access positions for any particular service will be determined by the undertaking or board concerned. Most services are laid under the footpath or verge so that repairs will cause the minimum of disturbance and to take advantage of the fact that the reinstatement of a footpath is usually cheaper and easier than that of the carriageway.

Services that can be grouped together are very often laid in a common trench, commencing with the laying of the lowest service and backfilling until the next service depth is reached and then repeating the procedure until all the required services have been laid. The selected granular backfilling materials should be placed in 200 to 250 mm well-compacted layers. All services should be kept at least 1.500 m clear of tree trunks, and any small tree roots should be cut, square-trimmed and tarred. Typical common service trench details are shown in Fig. 14.1.9.

■■■ INTERNAL SLABS AND PAVINGS

The design and construction of a large paved area to act as a ground floor slab is similar to that already described in the context of rigid pavings. The investigation and preparation of the subgrade is similar in all aspects to the preparatory works for roads. A sub-base of well-compacted graded granular material to a thickness of 150 mm for weak subgrades and 80 mm for normal subgrades should be laid over the formation level. If the sub-base is to be laid before the roof covering has been completed, the above thicknesses should be increased by 75 mm to counteract the effects of any adverse weather conditions. A slip membrane of 500 gauge plastic sheeting or similar material should be laid over the sub-base as described for rigid pavings for roads.

The *in-situ* concrete slab thickness is related to the load intensity, the required life of the floor and the classification of the subgrade. Floors with light loadings of up to 5 kN/m^2 would require a minimum thickness of 175 mm for weak subgrades and 150 mm for normal subgrades, whereas heavier slab loadings would require minimum slab thicknesses of 200 mm for weak subgrades and 175 mm for normal subgrades. The grade of concrete specified would range from 20 N/mm^2 for offices and shops to 40 N/mm^2 for heavy industrial usages. The need, design, detail and spacing of the various types of joints are as described for rigid pavings and detailed in Fig. 14.1.6.

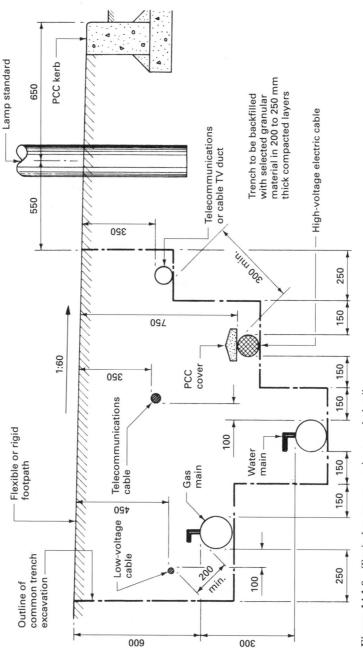

Figure 14.1.9 Typical common service trench details

Large internal pavings or slabs are laid in bays to control the tensile stresses due to the thermal movement and contraction of the slab. Laying large areas of paving in bays is also a convenient method of dividing the work into practicable sizes. Bay widths of 4.500 m will enable a two-operative tamping beam to be used, and will facilitate the easy placing and finishing of the slab. This is also an ideal width for a standard sheet of welded steel fabric reinforcement. Two basic bay layouts are possible: chequered board and long strip. **Chequered board** layout is where alternate bays in all directions are cast with joints formed to their perimeters. The intermediate bays are laid some 7 days later to allow for some shrinkage to take place. Although this is the traditional method it has two major disadvantages, in that access to the intermediate or infill bays is poor and control over shrinkage is suspect.

The **long strip** method of internal paving construction provides an excellent alternative and is based on established rigid road-paving techniques. The 4.500 m wide strips are cast alternately, which gives good access; expansion joints if required are formed along the edges; transverse contraction joints can be formed as the work proceeds, or the surface groove can be sawn in at a later time. With this method it may be necessary to cast narrow edge strips some 600 mm to 1.000 m wide at the commencement of the operation if perimeter columns or walls would hinder the use of a tamping board or beam across the full width of the first strip.

The concrete paving must be fully compacted during laying, and this can be achieved by using a single or double tamping beam fitted with a suitable vibration unit. The finish required to the upper surface will depend upon the usage of the building and/or any applied finishes that are to be laid. Common methods employed are hand trowelling using steel trowels, power floating and power trowelling. The latter methods will achieve a similar result to hand trowelling but in a much shorter space of time. None of the above finishing methods can be commenced until the concrete has cured sufficiently to accept the method being used.

Vacuum dewatering is a method of reducing the time delay before power floating can take place. The object is to remove the excess water from the slab immediately after the initial compaction and levelling has taken place. The slab is covered with a fine filter sheet and a rigid or flexible suction mat to which is connected a transparent flexible plastic pipe attached to a vacuum generator. The vacuum created will compress the concrete slab and force the water to flow out up to a depth of 300 mm. The dewatering process will cause a reduction of about 2% in the slab depth, and therefore a surcharge should be provided by means of packing strips on the side forms or at the ends of the tamping boards. The filter sheet will ensure that very little of the cement fines of the mix is carried along in suspension by the water being removed. The vacuum should be applied for about 3 minutes for every 25 mm of concrete depth, which will generally mean that within approximately 20 minutes of casting the mats can be removed and the initial power floating operation commenced, followed by the final power trowelling operation. This method enables the laying of long strips of paving in a continuous operation. It is possible for a team consisting of six operatives and a supervisor to complete 200 m^2 of paving per day using this method.

PART 15

ACCESS AND FACILITIES FOR DISABLED PEOPLE: BUILDINGS OTHER THAN DWELLINGS

▨ ▇ ▇ INTRODUCTION

Much of the guidance in the Building Regulations, Approved Document M: *Access to and use of buildings* is complementary to the recommendations provided in BS 8300: *Design of buildings and their approaches to meet the needs of disabled people – Code of practice*. The British Standard contains additional material not considered appropriate for Building Regulation guidance, such as signposting and other management procedures. It also contains some variations from the Approved Document. Therefore, although the two publications have similar objectives, and compliance with the BS will ensure good practice, it is not necessarily equivalent with the guidance in the Approved Document.

The Disability Discrimination Act differs from Approved Document M and BS 8300 in that it is not directly about the design and construction of buildings. It is concerned with the provision of goods, services and premises with non-discriminatory facilities. Therefore the facilities available in buildings must be favourable for *all* users, including the physically less able.

All three documents apply to newly constructed buildings as well as to newly extended buildings and those undergoing material alteration or change of use. The Act has particular reference to all buildings that function as a service provider. Owners and business operators are required to remove and change features that prevent access to that service or facility. For example, in retail outlets, displays need to be at an accessible height, aisle widths adequate for wheelchairs, lighting levels increased, accesses ramped, etc.

Chapters 15.1 to 15.5 of this book provide a general guidance for some of the more common provisions and facilities for the less ambulant using non-domestic buildings. Although the objectives are to show how the less able can be assisted, it is noteworthy that the title of the Building Regulations Approved Document does not include specific reference to disabled people. This emphasises the need for buildings to be convenient for all users, including those carrying heavy or large packages or luggage as well as people with pushchairs. Furthermore, note that particular purpose groups and building functions, e.g. theatres and sports centres, will have provisions specifically detailed for that activity, and appropriate references should be sought.

15.1

ACCESS TO BUILDINGS

The objective is to provide a suitable means for accessing entrances to buildings from the point of site entry and from the car parking areas. It is also important to provide suitable means for gaining access between buildings on the same site. Car parking designated for disabled persons should be planned to have a priority for closeness to the building entrance.

■■■■ CAR PARKING

Parking provisions should be made available for visitors, customers and members of staff needing to access a building. The main entrance and staff entrances should be reasonably accessible on foot or by wheelchair from any of the allocated car parking spaces. Priority parking for the disabled should be suitably designated and given larger parking bays to allow people with limited mobility to access their vehicles unimpeded: see Fig. 15.1.1.

- Location of accessible bays clearly signed at the car park entrance.
- Designated parking for disabled drivers or passengers clearly identified.
- Firm, durable and slip-resistant surface on level ground (no undulations greater than 5 mm).
- Parking space for the disabled within 50.000 m of the building main entrance or other facility served.
- Parking bays for use by the disabled to have adequate width to allow car doors to be fully opened for wheelchair space and passenger transfer.
- Dropped kerbs for wheelchair access.
- Parking ticket machine in close proximity to designated parking with controls between 750 and 1,200 mm above adjacent ground.

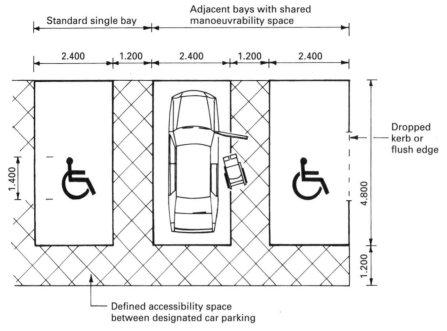

Figure 15.1.1 Designated parking bays

▦◼◼ GENERAL PROVISIONS

- Level or slightly graded foot or wheelchair approach to building entrance.
- Approach surface to be firm, durable, trip free, slip resistant and unobstructed to a height of 2.100 m.
- Where the surface comprises paving slabs, differentials between slab levels to be no more than 5 mm, with joints no deeper than 5 mm. Preferably not exceeding 3 mm under a 1.000 m level.
- Tactile (blistered) paved surfaces to be located to indicate deviation from the pedestrian access.
- Surface width to be at least 1.500 m, with established passing places.
- Passing places to be at least 1.800 m wide × 2.000 m long, at a maximum distance of 50.000 m apart.
- Main access route to be adequately signed.

▦◼◼ GRADIENTS

- Formed as a gentle continuous slope (max. gradient 1 in 60).
- May include a series of shorter gradients.
- Maximum gradient 1 in 20 in any part, with level landings at each 500 mm rise.
- Crossfalls, maximum gradient 1 in 40.

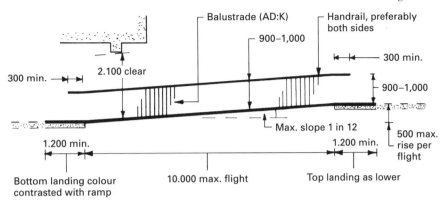

Figure 15.1.2   Ramped access

■■■ RAMPED ACCESS

Required where site conditions create a gradient greater than 1 in 20. Ramps are essential for wheelchair users and parents with pushchairs to overcome changes in ground levels. See Fig. 15.1.2 for outline requirements.

- Must be clearly signed.
- Maximum going of flight, 10.000 m.
- Maximum rise of flight, 0.500 m.
- An alternative access is required where the total rise exceeds 2.000 m, e.g. a platform lift.
- Slip-resistant surface to ramp and landings.
- Ramp surface colour to contrast with that of the landings.
- Surface width, minimum 1.500 m.
- Landings provided at the top and bottom of ramp, minimum 1.200 m long and clear of opening doors and other possible obstructions.
- Intermediate landings at least 1.800 m long and 1.800 m wide to function as passing places.
- Landings level, subject to maximum gradient 1 in 60 along their length, with maximum cross-gradient of 1 in 40.
- Handrails to both sides; one side only is acceptable if there is another means for accessing the main entrance to a building.
- Handrail to be accompanied by balustrade or other guarding (see guidance in Building Regulations Approved Document K). If there is an open side, the adjacent ground level to be with the ramp surface or a visually contrasting kerb of at least 100 mm height to be provided.
- Grading of ramps:
  - Flight going up to 2.000 m, maximum gradient is 1 in 12.
  - Flight going of 10.000 m, maximum gradient is 1 in 20.

- Maximum flight gradient in between 2.000 and 10.000 m going can be interpolated directly:
  e.g. A going of 6.000 m has a maximum gradient of 1 in 16.
       A going of 8.500 m has a maximum gradient of 1 in 18.5.

▪▪▪ STEPPED ACCESS

Ramps should be accompanied by a stepped access, because many ambulant disabled people may find them more negotiable.

- **Landings**   Approach identified by an 800 mm deep tactile (corduroy) warning surface. Intermediate landings provided with a tactile warning surface where the landing is accessed other than from the steps. Top and bottom of flights to have a level landing not less than 1.200 m long. Landings to be clear of door swings and other potential obstructions.
- **Steps and flights**
  - Unobstructed surface width not less than 1.200 m.
  - Maximum 12 risers between landings for going less than 350 mm.
  - Maximum 18 risers between landings for going 350 mm or more.
  - No open risers.
  - Nosings integral (not projecting) and colour-contrasted 55 mm on tread and riser.
  - Nosing projection a maximum of 25 mm over tread below.
  - Risers between 150 and 170 mm and consistent.
  - Going between 280 and 425 mm and consistent.
- **Handrail**
  - Continuous to both sides, including landing.
  - Central handrail provided where unobstructed stair width exceeds 1.800 m.
  - Height between 900 and 1,000 mm above the pitch line.
  - Height between 900 and 1,100 mm above landings.
  - Extends at least 300 mm beyond ramps or steps.
  - Contrasts visually with the background.
  - Suggested profile 40–50 mm diameter and 50–60 mm from wall.
  - Projection from wall preferably not more than 100 mm into stair width.
  - Terminates in a non-hazardous manner, generally by returning to the wall to prevent clothing being caught.

Note: Some variations may be more practical and safer in special circumstances, e.g. the given riser, going and handrail dimensions may not be considered appropriate for use in schools designed specifically for young children, e.g. rise 150 mm and going 280 mm.

Figure 15.1.3 shows the general provisions for stair flights, landings and handrails.

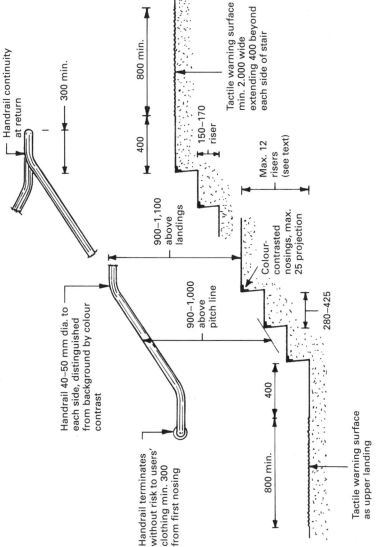

Handrail continuity
at return

300 min.

800 min.

400

150–170
riser

Tactile warning surface
min. 2.000 wide
extending 400 beyond
each side of stair

Max. 12
risers
(see text)

Handrail 40–50 mm dia. to
each side, distinguished
from background by colour
contrast

900–1,100
above
landings

900–1,000
above
pitch line

Colour-
contrasted
nosings, max.
25 projection

280–425

Handrail terminates
without risk to users'
clothing min. 300
from first nosing

400

800 min.

Tactile warning surface
as upper landing

Note: Top, intermediate and lower landings min. 1.200 long clear of opening doors

Figure 15.1.3   Stepped access

ACCESS INTO BUILDINGS: ENTRANCES

The disposition of entrances to buildings should have regard for the routes that service them. They should be prominent, and easily distinguished from the general facade.

Further considerations:

- Entrance to be clearly identified with adequate signage.
- If an alternative access for the disabled is provided, it should link directly with the area served by the principal entrance.
- Architectural featurework or structural supports, e.g. columns to a canopy, to be colour contrasted and of a rounded finish to minimise any hazard to the visually impaired.
- Threshold to have a level approach at least 1.500 × 1.500 m clear of swing doors.
- Threshold to be flush and trip free for easy transit with a wheelchair (see Fig. 3.5.9 in *Construction Technology*). A 15 mm maximum upstand is acceptable, provided it has a chamfered or rounded finish.
- Floor finishes in the immediate threshold area (inside and out), of materials not likely to impede movement.
- Entrance doors ideally activated automatically by motion sensor, not to swing outwards (horizontal sliding preferred), and to remain open long enough for a slow-moving person to progress through.
- If not automatic, entrance doors to have large push button controls at a suitable height for all users (750–1,000 mm above ground finish).
- Revolving doors are acceptable for wheelchair access, if they are slow moving and enclose a 1.500 m × 1.200 m footprint, and there is an alternative suitably furnished single-leaf side-hung door available: see Fig. 15.2.1.
- Immediately inside the entrance, the area should be sympathetically lit for transition and adjustment by the visually impaired, when contrasting with outside lighting levels.

Door furniture and visibility provisions

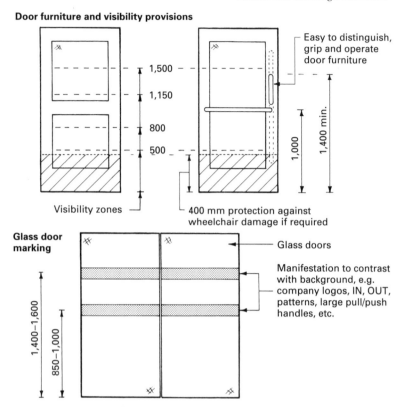

Figure 15.2.1   Entrance doors

- Inside the entrance, location of facilities, e.g. WC, lift, reception, to be clearly signed.
- Entrance lobby to be of adequate dimensions to allow a wheelchair user to move sufficiently clear of the first door before negotiating the next: see dimensionally enhanced diagrams provided in Section 2 of Approved Document M to the Building Regulations.
- Door opening widths as Fig. 15.2.2.
- Glass doors to be defined by manifestation (lettering or patterns) at 0.850 to 1.000 m and 1.400 to 1.600 m above floor level. Manifestation to contrast with background: see Fig. 15.2.1.

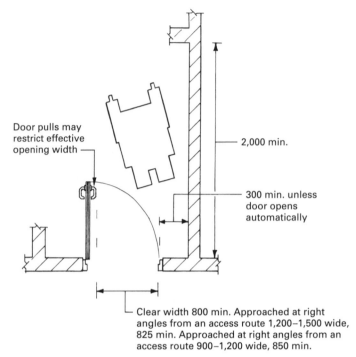

Door pulls may
restrict effective
opening width ─

2,000 min.

300 min. unless
door opens
automatically

Clear width 800 min. Approached at right
angles from an access route 1,200–1,500 wide,
825 min. Approached at right angles from an
access route 900–1,200 wide, 850 min.

Figure 15.2.2   Door opening widths

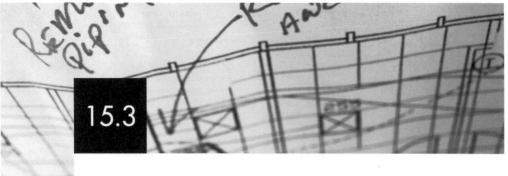

15.3

ACCESSIBILITY WITHIN BUILDINGS

Circulation within buildings applies to both the horizontal and the vertical dimension. The facility should provide for unimpeded independent circulation by users of wheelchairs and for those with hearing and/or visual impairments.

◼◼◼ HORIZONTAL CIRCULATION

Corridors and passageways have the dual function of connecting spaces and providing a means for emergency escape. With particular regard to escape, they must be easy and safe to negotiate by the less ambulant. Information about the building (fire escape routes, location of lavatories, function rooms, etc.) should be adequately signed to assist circulation.

◼◼◼ ENTRANCE AND RECEPTION AREA

- Reception point to be identified and accessible from the entrance door by a wheelchair user.
- Reception to be sufficiently clear of the building entrance to be unimpaired by external noise.
- The minimum space for wheelchair manoeuvring in front of reception is 1,200 mm deep × 1,800 mm wide where there is a 500 mm minimum knee recess under the counter. With no recess, a 1,400 mm × 2,200 mm space is required.
- Part of the reception counter should be at a lower level. Maximum surface height 760 mm with at least 700 mm clear space above floor finish: see Fig. 15.3.1.

◼◼◼ DOORS

All doors need to perform well in containing a fire, yet must still be easily operable by all users, including those of limited strength. The maximum door closer pressure

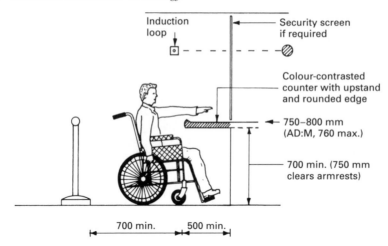

Figure 15.3.1   Counters and checkouts

by powered operation, as specified in Approved Document M to the Building Regulations, relates directly to a minimum opening force for manual operation of a door. The exact relationship between closing and opening operating pressure depends on the efficiency of the closing device: see BS EN 1154: *Building hardware. Controlled door closing devices. Requirements and test methods.*

Doors hinder circulation, particularly for disabled people. Where specified as 'fire doors' (Building Regulations Part B: *Fire safety*) they are normally closed, but may be held open by an electromagnetic device. Such doors must self-close when activated by a smoke detector, a fire alarm system, a power failure or a hand operated device, but must still open manually. BS 8300 has guidance on the use of electromagnetic door retention devices. Powered swing doors must also be capable of manual operation. Other considerations include:

- Clear opening width 800 mm minimum with a 300 mm minimum access space for convenience of a wheelchair user. Situations exceeding 800 mm as shown in Fig. 15.2.2.
- Door latch to be lever operated in preference to a rounded knob.
- Door furniture to colour contrast with door.
- Door surround (frame and/or architrave) to colour contrast with the adjacent wall.
- Door leading edge to colour contrast with door surface and surrounds.
- Vision panels to be provided at a height between 500 and 1,500 mm above floor level. If necessary, interrupted between 800 and 1,150 mm for an intermediate rail: see Fig. 15.2.1.
- Glass doors to be suitably manifested as Fig. 15.2.1, and differentiated from any adjacent glazed wall system.

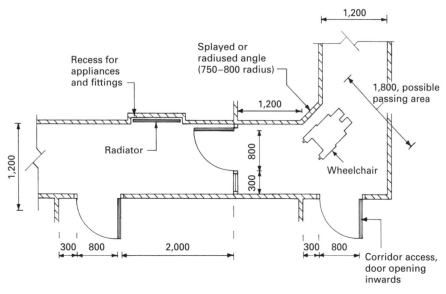

Figure 15.3.2 Corridors and passageways, minimum dimensions (mm)

▨▪■ CORRIDORS

See Fig. 15.3.2.

■ Unobstructed width of at least 1,200 mm.
■ Structural elements, e.g. columns and piers, are not to project into the passageway.
■ Unavoidable wall fixings, e.g. radiators, grills, fire hoses, to be fully recessed.
  Note: If projections are unavoidable, a means such as a colour-contrasted guard rail may be acceptable.
■ Other hazard potential projections into a corridor, such as outward-opening doors, are unacceptable.
■ Monochromatic decorative schemes to be avoided. Use of colour contrasts between elements preferred.
■ Lighting design to avoid glare and silhouettes.
■ Surface finishes: patterns to be avoided as they can be confused with components or facilities by those with impaired vision. Light reflecting surfaces can also be confusing.
■ Special facilities:
  ■ Splayed or radiused turns to ease wheelchair transport.
  ■ Occasional 1,800 mm widths to facilitate wheelchairs to pass.
■ Floor to be level. If a gradient is unavoidable, it should not exceed 1 in 20 and rise no more than 500 mm without a horizontal landing or rest area of at least 1,500 mm length.
■ Any irregularity in a floor finish to be clearly identified by visual colour contrasts.

▪▪▪ VERTICAL CIRCULATION

Lifts are usually the most convenient means for travelling between and accessing different levels in a building. Some buildings, particularly those subject to modernisation, may not have sufficient space to accommodate a lift. In these situations a platform lift or a stair lift may be considered for use between two levels. Platform and stair lifts may be used only if they do not interfere with the requirements for fire escape on a stairway. A ramp in place of a step may be used for access limited to a 300 mm rise, and as described in Chapter 15.1. In excess of a 300 mm rise, steps are also to be provided. Steps or internal stairs are as described in Chapter 15.1 and limited to 12 risers (up to 16 risers may be used in small premises where space for landings is restricted) with 900 to 1,100 mm handrail heights at landings. An internal stair must always be provided as an alternative to vertical access by any form of lift. Stairs are suitably designed for use by the ambulant disabled and the visually impaired.

PASSENGER LIFT

See Fig. 15.3.3.

- Landing space in front of lift doors should have enough area (min. 1,500 mm × 1,500 mm) for a wheelchair user to turn and reverse into the lift car.
- Where a building design allows enough space, opposing doors are beneficial so that a wheelchair user can enter one side and leave via the other side without turning around.
- Lift doors to remain open a sufficient time for a wheelchair user and assistant to enter the lift car unimpeded. Door-reactivating device essential.
- Lift doors colour-contrasting and distinguishable from surroundings.
- Landing and in-car call panel to be easily distinguishable from its wall mounting. Located between 900 and 1,100 mm from the floor and within 400 mm of the wall return.
- Lift control buttons to be embossed for tactile identification and to contrast visually with their background.
- Landing and in-car audible announcements and visual/illuminated displays of lift arrival, travel direction and floor reached.
- Emergency telephone in lift car to have inductive couplers for use by people with hearing difficulties.
- Alarm buttons to have a visual display acknowledging that an alert has been sounded.
- Lift car floor finish to be resistant to slippage, particularly when wet.
- Support rail in lift car at 900 mm top height above floor level.
- Floor number displayed clearly on the wall opposite the lift doors.
- Independent power supply for emergency use.

Reference: BS EN 81: *Safety rules for the construction and installation of lifts.*

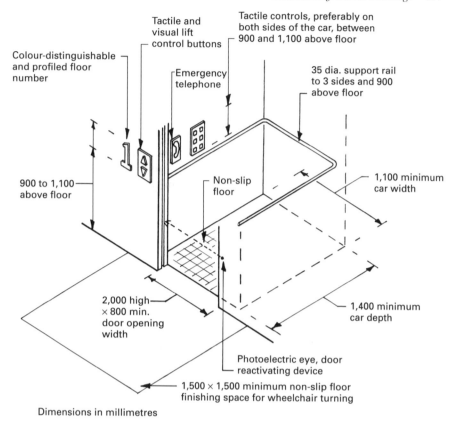

Tactile and visual lift control buttons

Tactile controls, preferably on both sides of the car, between 900 and 1,100 above floor

Colour-distinguishable and profiled floor number

Emergency telephone

35 dia. support rail to 3 sides and 900 above floor

900 to 1,100 above floor

Non-slip floor

1,100 minimum car width

2,000 high × 800 min. door opening width

1,400 minimum car depth

Photoelectric eye, door reactivating device

1,500 × 1,500 minimum non-slip floor finishing space for wheelchair turning

Dimensions in millimetres

Figure 15.3.3 Lift provisions

PLATFORM LIFT

Platform lifts have evolved from industrial applications for movement of goods between different levels within floors. For people with mobility impairments, platform lifts have been adapted to provide safe vertical transport. The main differences from conventional passenger lifts are a relatively small floor area, limitations on speed, and operation by continuous pressure controls by push button or joystick. Section 3 of Approved Document M contains design and installation provisions deemed to satisfy the Building Regulations. Other guidance is found in:

- BS 6440: *Powered lifting platforms for use by disabled persons. Code of practice.*
- BS 8300: *Design of buildings and their approaches to meet the needs of disabled people. Code of practice.*
- ISO/DIS 9386-1: *Power operated lifting platforms for persons with impaired mobility. Rules for safety, dimensions and functional operations. Part 1. Vertical lifting platforms.*

STAIR LIFT

These have evolved from special applications used in hospitals and homes for the elderly. More recently they have become standardised in terms of components for use in dwellings and now have a place in Approved Document M to the Building Regulations. All newly created dwelling-houses are required to be suitably structured to support and accommodate a retrospective installation. The objective is to allow people with limited disabilities to continue the enjoyment of their home without the need to move. For non-domestic applications, a stair lift should be considered only where it is impractical to install a conventional lift or platform lift. Where used, a stair lift must not restrict the use of a stair by other people, and there must be no intrusion on the required width of stair for emergency escape purposes.

References:
BS 5776: *Specification for powered stair lifts.*
ISO/DIS 9386-2: *Power operated lifting platforms for persons with impaired mobility. Rules for safety, dimensions and functional operations. Part 2. Powered stairlifts moving in an inclined plane for seated, standing and wheelchair users.*

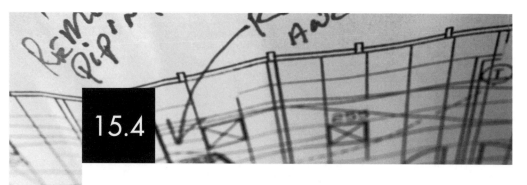

15.4

USE OF FACILITIES

Buildings provide a huge range of uses, functions and facilities. All people should be able to participate and enjoy the proceedings, the entertainment or the service that a particular building offers. Buildings should be designed to provide disabled people with a choice of unobstructed seating locations. This is particularly appropriate when designing a building for spectators and audiences. Refreshment bars, sanitary conveniences and other public services should be on a level and accessible path for the less ambulant. Different floor levels should have provision for access.

Buildings designed as commercial accommodation, e.g. hotels and hostels, should have some rooms arranged and serviced specifically for the convenience of disabled occupants. These are most likely to be situated at ground level; accommodation elsewhere should provide for the ambulant disabled.

Determining the requirements and the subsequent design of buildings with adequate facilities for the disabled depends considerably on a building's function. Functions vary: therefore the subject of special accommodation and facilities is diverse and not within the scope of this text. The reader should seek design manuals relating to particular purpose groups. Some outline guidance for audience seating to lecture theatres, conference centres, theatres, cinemas and sports stadia is provided in the Building Regulations, Approved Document M, Section 4: *Facilities in buildings other than dwellings*. This Section also considers accessibility to and the use of refreshment facilities, and the design considerations for sleeping accommodation, with particular regard for wheelchair accessibility to bedrooms and bathrooms.

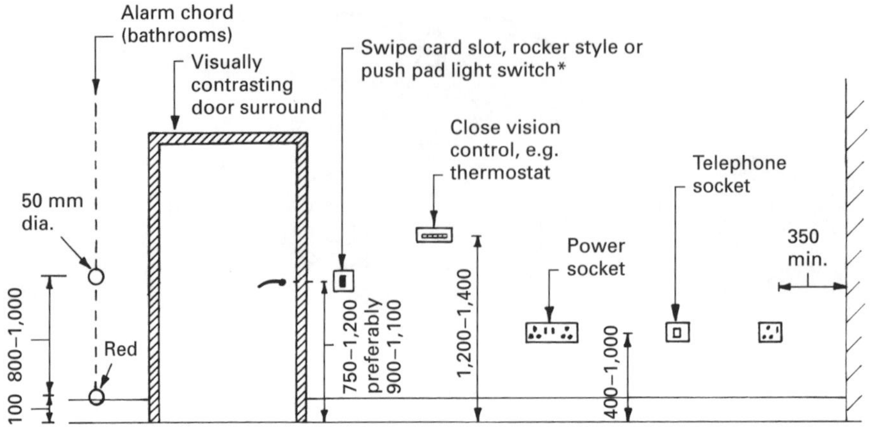

*Where these switch types are impractical, a pull chord with a 50 mm dia. bangle may be used, set between 900 and 1,100 above the floor. Bangle to be colour-differentiated from an alarm pull.

Dimensions in millimetres

Figure 15.4.1   Switch and socket location

■■■ MANUAL LIGHTING CONTROLS AND SOCKET OUTLETS

See Fig. 15.4.1.

Provision and disposition of controls and connections is common to most buildings, and should feature the following:

- Switches and sockets with colour and tonal contrast to distinguish visually from background.
- Switch location 750 to 1,200 mm above floor.
- Switches preferably aligning with door handles at 900 to 1,100 mm above floor.
- Switches of simple design, preferably of push pad type or large rocker style for operating with one hand.
- Switches and thermostats that require hand movement other than a push/pull function, 750 to 1,000 mm above floor.
- Alarm pull chords in sanitary accommodation and other private areas, coloured red, located close to the wall with two 50 mm diameter red bangles for easy grasp at 100 mm and between 800 to 1,000 mm above floor.
- Printouts or metered indicators at 1,200 to 1,400 mm above floor.
- Socket location 400 to 1,000 mm above floor.
- Socket outlets (power or telecommunications) not closer than 350 mm to corners.
- Switched sockets and circuit isolators to clearly indicate ON or OFF.

SANITARY ACCOMMODATION

Accessibility to and within sanitary accommodation is essential for the freedom of people with disabilities. This generally requires the provision of supplementary sanitary accommodation for wheelchair users. Special facilities should also be provided for people of either sex with babies and small children, and in certain buildings such as hotels and airport terminals for people encumbered with luggage.

The design and physical layout of sanitary facilities especially for the disabled is a specialised subject, covering provisions for:

- WC compartments for ambulant disabled people within standard separate-sex toilets;
- unisex facilities that are accessible for a wheelchair user with an assistant, and that allow space for lateral transfer from 'chair to WC pan';
- enlarged cubicle space in separate-sex toilets.

In buildings designed for sports participation, accessibility to changing facilities, showers and bathrooms is also a priority. The Building Regulations, Approved Document M, Section 5: *Sanitary accommodation in buildings other than dwellings* in conjunction with BS 8300 provides a basis for the design of these requirements. Both publications include applications to cinemas, theatres, sports stadia, office buildings and recreational facilities. The subject is both extensive and specialised, with specific provisions for individual buildings. As a guide, the following general considerations for accessibility to, and use of facilities within WC compartments, and those shown in Figs 15.5.1 and 15.5.2, may be applied to most non-domestic buildings:

- Passageways, minimum 1,200 mm clear width.
- Door openings in passageways preferably 900 mm wide, 1,000 mm in public buildings.
- Minimum floor dimensions 1,500 mm × 2,000 mm, to accommodate a wheelchair in clear turning space and people with impaired leg movement and those with crutches.

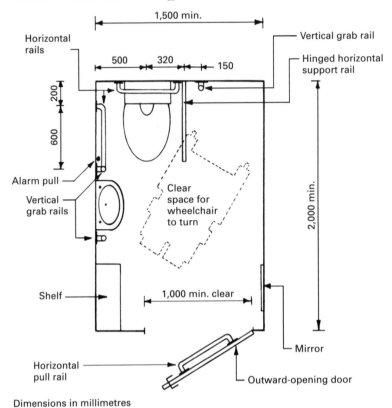

Dimensions in millimetres

Figure 15.5.1   Disposition of facilities in a WC compartment

- Door opening into a compartment, minimum 1,000 mm clear width.
- Compartment door to open outwards and to be provided with an emergency release device that may be operated from the outside. A horizontal bar for closing is fitted to the inside.
- Support or hand grab rails of 50 mm diameter located each side of a wash basin.
- Support rail of the hinged drop-down type positioned on the exposed side of the WC pan. Minimum length of rail 300 mm.
- WC pan located at least 500 mm from the adjacent wall to allow for unimpeded transfer from a wheelchair.
- WC pan seat set at a 480 mm height above the floor. Seat to be of a robust and rigid composition.
- WC cistern flush handle or push button at 1,200 mm maximum above the floor.
- Toilet paper dispenser to be on the nearest adjacent wall and within easy reach.
- Wash basin rim set between 720 and 740 mm above the floor and within reachable distance of a user seated on the WC pan.
- A towel facility is acceptable, but an automatic warm air hand dryer is preferable, with the air temperature set at 35 °C maximum. Height above floor, 800 to 1,200 mm.

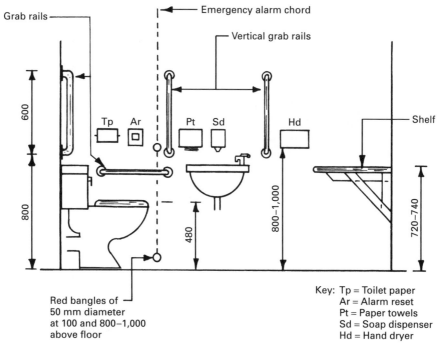

Key: Tp = Toilet paper
Ar = Alarm reset
Pt = Paper towels
Sd = Soap dispenser
Hd = Hand dryer

Red bangles of
50 mm diameter
at 100 and 800–1,000
above floor

Dimensions in millimetres

Figure 15.5.2   Facilities in a typical unisex WC compartment

- Basin taps of the quarter-turn lever type or of photoelectric cell automatic operation. Water temperature controlled through a thermostatic mixing valve to 35 °C maximum.
- Emergency alarm chord suspended from the ceiling, close to the wall and unobstructed. Two distinctive red bangles of 50 mm diameter fitted at 100 mm and between 800 and 1,000 mm above the floor.

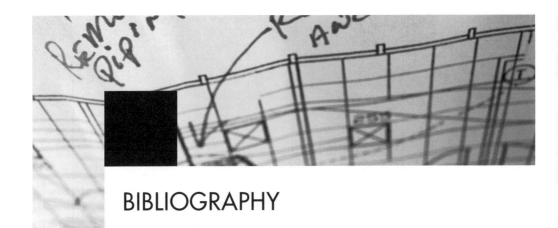

BIBLIOGRAPHY

Barry, R. *The Construction of Buildings*. Blackwell Scientific.

Boughton, B. *Reinforced Concrete Detailers Manual*. Blackwell Scientific.

Fisher Cassie, W. and Napper, J. H. *Structure in Buildings*. The Architectural Press.

Holmes, R. *Introduction to Civil Engineering Construction*. College of Estate Management.

Handisyde, Cecil C. *Building Materials*. The Architectural Press.

Leech, L. V. *Structural Steelwork for Students*. Butterworth/Heinemann.

Llewelyn Davies, R. and Petty, D. J. *Building Elements*. The Architectural Press.

McKay, W. B. *Building Construction*, Vols 1 to 4. Pearson Education.

Smith, G. N. *Elements of Soil Mechanics for Civil and Mining Engineers*. Blackwell Scientific.

West, A. S. *Piling Practice*. Butterworth/Heinemann.

Whitaker, T. *The Design of Piled Foundations*. Elsevier Science.

GENERAL

A Guide to Scaffolding Construction and Use. Scaffolding (GB) Ltd.

BSP. Pocket Book. The British Steel Piling Co. Ltd.

Building Regulations, Approved Documents. ODPM.

Construction Safety. Construction Industry Publications Ltd.

Data sheets of the British Precast Concrete Federation.

Design Manual for Roads and Bridges, Vol. 7. The Stationery Office Ltd.

DoE Construction Issues: 1 to 17. The Stationery Office Ltd.

Drained Joints in Precast Concrete Cladding. The National Buildings Agency.

Glass. Pilkington UK Ltd.

Handbook on Structural Steelwork. The British Constructional Steelwork Association Ltd.

Lifting Operations and Lifting Equipment Regulations 1998. The Stationery Office Ltd.

Lighting for Building Sites. Electricity Association Services Ltd.

Loss Prevention Standards. Loss Prevention Certification Board.

Mitchell's Building Series. Pearson Education.

The Confined Spaces Regulations 1997. The Stationery Office Ltd.

The Construction (Health, Safety and Welfare) Regulations 1996. The Stationery Office Ltd.

The Fire Precautions (Workplace) Regulations 1997. The Stationery Office Ltd.

The Work at Height Regulations 2005. The Stationery Office Ltd.

OTHER READING

Relevant advisory leaflets. The Stationery Office Ltd.

Relevant AJ Handbooks. The Architectural Press.

Relevant BRE Digests. Construction Research Communications Ltd.

Relevant British Standards, Codes of Practice and EuroNorms. British Standards Institution.

Relevant manufacturers' catalogues and technical guides contained in the Barbour Index and Building Products Index libraries.

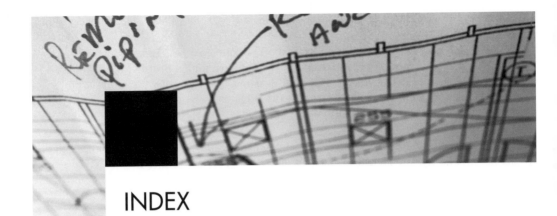

INDEX